II–VI Blue/Green Light Emitters: Device Physics and Epitaxial Growth

SEMICONDUCTORS
AND SEMIMETALS
Volume 44

Semiconductors and Semimetals

A Treatise

Edited by R. K. Willardson
CONSULTING PHYSICIST
SPOKANE, WASHINGTON

Albert C. Beer
CONSULTING PHYSICIST
COLUMBUS, OHIO
FOUNDING EDITOR, EDITOR EMERITUS

Eicke R. Weber
DEPARTMENT OF MATERIALS SCIENCE
AND MINERAL ENGINEERING
UNIVERSITY OF CALIFORNIA AT
BERKELEY

II–VI Blue/Green Light Emitters: Device Physics and Epitaxial Growth

SEMICONDUCTORS
AND SEMIMETALS

Volume 44

Volume Editors

R. L. GUNSHOR

SCHOOL OF ELECTRICAL AND COMPUTER ENGINEERING
PURDUE UNIVERSITY
WEST LAFAYETTE, INDIANA

A. V. NURMIKKO

DIVISION OF ENGINEERING
BROWN UNIVERSITY
PROVIDENCE, RHODE ISLAND

ACADEMIC PRESS

San Diego   London   Boston
New York   Sydney   Tokyo   Toronto

In Chapter 2, "Growth and Characterization of ZnSe-based II–VI Semiconductors by MOVPE," by Sz. Fujita/ Sg. Fujita, reprinted with the kind permission of Elsevier Science–NL, Sara Burgerhartstraat 25, 1055 KV Amsterdam, The Netherlands, are: Fig. 1, Toda, A., Asano, T., Funato, K., Nakamura, F., Mori, Y. (1994), *J. Cryst. Growth* **145**, 537–540; Fig. 2, Kukimoto, H. (1990), *J. Cryst. Growth* **101**, 953–957; Fig. 3, Danek, M., Huh, J. S., Foley, L., Jensen, K. F. (1994), *J. Cryst. Growth* **145**, 530–536; Fig. 4, Fujita, Sz., Fujita, Sg. (1995), *Appl. Surf. Sci.* **86**, 431–436; Fig. 5, Fujita, Sz., Fujita, Sg. (1994), *Appl. Surf. Sci.* **79/80**, 41–46; Fig. 10, Ogata, K., Kawaguchi, D., Kera, T., Fujita, Sz., Fujita, Sg. (1996), *J. Cryst. Growth* **159**, 312–316. Reprinted with the kind permission of *Japanese Journal of Applied Physics*, are: Fig. 6, Ichimura, M., Wada, T., Fujita, Sz., Fujita, Sg. (1991), *Jpn. J. Appl. Phys.* **30**, 3475–3481; Fig. 8, Yoshikawa, A., Nomura, H., Yamaga, S., Kasai, H. (1988), *Jpn. J. Appl. Phys.* **27**, L1948–L1951. Fig. 7 reprinted with the kind permission of TMS, Fujita, Sz., Matsumoto, S., Fujita, Sg. (1993), *J. Electron. Mater.* **22**, 521–527. Fig. 9 reprinted with the kind permission of *Japanese Journal of Applied Physics*, Fujita, Sz., Asano, T., Maehara, K., Fujita, Sg. (1993). *Jpn. J. Appl. Phys.* **32**, L1153–L1156; Elsevier Science–NL, Fujita, Sz., Fujita, Sg. (1994), *J. Cryst. Growth* **145**, 552–556; TMS, Fujita, Sz., Tojyo, T., Yoshizawa, T., Fujita, Sg. (1995), *J. Electron. Mater.* **24**, 137–141; American Institute of Physics, Hauksson, I. S., Simpson, J., Wang, S. Y., Prior, K. A., Cavenett, B. C. (1992), *Appl. Phys. Lett.* **61**, 2208–2210. Fig. 11 reprinted with the kind permission of Institution of Electrical Engineers, Toda, A., Margalith, T., Imanishi, D., Yanashima, K., Ishibashi, A. (1995), *Electron. Lett.* **31**, 1921–1922.

In Chapter 3, "Gaseous Source UHV Epitaxy Technologies for Wide Bandgap II–VI Semiconductors," by E. Ho/ L. A. Kolodziejski, reprinted with the kind permission of TMS are: Figs. 4, 5, 7, Ho, E., Coronado, C. A., Kolodziejski, L. A. (1993), *J. Electron. Mater.* **22**, 473–478; Figs. 9, 12, Coronado, C. A., Ho, E., Fisher, P. A., House, J. L., Lu, K., Petrich, G. S., Kolodziejski, L. A. (1994), *J. Electron. Mater.* **23**, 269–273. Fig. 8 reprinted with the kind permission of American Vacuum Society, Lu, K., Fisher, P. A., House, J. L., Ho, E., Coronado, C. A., Petrich, G. S., Kolodziejski, L. (1994), *J. Vac. Sci. Technol.* **B 12**, 1153. Figs. 10, 11 reprinted with the kind permission of Materials Research Society, Fisher, P. A., Ho, E., House, J. L., Petrich, G. S., Kolodziejski, L., Brandt, M. S., Johnson, N. M. (1994), *Materials Research Society Symposium Proceedings* **340**, 451–456. Fig. 14 reprinted with the kind permission of American Institute of Physics, Ho, E., Fisher, P. A., House, J. L., Petrich, G. S., Kolodziejski, L. A., Walker, J., Johnson, N. M. (1995), *Appl. Phys. Lett.* **66**, 1062–1064. Reprinted with the kind permission of Elsevier Science–NL, Sara Burgerhartstraat 25, 1055 KV Amsterdam, The Netherlands, are: Fig. 15, Fisher, P. A., Ho, E., House, J. L., Petrich, G. S., Kolodziejski, L. A., Walker, J., Johnson, N. M. (1995), *J. Cryst. Growth* **150**, 731; Fig. 17, Ho, E., Petrich, G. S., Kolodziejski, L. A. (1996), *J. Cryst. Growth* **159**, 269.

In Chapter 6, "II–VI Diode Lasers: A Current View of Device Performance and Issues," by A. V. Nurmikko/A. Ishibashi, reprinted with the kind permission of American Institute of Physics are: Fig. 5, Gaines, J. M., Drenten, R. R., Haberen, K. W., Marshall, T., Mensz, P., Petruzzello, J. (1993), *Appl. Phys. Lett.* **62**, 2462–2464; Fig. 10, Buijs, M., Shahzad, K., Flamholtz, S., Haberen, K., Gaines, J. (1995), *Appl. Phys. Lett.* **67**, 1987. Fig. 6 reprinted with the kind permission of *Japanese Journal of Applied Physics*, Itoh, S., Nakayama. N., Matsumoto, S., Nagai, N., Nakano, K., Ozawa, M., Okuyama, H., Tomiya, S., Ohata, T., Ikeda, M., Ishibashi, A., Mori, Y. (1994), *Jpn. J. Appl. Phys.* **33**, L938. Reprinted with the kind permission of Institution of Electrical Engineers are: Fig. 9, Grillo, D. C., Han, J., Ringle, M., Hua, G., Gunshor, R. L., Kelkar, P., Kozlov, V., Jeon, H., Nurmikko, A. V. (1994), *Electron. Lett.* **30**, 2131; Fig. 11, Kawasumi, T., Nakayama, N., Ishibashi, A., Mori, Y. (1995), *Electron. Lett.* **31**, 1667; Fig. 13, Jeon, H., Kozlov, V., Kelkar, P., Nurmikko, A. V., Grillo, D. C., Han, J., Ringle, M., Gunshor, R. L. (1995), *Electron. Lett.* **31**, 106. Reprinted with the kind permission of The American Physical Society are: Figs. 16, 17, Kozlov, V., Kelkar, P., Nurmikko, A. V., Grillo, D. C., Han, J., Gunshor, R. L. (1996), *Phys. Rev.* **B53**, 10837; Figs. 18, 20, Ding, J., Hagerott, M., Kelkar, P., Nurmikko, A. V., Grillo, D. C., He, L., Han, J., Gunshor, R. L. (1994), *Phys. Rev.* **50**, 5787.

This book is printed on acid-free paper.

COPYRIGHT © 1997 BY ACADEMIC PRESS

ALL RIGHTS RESERVED.
NO PART OF THIS PUBLICATION MAY BE REPRODUCED OR TRANSMITTED IN ANY FORM OR BY ANY MEANS, ELECTRONIC OR MECHANICAL, INCLUDING PHOTOCOPY, RECORDING, OR ANY INFORMATION STORAGE AND RETRIEVAL SYSTEM, WITHOUT PERMISSION IN WRITING FROM THE PUBLISHER.

The appearance of the code at the bottom of the first page of a chapter in this book indicates the Publisher's consent that copies of the chapter may be made for personal or internal use, or for the personal or internal use of specific clients. This consent is given on the condition, however, that the copier pay the stated per copy fee through the Copyright Clearance Center, Inc. (222 Rosewood Drive, Danvers, Massachusetts 01923), for copying beyond that permitted by Sections 107 or 108 of the U.S. Copyright Law. This consent does not extend to other kinds of copying, such as copying for general distribution, for advertising or promotional purposes, for creating new collective works, or for resale. Copy fees for pre-1997 chapters are as shown on the chapter title pages; if no fee code appears on the chapter title page, the copy fee is the same as for current chapters. 0080-8784/97 $25.00

ACADEMIC PRESS
525 B Street, Suite 1900, San Diego, CA 92101-4495, USA
1300 Boylston Street, Chestnut Hill, Massachusetts 02167, USA
http://www.apnet.com

ACADEMIC PRESS LIMITED
24–28 Oval Road, London NW1 7DX, UK
http://www.hbuk.co.uk/ap/

International Standard Book Number: 0-12-752144-5

Printed in the United States of America
96 97 98 99 00 BB 9 8 7 6 5 4 3 2 1

Contents

LIST OF CONTRIBUTORS ix
FOREWORD xi

Chapter 1 MBE Growth and Electrical Properties of Wide Bandgap ZnSe-based II–VI Semiconductors
J. Han and R. L. Gunshor

I. Introduction	1
II. MBE Growth of ZnSe-based alloys	3
1. Review of MBE Concepts	3
2. Control of Growth Parameters	5
3. Growth of ZnMgSSe (Strains and Morphology)	13
4. Correlation with Device Performance	19
III. The ZnSe/GaAs Heterovalent Interface	22
1. Control of Interface States	23
2. Band Offset	25
3. Vertical Transport across the ZnSe/GaAs Interface	26
4. Nucleation of ZnSe and Its Alloys on GaAs	28
IV. Electrical Contact to p-ZnSe	30
1. Zn(Se,Te) Graded Contact	31
2. Other Techniques for Forming Contacts to p-ZnSe	37
V. p-Type Doping of ZnSe-based Alloys	38
1. Background (P, As, O, and Li Doping)	38
2. Nitrogen-doping of p-ZnSe	39
3. Comparative Study of Nitrogen Doping for ZnSe and ZnTe	45
4. Nitrogen-doped Wide Bandgap Alloys of ZnSe: AX Centers?	47
VI. Conclusion	53
References	54

Chapter 2 Growth and Characterization of ZnSe-based II–VI Semiconductors by MOVPE
Shizuo Fujita and Shigeo Fujita

I. Introduction	59
II. Brief Review of Pioneering Works	60
III. Recent Growth Techniques	61
1. New Precursors	61
2. Photo-Assisted Growth	65
IV. Doping and Devices	72
1. n-Type Doping	72
2. p-Type Doping	74
3. Present Status of Device Applications	77
References	79

Chapter 3 Gaseous Source UHV Epitaxy Technologies for Wide Bandgap II–VI Semiconductors
Easen Ho and Leslie A. Kolodziejski

I. Introduction	83
II. Metalorganic Molecular Beam Epitaxy of ZnSe	85
1. Advantages and Disadvantages of MOMBE	85
2. Experimental Setup	86
3. Growth Rate Limitations Due to Surface Blockage	88
4. Beam-assisted Growth	91
III. Gas Source Molecular Beam Epitaxy	97
1. Advantages and Disadvantages of GSMBE	97
2. Experimental Details	99
3. n-Type Doping	102
4. p-Type Doping	104
5. Hydrogen Passivation	108
IV. Concluding Remarks	114
References	116

Chapter 4 Doping of Wide-band-gap II–VI Compounds—Theory
Chris G. Van de Walle

I. Introduction	122
II. Mechanisms that Limit Doping	122
1. Self-compensation by Native Defects	122
2. Compensation by Other Configurations of the Impurity	124
3. Formation of Complexes	125
4. Solubility Limits	125
5. Compensation by Foreign Impurities	126
II. Formalism for Calculating Doping Levels	126
1. Formation Energies	126
2. Chemical Potentials	128

	3. General Expression for Formation Energy	130
	4. Thermodynamic Equilibrium	131
	5. Charge Neutrality — Self-consistent Solutions	132
	6. First-principles Calculations	133
IV.	Native Defects	135
	1. First-principles Investigations of Native Defects	135
	2. The Zn Interstitial	136
	3. Antisite Defects	137
	4. The Se Vacancy	137
	5. Self-activated Centers	137
	6. Broken-bond Defects	138
	7. Critical Examination of Native Defect Compensation as a Generic Compensation Mechanism	138
V.	p-Type Doping	140
	1. Lithium in ZnSe	140
	2. Sodium in ZnSe	147
	3. Phosphorus and Arsenic in ZnSe	147
	4. Nitrogen in ZnSe and ZnTe	148
	5. Oxygen in ZnSe	151
VI.	n-Type Doping	151
	1. Al and Ga in ZnSe and ZnTe	151
	2. Cl in ZnSe and ZnTe	153
VII.	Comparison between Theory and Experiment	153
	1. Compensation Due to Native Defects and Native-defect Complexes	153
	2. Discussion of Doping Saturation Effects	156
	3. Nucleation of Misfit Dislocations	157
	4. Comparison Between Different II–IV Materials	157
	5. Solubility-limiting Phases	158
	6. Effect of N Incorporation on the Lattice Constant	158
VIII.	Conclusions and Future Directions	159
	References	160

Chapter 5 Optical Properties of Excitons in ZnSe-based Quantum Well Heterostructures

Roberto Cingolani

I.	Introduction	163
II.	Modeling Excitonic States in II–VI Quantum Wells	164
III.	Linear Optical Properties of Quasi-two-dimensional Exciton	169
	1. Optical Absorption	169
	2. Excitons at the Dimensionality Cross-over	185
	3. Phototransport Processes	189
	4. Temporal Evolution of the Excitonic Transitions	195
IV.	Nonlinear Excitonic Properties	202
	1. Basic Theoretical Concepts	202
	2. Excitons and the One-component Electron Plasma	205
	3. Excitons and the Electron–Hole Plasma	211
V.	Role of Excitons in the Lasing of ZnSe-based Quantum Wells	215

VI. Conclusions 223
References 223

Chapter 6 II–VI Diode Lasers: A Current View of Device Performance and Issues

A. V. Nurmikko and A. Ishibashi

I. Introduction 227
II. Designs Considerations 228
   1. Electronic Confinement: Bandoffsets and Quantum Wells 228
   2. Electrical Contacts and Vertical Transport 236
III. Diode Laser Performance and Characteristics 239
   1. Evolution of Diode Laser Design 240
   2. Diode Laser Characteristics 242
   3. Diode Laser Degradation and Reliability 252
IV. Physics of Gain and Stimulated Emission in ZnSe-based Quantum Well Lasers 256
   1. Excitonic Molecules and Lasing in ZnSe Quantum Wells 258
   2. Gain Spectroscopy of Blue-Green Diode Lasers at Room Temperature . . . 262
V. Summary 267
References 268

Chapter 7 Defects and Degradation in Wide-gap II–VI-based Structures and Light-emitting Devices

Supratik Guha and John Petruzzello

I. Introduction 271
II. Structural Characteristics 272
   1. Single Epilayer Structures 272
   2. Laser Structures 288
III. Degradation Effects 298
   1. Experimental Techniques 299
   2. Degradation in Light-emitting Diodes and Lasers 301
References 317

INDEX 319
CONTENTS OF VOLUMES IN THIS SERIES 325

List of Contributors

Numbers in parenthesis indicate the pages on which the authors' contributions begin.

ROBERTO CINGOLANI (163), *Departimento de Scienza dei Materiali, University of Lecce, 1-73100 Leece, Italy*

SHIGEO FUJITA (59), *Department of Electronic Science and Engineering, Kyoto University, Kyoto 606-01 Japan*

SHIZUO FUJITA (59), *Department of Electronic Science and Engineering, Kyoto University, Kyoto 606-01 Japan*

SUPRATIK GUHA (271), *3M Corporate Research Labs, St. Paul, Minnesota 55144.* Present address: *IBM T. J. Watson Research Center, Yorktown Heights, New York, 10598*

R. L. GUNSHOR (1), *School of Electrical and Computer Engineering, Purdue University, West Lafayette, Indiana 47907*

J. HAN (1), *School of Electrical and Computer Engineering, Purdue University, West Lafayette, Indiana 47907*

EASEN HO (83), *Department of Electrical Engineering and Computer Science, Research Laboratory of Electronics, Massachusetts Institute of Technology, Cambridge, Massachusetts 01238*

A. ISHIBASHI (227), *Sony Corporation Research Center, Yokohama, 240 Japan*

LESLIE A. KOLODZIEJSKI (83), *Department of Electrical Engineering and Computer Science, Research Laboratory of Electronics, Massachusetts Institute of Technology, Cambridge, Massachusetts 02138*

A. V. NURMIKKO (227), *Brown University, Providence, Rhode Island 02912*

JOHN PETRUZZELLO (271), *Philips Laboratories, Philips Electronics North American Corporation, Briarcliff Manor, New York 10510*

CHRIS G. VAN DE WALLE (121), *Xerox Palo Alto Research Center, Palo Alto, California 94304*

Foreword

The challenging search for blue/green injection laser diodes and light emitting diodes (LEDs) has been ongoing for roughly 30 years. While the GaAs-based devices dominated the longer wavelength light emitters, the members of the II–VI family of compound semiconductors have long been attractive as candidates for the realization of blue/green laser diodes and LEDs. The primary reason for the interest in the II–VIs was their range of bandgap energies, encompassing the entire visible spectrum while at the same time providing direct bandgap transitions. The various difficulties encountered over the years are well documented in the past literature. The heart of the problem seemed associated with materials issues, primarily the tendency for defects to form under equilibrium growth conditions and the lack of doping control. The difficulty in achieving both n- and p-type conduction in a single wide bandgap II–VI host compound prevented the formation of a p-n junction, the structure at the heart of an electrical injection device.

The current "era" in the course of development of the wide bandgap II–VIs began in the early 1980s with the application of nonequilibrium growth techniques such as molecular beam epitaxy and metalo-organic vapor phase epitaxy. These growth techniques enabled not only greatly improved material quality in metastable structures, but also an opportunity to extend the quantum confinement revolution, begun in the GaAs family, to the II–VIs. In 1990 two groups working independently, one headed by Robert Park at the University of Florida, and the other at the Central Research Laboratories of Matsushita, demonstrated that an rf plasma source manufactured by Oxford Research in the UK could incorporate sufficient nitrogen into ZnSe to produce useful p-doping levels. Almost simultaneously a p-n junction LED was demonstrated, and the following year (summer of 1991) brought the reports of the first p-n junction laser diodes, structures which incorporated the heterostructure developments which had paralleled the search for suitable p-dopants. Since the first laser diodes and LEDs were demonstrated, efforts have been ongoing to improve the performance and lifetime by addressing a variety of problem areas including doping, contacts, epitaxial growth, and device processing. At the same time, there appeared exciting and controversial developments in the device physics, specifically concerning the role of electron hole correlation (excitons) in the laser gain mechanism.

The current progress towards the development of commercially viable devices, especially as regards the challenge of extending the device lifetime beyond the 101 hr room temperature cw device reported by Sony in early 1996, reveals echoes of the experience of the GaAs-based laser diodes. In the same timeframe the emergence of the group III nitride compounds, both in the form of commercial green and blue LEDs (1993) and the report of a room temperature pulsed laser diode at the end of 1995 (both developments by Nichia Chemical Corporation), has created an atmosphere of intense competition between the two materials systems to see which ultimately will dominate the commercial market.

In this volume of the Semiconductors and Semimetals series we have attempted to provide an overview of the current issues relating to the development of both blue/green laser diodes and LEDs based on the II–VI semiconductors, though with a specific focus on the somewhat more technically challenging of the two devices, the laser diode. The first three chapters primarily deal with the growth aspects of the development of blue/green light emitting devices. The chapter by Han and Gunshor outlines materials issues as growth is approached by the MBE technique, and includes discussions of doping, contacts, and techniques for crystalline defect reduction. The chapter by Fujita and Fujita describes the progress of the MOVPE growth techniques, and includes interesting experiments with photo-assisted growth, while presenting the aspects of doping particular to the MOVPE method of growth. The chapter by Ho and Kolodziejski presents those issues relating to the use of gas sources in the UHV context. Details of the growth of ZnSe by both MOMBE and GSMBE are discussed in terms of the chemistry of the surface reactions. The crucial aspects of the role of hydrogen passivation on p-doping are addressed in the context of the use of gas sources.

Wide bandgap semiconductors traditionally exhibit more of a challenge to control of electrical properties. The fourth chapter by Van de Walle provides insight into the physics behind the doping issues particularly associated with the wide bandgap II–VI compounds. The behavior of a variety of substitutional elements in host ZnSe is discussed in detail. In this chapter one can follow the evolution of the physical view into the origins of the doping limitations occurring in these compounds.

The II–VI compounds not only exhibit direct bandgap transitions over the visible spectrum, but also have strong electron hole correlation leading to high exciton binding energies. The chapter by Cingolani discusses the exciton-dominated optical properties of the ZnSe-based quantum well structures which form the active gain region of the light emitting devices.

The chapter by Nurmikko and Ishibashi focuses on the laser diode devices themselves with emphasis on fabrication issues, physical aspects such as optical gain that are peculiar to the ZnSe-based alloy devices, and a discussion of current and anticipated performance characteristics as they relate to the fundamental characteristics of the II–VI family of compounds. The device physics of both edge and surface emitting device configurations are presented.

In the final chapter by Guha and Petruzzello one can see the extent to which the ZnSe-based laser and LED device development echoes the experience with the (Al,Ga)As system in aspects of the role of defects in limiting the device lifetime. Transmission electron microscopy is applied to relate the microstructure of degraded devices to the presence of extended defects in as-grown samples. The origins of the extended defects and point defects are explored, as well as their influence on device degradation.

R. L. Gunshor
A. V. Nurmikko

CHAPTER 1

MBE Growth and Electrical Properties of Wide Bandgap ZnSe-based II–VI Semiconductors

J. Han and R. L. Gunshor

SCHOOL OF ELECTRICAL AND COMPUTER ENGINEERING
PURDUE UNIVERSITY
WEST LAFAYETTE, INDIANA

I. INTRODUCTION	1
II. MBE GROWTH OF ZnSe-BASED ALLOYS	3
1. *Review of MBE Concepts*	3
2. *Control of Growth Parameters*	5
3. *Growth of ZnMgSSe (Strains and Morphology)*	13
4. *Correlation with Device Performance*	19
III. THE ZnSe/GaAs HETEROVALENT INTERFACE	22
1. *Control of Interface States*	23
2. *Band Offset*	25
3. *Vertical Transport across the ZnSe/GaAs Interface*	26
4. *Nucleation of ZnSe and Its Alloys on GaAs*	28
IV. ELECTRICAL CONTACT TO p-ZnSe	30
1. *Zn(Se, Te) Graded Contact*	31
2. *Other Techniques for Forming Contacts to p-ZnSe*	37
V. P-TYPE DOPING OF ZnSe-BASED ALLOYS	38
1. *Background (P, As, O, and Li Doping)*	38
2. *Nitrogen-doping of p-ZnSe*	39
3. *Comparative Study of Nitrogen Doping for ZnSe and ZnTe*	45
4. *Nitrogen-doped Wide Bandgap Alloys of ZnSe: AX Centers?*	47
VI. CONCLUSION	53
REFERENCES	54

I. Introduction

The efforts to develop blue-green visible light-emitting devices based on the II–VI semiconductors commenced at the same time as the longer wavelength III–V devices. While solutions to the materials problems encountered in the development of III–V based optoelectronic devices were found, major barriers emerged early with the II–VI semiconductors. The problems with the II–VIs could be broadly defined in terms of material quality originating from the intrinsic physical make-up (primarily the strong ionicity) of wide bandgap semiconductors. While the past two decades have witnessed a fruitful development of commercial III–V infrared ($\lambda > 780$ nm), and recently red (650 nm), light-emitting diodes (LED) and laser diodes (LD), the potential applications of wide bandgap II–VI semiconductors for short wavelength emitters, which range from high-density optical information storage/processing to multicolor (red/green/blue) flat-panel displays, remained either unfulfilled or were pursued along other technological paths.

In the early 1980s a small number of research efforts were begun to apply the modern nonequilibrium epitaxial growth methods, such as molecular beam epitaxy (MBE), to the synthesis of II–VI compound semiconductors and their heterostructures (Gunshor et al., 1991). The initiation of the epitaxial growth efforts were motivated by several considerations: (a) to provide the II–VI compounds with the reduced-dimensional optoelectronic features associated with the fabrication of quantum-sized heterostructures in the III–V compounds, (b) in an attempt to circumvent the traditional doping problems (Mandel, 1964a; Mandel et al., 1964b; Kroger, 1965; Aven, 1967; Watts et al., 1971) and defect generation/compensation which had plagued the more conventional equilibrium growth techniques, such as those employed for the growth of bulk crystals, and (c) to fabricate II–VI device structures in the planar configurations which are compatible with modern semiconductor device technology. By the end of the decade, substantial progress had been made in research on II–VI multilayer structures such as quantum wells and superlattices, including demonstrations of lasing action (Bylsma et al., 1985; Cammack et al., 1987) and proposals for potential heterostructures for efficient light-emitting applications (Ding et al., 1990); at the same time major advances occurred in the control of material purity and defects.

As a direct consequence of the improved material quality provided by MBE growth, advances were made in the conductivity control (doping) of ZnSe; it is primarily this compound and its alloys which form the basis for today's II–VI heterostructure research (Fig. 1). The attaining of p-type doping using nitrogen in particular, first demonstrated in 1990 (Park et al., 1990; Ohkawa et al., 1991), when combined with the knowledge accumulated from research into the physical properties of II–VI heterostructures and the materials advances associated with MBE process, provided an opportunity for attempting the realization of a blue-green diode laser. During the summer of 1991 a group of 3M Corporate Research Laboratories (Haase et al., 1991) followed by the Brown and Purdue Universities collaboration (Jeon et al., 1991) announced the demonstration of such diode lasers, first under pulsed excitation at cryogenic temperatures, thus ending a three-decade quest. Following rapid progress, the first demonstra-

1 MBE GROWTH AND ELECTRICAL PROPERTIES OF SEMICONDUCTORS 3

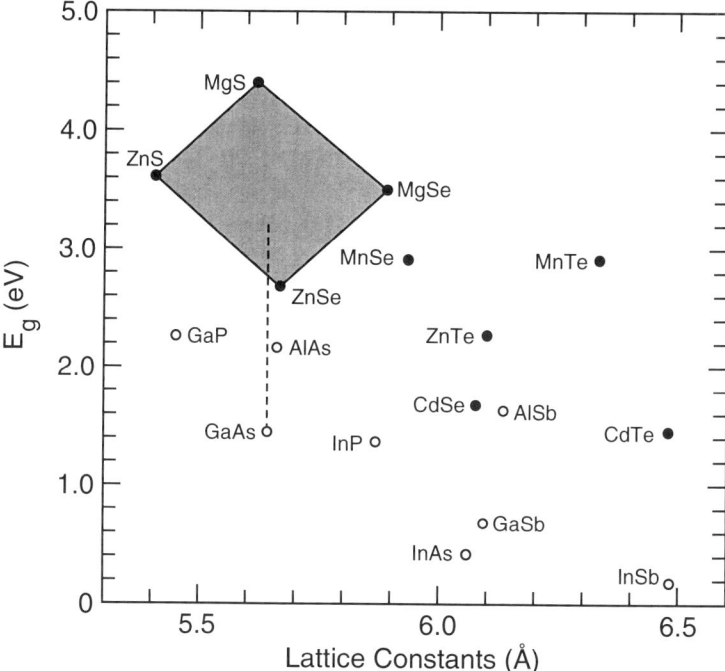

FIG. 1. Bandgap (eV) vs. lattice parameter (Å) for cubic II–VI wide bandgap semiconductors. The connection by the solid lines indicates some of the strained heterostructures studied. The shaded quadrangle (which covers the green, blue, and UV spectrum) is particularly relevant to present laser device structures. Selected choices for III–V buffer layers and substrates are indicated.

tions of room temperature continuous-wave (cw) operation of the diode laser were made (Nakayama *et al.*, 1993; Salokatve *et al.*, 1993) in the summer of 1993.

The chapter will provide a sense of the role played by MBE in the development of the blue/green light emitters, as well as insight into the problems, including both the sample growth and characterization, associated with the MBE of the wide bandgap II–VI compound semiconductors.

II. MBE Growth of ZnSe-based Alloys

1. REVIEW OF MBE CONCEPTS

Crystal growth using molecular beam epitaxy (MBE) is fundamentally different from bulk crystal growth in that the atom elements on the growth surface do not necessarily incorporate in a manner corresponding to the lowest energy configurations available. The growth by MBE represents a nonequilibrium process where the surface kinetics, instead of the thermodynamic equilibrium of the elements on the growing surface of the crystal, determines the configuration adopted by the condensing molecules/atoms. Within the ultra-high vacuum chamber, the constituent elements, normally generated thermally from heated crucibles in the case of solid source MBE, are made to impinge upon a heated substrate. The atoms/molecules tend to stick to the growth surface by a combination of physisorption (such as associated with van der Walls forces between the atoms and the surface) and/or chemisorption where chemical bonding to the surface dangling bonds is formed under the influence of several factors. The process depends on the finite surface dwelling time (physisorption), surface mobility, and a spatial configuration of the available bonds from an underlying crystal structure such that the incorporated take on a local energy minimum configuration. The desorbed species from the heated substrate surface enter the (UHV) vapor background of the growth chamber until condensed out on the walls of the chamber which are usually cooled to cryogenic temperatures. A characteristic of successful MBE is a low background pressure, of the order of 10^{-10} to 10^{-11} torr, a feature required to maximize incorporation of only the desired species directly into the growing crystal.

An interesting demonstration of such a nonequilibrium mode of crystal growth was the synthesis of zincblende MnTe (Durbin et al., 1989; Han et al., 1991). Bulk MnTe crystallizes in the NiAs structure which has a hexagonal symmetry; the optical band gap of the hexagonal MnTe lies in the infrared range (~ 1.3 eV). Yet when appropriate zincblende (cubic) substrates (CdTe, for example) are employed during MBE growth, the bonding and stacking of Mn and Te atoms are arranged so as to preserve the spatial configuration of the underlying zincblende substrates. Zincblende MnTe prepared in this manner has distinctly different material properties, including a band gap energy in UV range (3.2 eV).

Being a growth process far from equilibrium, crystal growth by MBE often takes place at a temperature well below the melting point (the temperature where equilibrium between liquid phase and solid phase can be maintained) of the compound. For example, ZnSe, which has a melting point of 1250°C, can easily be prepared by MBE at temperatures as low as 200°C, although the minimum point defect densities are likely obtained at growth temperatures closer to or somewhat above 300°C. The upper range of the growth temperature, approaching 450°C for ZnSe, is primarily limited by the surface desorption of the relatively high vapor pressure constituents of the II–VI compounds.

It was the potential metastable nature of the structure which suggested that the MBE process could act to avoid the defect formation events predicted to occur under equilibrium growth conditions (Aven, 1967), and hence many of the difficulties associated with the early bulk growth of II–VI compounds could conceivably be circumvented.

In many respects the MBE of the wide bandgap II–VI compounds is similar to the growth of the III–V materials, however there tend to be some significant differences. One determining factor is the use of columns II(A&B) and VIB elemental sources having a relatively high vapor pressure which affects various growth parameters in a profound way. The source ovens tend to have lower operating temperatures than for typical III–V growth; the difficulty then arises in connection with maintaining and controlling a constant source temperature in the presence of a reduced radiation cooling. There can also be a problem with material condensing near the the output orifice of the cell. (To some extent the source temperature control problem can be alleviated by the use of compound sources, for example ZnSe.) The high vapor pressures of the constituent elements also restrict the growth temperature and therefore the mobility of surface atoms. An additional difference when compared to the MBE of, for example (In, Ga, Al)As is the fact that essentially none of the elements of II–VI alloy has unity sticking coefficient as do the group III elements. It is thus somewhat more demanding (as discussed below) to control the alloy fraction of a II–VI compound during MBE growth.

2. Control of Growth Parameters

a. Sticking Coefficients

In order to illustrate some of the MBE growth issues, we have chosen specific examples from the experience in our laboratory. The films of these investigations were grown in a multigrowth chamber facility (Perkin Elmer 430 MBE). A particularly demanding yet important alloy system is the wide bandgap ZnMgSSe quaternary first reported by Okuyama et al. (1991), and since incorporated in most II–VI blue-green laser device configurations. By adding both S and Mg to ZnSe one gains access to an alloy for which both the bandgap and lattice constant can be varied over a wide range (shaded area in Figure 1). Specifically, the capability of preserving lattice matching to the commercially available GaAs substrates, while having bandgap tunability to blue and UV, enables the growth of a broad range of heterostructures. In our laboratory the ZnMgSSe epilayers were grown on GaAs homoepilayers which had been grown in a separate III–V growth chamber connected to the II–VI chamber by an ultra-high vacuum transfer tube. The ZnMgSSe layers were grown using Zn(6N), Se(6N), ZnS(6N), and Mg in solid form. Before loading Mg source into the MBE chamber, the commercial-grade (4N purity) Mg was further refined by a two-step vacuum distillation process (Miotkowski and Ramdas, 1991) followed by extensive outgassing in a separate vacuum chamber.

Due to the less-than-unity sticking coefficients of II–VI elements, one has to know both the intensity of fluxes coming from the ovens and the amount that actually is incorporated. In order to measure the flux from each source, the chamber is equipped with a quartz crystal microbalance which can be placed at approximately the growth position; the deposition of molecules onto the water-cooled crystal changes its oscillating frequency. The supplied fluxes can be deduced from the change rate of the frequency. The actual rate of cation or anion incorporation (J_{inc}) can be calculated by the following equation provided that the epilayer is reasonably stoichiometric and the crystal takes a zincblende structure

$$J_{inc} = 4R/(a^3) \qquad (1)$$

where R is the growth rate and a is the lattice constant. The growth rate can be derived through the measurement of sample thickness by transmission electron microscopy (TEM), scanning electron microscopy (SEM), selective etching/step profiling, or by *in situ* monitoring of the pyrometer oscillations (Ringle *et al.*, 1994b). The sticking coefficient is defined as the fraction of the flux that is incorporated into the film. In the case of sulfur (K_s) in the quaternary ZnMgSSe

$$K_s = (J_{inc}/J_s)\,x \qquad (2)$$

J_s is the ZnS (in the case of a compound source) flux measured by the quartz crystal microbalance. The sulfur fraction x can be determined by either an electron microprobe (Han *et al.*, 1994a) or a combination of the knowledge of bandgap (from photoluminescence) and lattice constant (from x-ray diffraction rocking curves) (Okuyama *et al.*, 1991). It is expected that the sticking coefficients vary as a fraction of substrate temperatures and cation to anion flux ratios. A compilation of sulfur sticking coefficients, determined over many independent film growths, as a function of substrate temperatures is shown in Figure 2; the line is a guide to the eye. (It is worth noting that the surface stoichiometry during growth was maintained to approximately the same condition, namely slightly on the anion rich side of the transition between cation and anion stabilized growth as determined by reflection high-energy electron diffraction, RHEED.) In spite of the scattered data points, one can nevertheless see a clear trend showing that the sulfur incorporation decreases as the substrate temperature increases. As the growth temperature drops from 300 to 260°C, the sulfur sticking coefficient increases by a factor of 2 from 0.05 to 0.1. From 260 to 230°C, the sticking coefficient increases by another factor of 2 from 0.1 to 0.2. From this data, one can conclude that a 5°C fluctuation in substrate temperature during

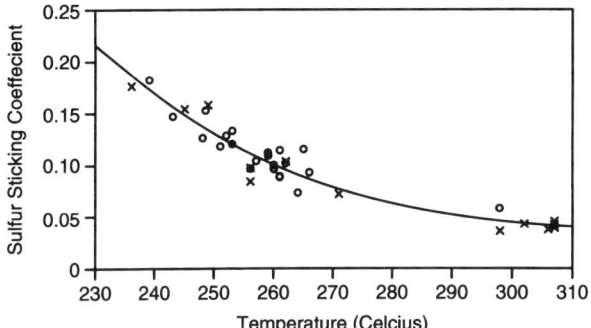

FIG. 2. Plot of measured sulfur sticking coefficients as a function of substrate temperature. (O) are ZnMgSSe growths, (X) are ZnSSe growths. The solid line represents a guide to the eye (from Ringle et al., 1994a).

growth will result in about a 10% change in both the sulfur sticking coefficient and the sulfur fraction (from $S = 7\%$ to $S = 7.7\%$, for instance). This particular study indeed highlights the fact that, in addition to a precise determination of fluxes from individual ovens, one has to monitor/control the substrate temperature both from run to run for reproducibility and during each growth to minimize the drifting of alloy fractions.

Before discussing the issue of temperature characterization, it is appropriate to comment on the rather scattered data points in Figure 2. The uncertainty in substrate temperatures (refer to the next section for details) can obviously shift points horizontally and change the presentation. There is also the fact that the quartz crystal microbalance is not exactly in the same location as the substrate growth position. Because the flux patterns from source ovens change with time as the charges are depleted and replenished, the measured flux will not always be the same as the flux at the exact growth position.

b. *Substrate Temperature*

The MBE surface kinetics (surface diffusion, desorption, etc.) are a function of temperature. It is well known that the standard provision of a thermocouple (in either contact or noncontact mode) at the back side of the sample holder gives a convenient yet frequently misleading calibration of growth temperature. On the Perkin-Elmer 430 system, the thermocouple is simply pressed against the back of the substrate holder. Such mechanical contact does not guarantee a consistent thermal contact, especially when the substrate is rotating, a necessary provision to ensure spatially uniform quaternary film growth. Another serious problem comes from direct heating of the unshielded thermocouple by the heating element. As the heating element radiates power to the substrate holder, the thermocouple is also directly heated, resulting in false readings. In the case of MBE growth of III–V AlGaAs compound semiconductors, the incorporation of column IIIB elements (Al, Ga, and In) is relatively insensitive to temperature variations; the temperature "window" for optimum growth tends to be quite wide and tolerant. In contrast, the generally high vapor pressures associated with columns II and VIB elements place stringent requirements on substrate temperature calibration. One has to ensure temperature reproducibility from run to run, as well as minimum fluctuations during each film growth. The growth of ZnMgSSe is especially vulnerable to, and revealing of, the occurrence of temperature drifting during a growth event. The alloy fraction, hence lattice parameter, is quite sensitive to the sulfur fraction/incorporation which in turn depends strongly on substrate temperature. Figure 3 shows a (004) four crystal x-ray diffraction rocking curve of a 2-μm-thick ZnMgSSe (Mg $\sim$ 8%, S $\sim$ 12%) epilayer. The appearance of multiple narrow peaks, rather than two peaks from the substrate and epilayer, respectively, suggests that drifting of lattice parameter occurs in spite of the pseudomorphic condition (Halliwell *et al.*, 1984). Further analysis of this diffraction curve will be provided after the following discussion of growth temperature characterization using optical pyrometry.

Optical pyrometry, in brevity the determination of temperatures based on the detection of black-body radiation from substrate/sample holder, can be used to directly probe the surface temperature provided that the radiation actually comes from the surface and the non-unity surface emissivity is taken into consideration. However, there is a complication associated with using an infrared optical pyrometer in the II–VI/GaAs heteroepitaxy. The low temperature (200 to 400°C) pyrometer has a detection window, in our case at around 2 μm where both the II–VI layers and GaAs substrates are transparent to the infrared radiation from the GaAs/sample holder. The

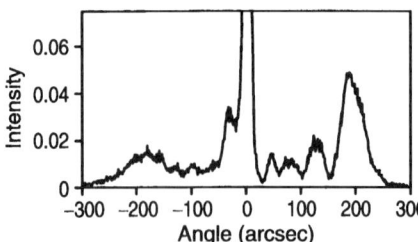

FIG. 3. (004) four crystal x-ray diffraction rocking curve of a 2-μm thickness ZnMgSSe (Mg $\sim$ 8%, S $\sim$ 12%) epilayer.

infrared radiation undergoes multiple reflections at both the II–VI epilayer surface and the II–VI/GaAs interface due to the large difference in optical refractive indices. The result is a time-modulated pyrometer reading originating in an interference effect as the epilayer thickness increases, analogous to the classical Fabry-Perot experiment with a varying cavity length.

The problem of analyzing the thin film interference from a single layer sandwiched between two layers of material of differing index of refraction can be solved in a manner almost identical to that of the Fabry-Perot interferometer problem (Born and Wolf, 1965) with the modification that the mirror separation is increasing with time. The tranmission coefficient (t) of optical power as a function of epilayer thickness (d) is expressed by the formula:

$$t(d) = (1 - r_1^2)(1 - r_2^2)/[1 + r_1^2 r_2^2 + 2r_1 r_2 \cos(4\pi n_2 d/\lambda)] \tag{3}$$

where

$$r_1 = (n_1 - n_2)/(n_1 + n_2), \tag{4}$$

and

$$r_2 = (n_2 - 1)/(n_2 + 1) \tag{5}$$

are the reflection coefficient at the II–VI/GaAs and the vacuum/II–VI interfaces, respectively. The refractive index of the (GaAs) substrate is n_1 and n_2 is the refractive index of the (II–VI) epilayer. λ is the radiation wavelength. The optical pyrometer converts a measured optical power (I) into a temperature reading $T(°K)$ based on the following relation (Gasiorowicz, 1974):

$$I \alpha \exp(-C/\lambda T) \tag{6}$$

where C = (Planck's constant) × (velocity of light)/(Boltzmann's constant) with a numerical value of 14,388 $\mu m°K$. The proportionality is eliminated by taking the ratio of optical power evaluated at two different thicknesses (where the real temperature is the same but the measured optical powers are different),

$$I_1/I_2 = t_1/t_2 = \exp[-C/\lambda(1/T_1 - 1/T_2)] \tag{7}$$

The ratio of I_1/I_2 is the ratio of the optical transmissions expressed in Eq (3) which depends only on optical indices. The cosine term in Eq (3) provides the oscillatory nature as the film thickness d (given by the product of growth rate and time) increases. The formula can predict the pyrometer readings for a constant growth temperature by fixing T_1, which in this case is the pyrometer reading at nucleation, and solving for T_2. Oscillations with rather large amplitude ($\Delta T \sim 15°$ C) have been observed in all films that we have grown: ZnSe, ZnSSe, ZnMgSSe, ZnCdSe, and ZnTe. The reason for this large oscillation amplitude is the large differences in refractive index between GaAs, ZnMgSSe, and vacuum which creates large reflections (refer to Eq (3)) at both the epilayer/substrate interface (r_1) and the epilayer surface (r_2). (The refractive index for GaAs is about 3.3 [Blakemore, 1982] in the 2.0 to 2.6 μm wavelength range used by the pyrometer, where ZnSe and related materials have a refractive index of about 2.3.) Other groups have used this effect to measure growth rate once the index of refraction is known (SpringThorpe and Majeed, 1990; Wright et al., 1990). We have had to *a priori* quantitatively predict these oscillations in order to accurately control the substrate temperature during a given film growth because of the large amplitude of the pyrometer oscillations. It is worth mentioning that the aforementioned model is not complete, in that the experimental pyrometer readings show damped oscillations. The observed damping is most likely caused by the dispersion due to the finite pass band (instead of a single wavelength) associated with the input selection of the pyrometer. After integrating the optical transmission over the finite bandwidth of wavelengths and then calculating the temperature reading, the theoretical result does indeed exhibit damping due to the phase difference of each wavelength resulting in a cancellation of the oscillation amplitude as the film grows thicker. The actual rate of damping depends on the spectral response of the pyrometer. Experimental observations of the damping of pyrometer oscillations were used to estimate the spectral response.

To understand the origin of the multiple peaks in the x-ray rocking curve associated with a nominally single epilayer, a computer simulation based on dynamical x-ray diffraction theory (analogous to the inverse Fourier transform) was employed. In principle, a variety of alloy fraction profiles can produce a specific x-ray rocking curve due to the absence of phase information in x-ray rocking curve measurements (which supply only the intensity of the reflected x-rays) (Segmuller et al., 1989). To remove this ambiguity, an iterative etching/x-ray diffraction measurement/simulation procedure was used to reconstruct the lattice constant variation responsible for the multiple peak x-ray diffraction rocking curve measurement shown in Figure 3. The simulated x-ray rocking curve measurement with a particular lattice constant profile (to be discussed later) is shown in Figure 4.

For the growth of quaternary ZnMgSSe epilayers, the change of lattice constant (alloy fraction) can be due to the compositional fluctuation of (at least) one of the four elements. As discussed in Section II.a, the use of a

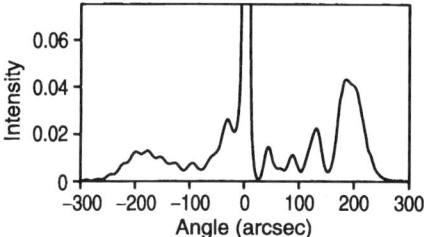

FIG. 4. Computer simulation (based on dynamical theory) of a ZnMgSSe epilayer described in Figure 3.

quartz crystal monitor to measure the fluxes before each film growth (along with alloy fraction and growth rate information obtained after growth) allows a calculation of the sticking coefficients of the constituent elements. The sticking coefficient data has indicated relatively constant sticking coefficients for Mg, Zn, and Se, but, as discussed above, the S sticking coefficient is found to be sensitive to the substrate temperature (Fig. 2). Thus if an alloy variation exists, one may assume that the variation would most likely occur in the sulfur composition. Figure 5a shows a proposed sulfur

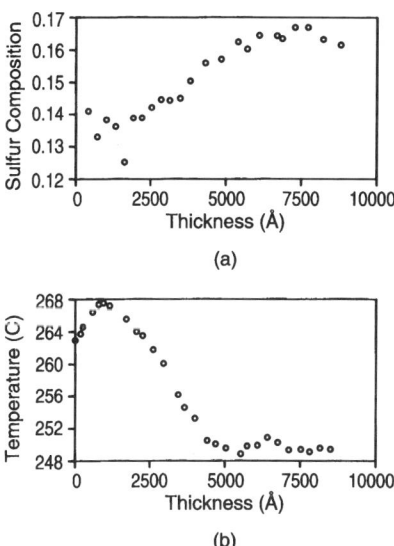

FIG. 5. (a) Proposed sulfur composition profile which could produce the simulated x-ray rocking curve shown in Figure 4. (b) Substrate temperature profile derived from the pyrometer data taken during the growth (ZMSS260-8); correction of pyrometer readings due to the interference effect is included.

composition profile as a function of film thickness (assuming a constant Mg fraction) which could produce the lattice constant variations consistent with the x-ray simulation for the ZnMgSSe film discussed (Fig. 4).

During the growth of this particular epilayer the pyrometric interferometery technique for temperature control discussed above (Eq (3)–(7)) was not utilized, although pyrometer data was recorded. However, by subsequently employing the pyrometric interferometry technique to interpret the pyrometer readings recorded during the film growth, the substrate temperature could be reconstructed and is plotted in Figure 5b. The visual correlation between Figure 5a and 5b suggests that the varying sulfur content was likely due to variations in substrate temperature. By converting the sulfur composition profile shown in Figure 5a to a sulfur sticking coefficient profile, an independent calculation of the sulfur sticking coefficient variation with temperature was made. This temperature variation of the sulfur sticking coefficient shows a trend similar to that seen from sulfur sticking coefficients calculated from many film growths (Fig. 2). The sulfur sticking coefficient determined from x-ray simulation data is 0.11 at 265°C and 0.145 at 250°C, which are in reasonable agreement with the data presented in Figure 2. The temperature-dependent variation in the sulfur concentration emphasizes the need to accurately control the substrate temperature during the growth of a sulfur containing epilayer.

We have also applied the pyrometer oscillation model to a film growth to confirm that improved substrate temperature control can (in the absence of drifting flux) result in constant epilayer composition. Figure 6 is a plot of the four crystal x-ray diffraction rocking curve of a ZnMgSSe (Mg $\sim$ 8%, S $\sim$ 12%) film grown with an improved temperature control; the power input to the growth manipulator was adjusted such that the apparent pyrometer readings follow the theoretical oscillation curve with a deviation of less than 2° throughout the film growth. The x-ray diffraction full width at half maximum (FWHM) is 24 arcseconds and there is essentially only one peak, consistent with a uniform composition profile.

It is possible to anticipate the two main processes within the growth chamber which change the amount of substrate heater power needed to maintain a constant temperature. The substrate temperature change due to changes in background thermal radiation field can be a major factor when using high-temperature effusion cells such as Ga, Al, and, in the case of II–VI growth, a compound ZnS source. For example, in the Perkin Elmer MBE, when the film is not being grown, the substrate holder is retracted away from the flux path of the ovens; the background thermal radiation seen by the wafer is essentially suppressed since it faces the liquid nitrogen cryoshrouds. When the substrate holder is brought back into the flux paths at nucleation, high temperature sources (such as ZnS at around 900°C)

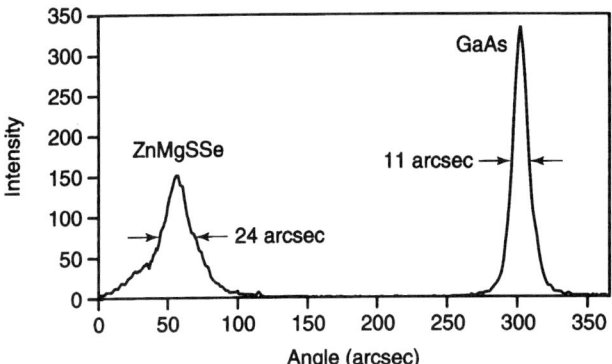

FIG. 6. Plot of the four crystal x-ray diffraction rocking curve of a ZnMgSe film grown with an improved temperature control scheme.

present a radiation background which can cause approximately a 10°C rise in substrate temperature for II–VI growth (growth temperature is normally 250 to 300°C), over a 5- to 10-min period. Therefore, adjustment has to be made to reduce the substrate heater power upon raising the sample block into the position for nucleation of the film. The second source comes from the change of surface emissivity during heteroepitaxy from either a bare molybdenum surface (using a bonded wafer holder) or a GaAs surface (In-free mounting scheme) to a II–VI surface. Additional power is needed to maintain a constant temperature as the emissivity changes during the deposition of the first 1 μm of epilayer.

Variation in substrate temperature is not the only cause of sulfur drifting. As mentioned earlier in reference to the flux measurements, the flux sources are not perfectly stable. There will be some drift in the flux from individual oven with time that will affect the composition of the films. However, this effect can be less severe than the observed drifting due to substrate temperature fluctuations.

3. Growth of ZnMgSSe (Strains and Morphology)

Unlike the AlGaAs–GaAs and the InP–AlInAs–GaInAs material systems where the flexibility in bandgap can be easily attained while preserving lattice coherence, (traditional) II–VI alloys such as ZnSSe, ZnCdSe, and ZnSeTe pose severe limitations on heteroepitaxial applications due to strong variations in lattice parameters as a function of alloy fraction. A vivid manifestation of this limitation is revealed in the first blue-green laser diodes reported in 1991 (Haase et al., 1991; Jeon et al., 1991) made of a ZnSSe–ZnSe–ZnCdSe separate confinement heterostructure (SCH) where only the thick ZnSSe (S ~ 6%) cladding can be lattice-matched to GaAs. These devices were not pseudomorphic, and arrays of misfit dislocations occurred at the interface between the ZnSSe cladding and the ZnSe waveguiding layers. To avoid the dislocations generated by the lattice mismatch inherent in these earlier structures, a multi-quantum well (MQW) diode scheme was designed and fabricated employing six (Zn, Cd)Se MQWs embedded in Zn(S, Se) (Xie et al., 1992). In this case, however, the advantage of pseudomorphic structural integrity was offset by the reduced electrical and optical confinement. The dilemma was resolved when Okuyama et al. (1991) reported the MBE growth of a Zn(Mg)SSe quaternary compound and subsequently (Okuyama et al., 1992) the demonstration of a six-well MQW laser similar to that described above, but replacing the ternary Zn(S, Se) with a wider bandgap quaternary (Zn, Mg)(S, Se) as barrier layers and using ZnSe as QWs to shift the emission wavelength to 447 nm (77 K). The addition of column IIA Mg, which possesses the properties of stronger chemical bonding (larger bandgap) and larger atomic radii for tetrahedral bonding (larger lattice parameter) (Kittel, 1976) than those of column IIB Zn, made possible a two-dimensional extension of the II–VI arena on the bandgap–lattice constant plot (Fig. 1). As a result, the (Zn, Mg)(S, Se) quaternary has become a currently important material, providing a capability both for lattice-matching to GaAs as well as increased bandgap energy. Following the first report of the six QW laser diodes, several groups described the incorporation of the (Zn, Mg)(S, Se) quaternary into pseudomorphic SCH diode laser structures (Gaines et al., 1993; Grillo et al., 1993; Itoh et al., 1993; Hasse et al., 1993).

Several important points relating to the MBE growth of the quaternary (Zn, Mg)(S, Se) of high crystalline quality still remain to be addressed. These issues include the suppression of stacking faults (Guha et al., 1993; Hua et al., 1994; Kuo et al., 1994; Han et al., 1995; Chu et al., 1996), the improvement of surface morphology, and the controllability of lattice parameters (fractions) sensitive to both sulfur and magnesium fractions, in order to maintain lattice coherence.

It was found in our earliest studies that the quaternary system tends to generate stacking faults. These are extended planar structural defects due to a deviation of the atomic stacking sequence along the [111] direction from the zincblende (ABCABCA...) to the wurtzite (ABABABA..) sequence. The generation of the stacking faults at the interface between the quaternary epilayer and a GaAs substrate appeared to occur to a greater degree than experienced when growing binary ZnSe. It has been shown that the presence of such stacking faults, and the associated partial dislocations, is the source of the currently observed degradation during laser operation.

The equilibrium crystalline structures for MgSe and MgS are both hexagonal, whereas ZnSe is cubic. The implication is that the stacking fault energy (i.e., the difference of formation energy between the cubic and hexagonal crystalline phases (Takeuchi et al., 1984; Yeh et al., 1992)) would tend to decrease as the S and Mg fractions increase from the binary ZnSe, eventually turning the MBE growth of zincblende ZnMgSSe into a metastable process. The tendency for the generation of stacking faults is related to the lattice-matching (strain) condition, surface treatment and cleaning of GaAs substrates, as well as the nucleation process employed during the initial II–VI growth on GaAs surface (Gaines et al., 1993). In order to avoid the complications from substrate surface preparation, homoepitaxial GaAs buffer layers are employed (Gunshor et al., 1987). In fact a dramatic reduction of stacking fault density was observed when nucleating the quaternary films on GaAs epilayers. Further discussion is given in Section III.4.

Figure 7 summarizes the x-ray measurement results on quaternary epilayers of different thicknesses ($1 \sim 2.8\,\mu$m). Significantly, it can be seen that the quaternary epilayer remain essentially pseudomorphic to GaAs under a wide range of strain from -0.225% tension to 0.137% compression at room temperature. This is evident by the observation that the parallel lattice parameters, as determined from the (400) and (511) reflections of x-ray diffraction rocking curves of the epilayers in the cited range, are almost identical with that of GaAs (Fig. 7a), and supported by the absence of misfit dislocations in TEM images. The asymmetry in the pseudomorphic ranges in compressive and tensile directions is partially the result of the difference in the thermal expansion coefficients between GaAs and the quaternary epilayer; this will be discussed below. When the strain increases in the direction of compression beyond some point, the parallel lattice parameter was observed to increased dramatically, indicating the epilayer starting to relax. This observation is consistent with the increase in the values of FWHM measured from (400) x-ray rocking curves shown in Figure 7b. In the strain range where the epilayers remains pseudomorphic, the FWHM is relatively narrow, ranging from 10 (multiply peaked interference patterns) to 44 arcsec. The FWHM increases as the film starts to relax, accompanied by the generation of misfit dislocations. In some epilayer samples, as well as laser structures, the x-ray diffraction peak splits up into several peaks, a phenomenon which simulations based on dynamical theory indicate is likely due to interference effects between multilayers and/or even modest composition drifting along the growth direction (refer to Section II.2 for details).

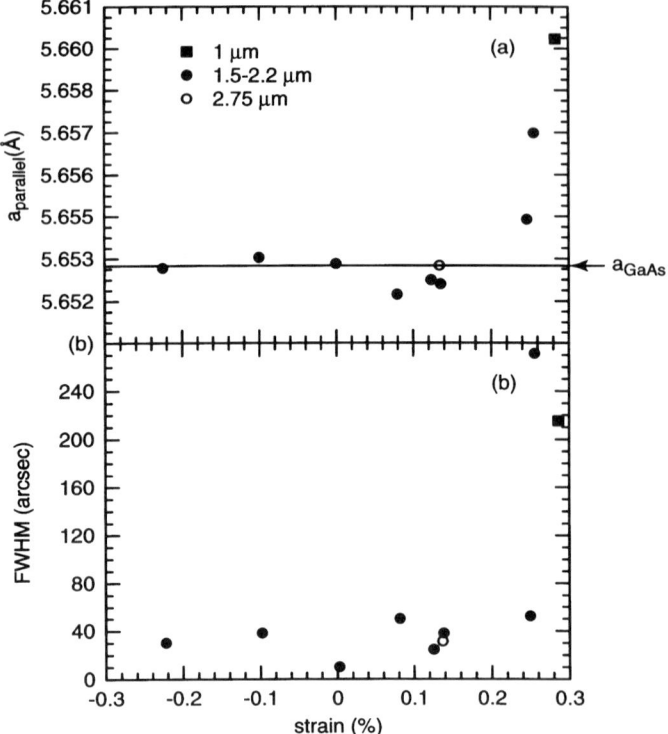

FIG. 7. (a) Parallel lattice parameters and (b) values of FWHM of diffraction peaks measured from (400) x-ray rocking curves vs. strain of the quaternary epilayers at room temperature. The solid line in (a) shows the lattice parameter of GaAs.

(The interference can, in a coherent structure, result in the very narrow FWHM sometimes observed.) For the calculations of lattice parameters and FWHM values shown in Figure 7, the most intense diffraction peak was selected from those samples exhibiting multiple peaks.

It is interesting to compare the results on the (Zn, Mg)(S, Se) quaternary shown in Figure 7a to that of a ZnSe binary epilayer grown on a GaAs substrate. The lattice mismatch between ZnSe and GaAs is 0.25%, and the pseudomorphic thickness range (critical thickness) for which a ZnSe epilayer can accommodate the compressive strain was reported to be approximately 0.15 μm (Yao et al., 1987; Petruzzello et al., 1988). Figure 7a shows that under a similar amount of compressive strain (0.248%) in the case of a quaternary epilayer of 1.65 μm thickness, the parallel lattice constant (5.6546 Å) of the quaternary begins to depart from that of GaAs (5.6535 Å), suggesting that the strain just starts to relax in this sample. A TEM plan-view study confirmed the result; misfit dislocations were observed at the quaternay/GaAs interface with an estimated average spacing between dislocations of $\sim 7000\,\text{Å}$. This value is nearly an order of magnitude larger as compared to the reported value of $830\,\text{Å}$ for a binary ZnSe epilayer grown on GaAs substrate having a similar thickness of $1.4\,\mu\text{m}$ (Petruzzello et al., 1988). The reason for the large discrepancy in the capability for accommodating strain between the quaternary (Zn, Mg)(S, Se) and binary ZnSe is not clear. However, one might speculate that the situation could be similar to the case of the (In, Ga)As ternary for which the addition of a small amount of In into GaAs is known to be effective in reducing dislocations. This phenomenon has been explained by various models involving the local strain effects induced by the bond length difference between In–As and Ga–As. Similarly, the strain introduced by the incorporation of Mg and/or S atoms having different atomic sizes, as compared to those of the host Zn and Se atoms, might cause the movement and multiplication of dislocations to be more difficult, thus enhancing the capability for accommodating strain.

Figure 8 shows examples of typical surface morphologies of quaternary films. A diffused cross-hatched pattern was observed when the film started to relax under a compressive strain, as shown in Figure 8a, where the film was partially relaxed and under a compressive strain of 0.257%. On the other hand, when the film was under a tensile strain, cracking was observed

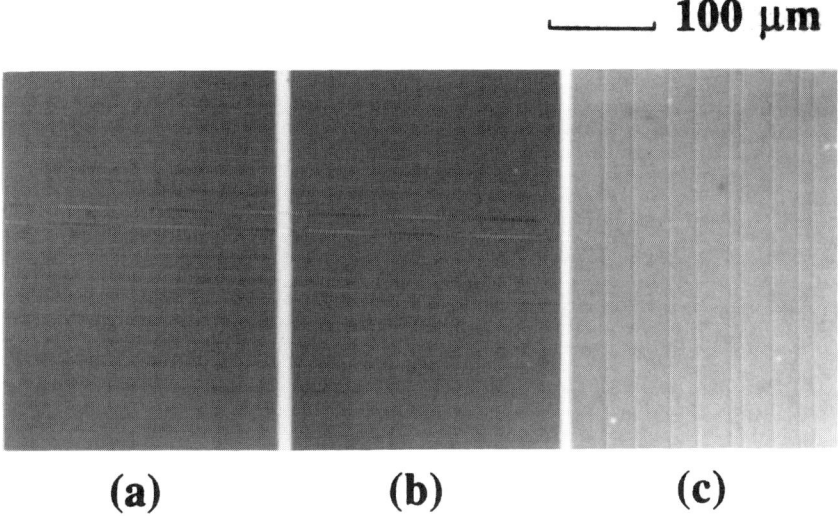

Fig. 8. Examples of typical surface morphology of ZnMgSSe quaternary epilayers; (a) diffused cross hatching, (b) smooth surface, and (c) cracks.

as shown in Figure 8c, where the film was under a strain of -0.225% and still essentially pseudomorphic as evident by x-ray measurements and TEM observation. (The identification of the surface features as cracks was confirmed by a cross-sectional observation using scanning electron microscopy.) The epilayers examined within a strain range from -0.10% (tension) to 0.25% (compression) exhibited few surface features; very diffused lines were visible in certain areas of some sample surfaces, possibly as a result of stress- (cleaving and indium mounting) induced slip lines. Paired oval-like defects were frequently observed on quaternary samples under compressive strain which have been correlated with the Schockly-type paired stacking faults (see next section for details). An example of a featureless surface (for the area shown) is shown in Figure 8b where the film was exactly lattice-matched to the GaAs epilayer.

It should be noted that all of the above mentioned strain values are those obtained at room temperature. Because of the lack of the thermal expansion coefficient data for the $(Zn, Mg)(S, Se)$ quaternary, the strain value at the growth temperature is unknown. However, it might be a reasonable ap-

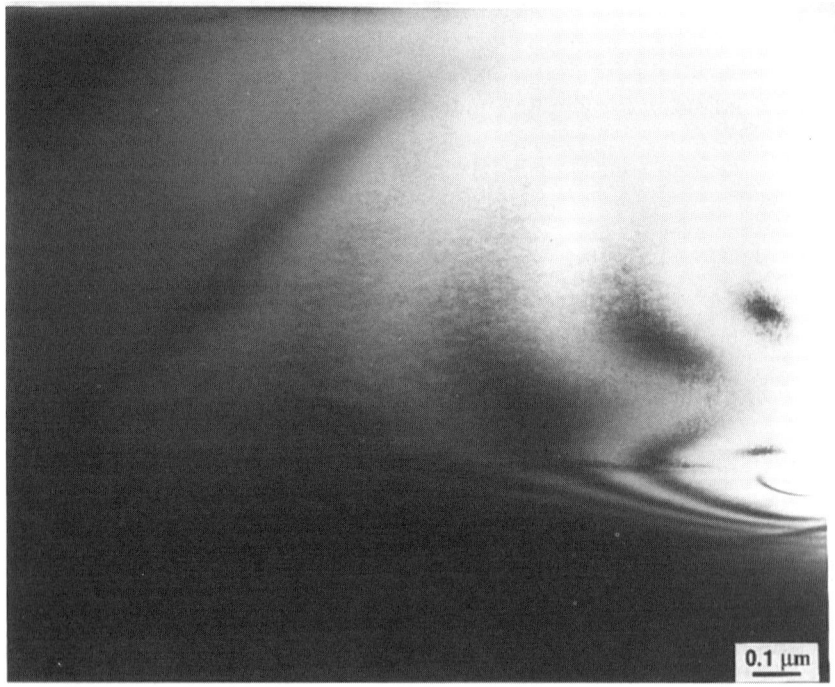

FIG. 9. Bright-field cross-sectional TEM of a pseudomorphic ZnMgSSE epilayer under a 0.136% compressive strain.

proximation to estimate that strain by using the thermal expansion coefficients of ZnSe (Ballard *et al.*, 1978) and GaAs (Blakemore *et al.*, 1982). A calculation shows that at the growth temperature of 260°C, the cracked epilayer (Fig. 8c) was under a tensile strain of -0.18%. In addition, at the growth temperature of 260°C the strain range over which the film remains pseudomorphic to GaAs is estimated to be -0.18% (tension) to 0.19% (compression). In terms of the surface morphology, observations show that the quaternary epilayer is more tolerant of accommodating compressive strain than tensile strain. The tensile strain causes epilayer cracking (striations). A similar behavior was also observed in the GaInP system (Bour and Shealy, 1988).

The essentially pseudomorphic conditions in each case were further confirmed by TEM cross-sectional observation. An example of a cross-sectional TEM bright field image is shown in Figure 9 for which the 2.8-μm-thick $(Zn, Mg)(S, Se)$ epilayer under study has an alloy fraction of $Mg = \sim 9\%$ and $S = \sim 12\%$; the compressive strain is 0.136% at room temperature. No stacking faults or dislocations could be observed in the imaged area, which is in agreement with the conclusions derived from the x-ray analysis.

4. Correlation with Device Performance

In order to characterize the degradation mechanisms that limit the lifetimes in the pseudomorphic SCH laser structures, plan-view transmission electron microscope (TEM) imaging was performed on degraded laser structures (Hua *et al.*, 1994). A plan-view TEM image showing the defect structures that are responsible for the degradation of a particular laser structure is shown in Figure 10a. A three-dimensional reconstruction of the defect structure is produced in Figure 10b through the use of TEM stereomicroscopy. A V-shaped pair of stacking faults which are nucleated at or near the II–VI/III–V interface is seen in Figure 10a. The depth in the structure over which the stacking faults extend was derived from the dimensions of the defects when observed from the exact [001] direction. These stacking faults, which are bounded by Shockley partial dislocations, propagate through the lower $(Zn, Mg)(S, Se)$ cladding layer. When the stacking faults reach the different strain field in the $Zn(S, Se)$ waveguiding layer, the partial dislocations unite and form a pair of perfect or slightly dissociated threading dislocations which continue to propagate to the surface of the structure. During laser operation, the points at which the threading dislocations intersect the quantum wells of this structure act as nucleation sites for the formation of dislocation dipoles. Through the climb process, these dislocation dipoles form extended dislocation networks which lie in the plane of the quantum wells. These dislocation networks, which appear as dark patches under electroluminescence (Hovinen et al., 1995), reduce the gain in the laser structures due to the increased nonradiative recombination rate in these regions.

It is known from studies of AlGaAs laser diodes that point defects (vacancies and interstitials, for example) are required in assisting the dislocation climb process (Petroff and Kimerling, 1976). A series of experiments was conducted which was aimed at reducing the point defect concentrations in the active region of the laser structures. (For instance, the density of Zn vacancies in ZnSe might be expected to be suppressed by an increase of the Zn/Se flux ratio during MBE growth.) The effectiveness of any modification in affecting the density of point defects can be gauged by temperature-dependent photoluminescence (PL) intensity measurements (Garbuzov and Khalfin, 1993) since thermally activated nonradiative recombination centers have been associated with point defects (Henry and

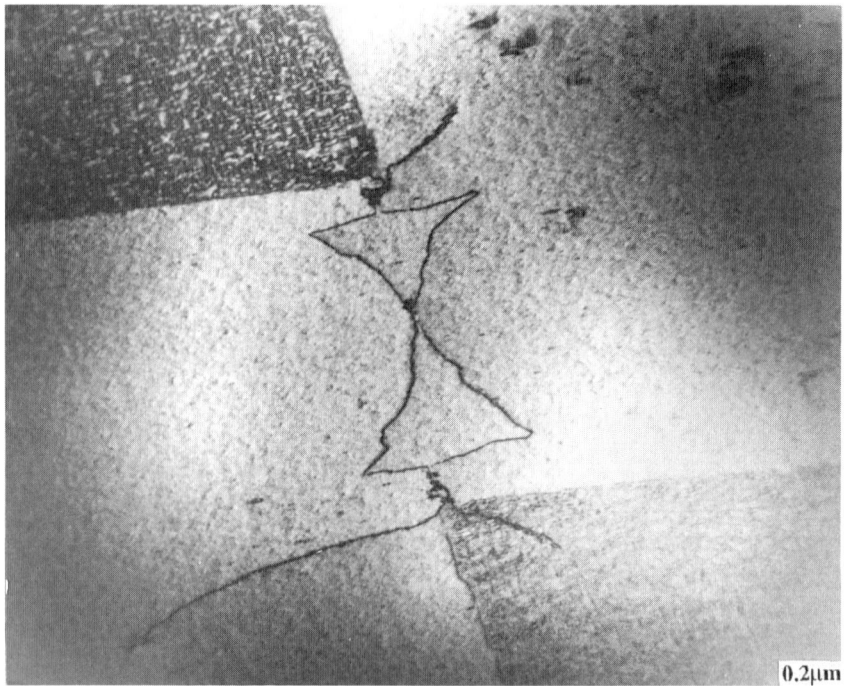

FIG. 10. (a) Plan-view TEM bright-field image showing a pair of V-shaped stacking faults. (b) Three-dimensional arrangement of the defects shown in (a) determined by using TEM stereomicroscopy.

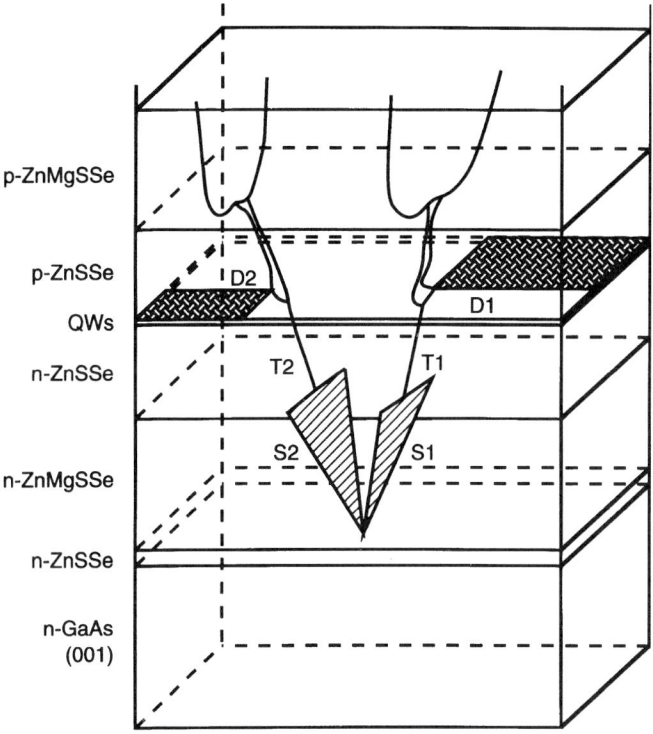

FIG. 10. (*Continued*)

Lang, 1977). It is not uncommon, in MBE grown ZnSe, for the integrated intensity from near band edge PL to drop about three orders of magnitude from low temperature to room temperature. By raising the growth temperature (from 260 to 300°C) while maintaining a barely selenium stabilized surface, it was found that the quenching was reduced to about two orders of magnitude. This result is shown in Figure 11 by the curve labeled "−5". (The same growth conditions used during the growth of sample "−5" were also used for the growth of the laser structure which demonstrated a room temperature CW lifetime of 37 sec [Salokatve *et al.*, 1993].) Although systematic studies are still in progress, further changes in the growth conditions, in this case the increase of the II to VI flux ratio of the ZnCdSe quantum well resulted in a structure which exhibits a PL intensity quenching of only a factor of four between low temperature and room temperature, as shown in Figure 11 by the data labeled "−8". (It should be noted that this factor of four may include a possible decrease in the carrier confinement in the quantum well with increasing temperature.) The reduced quenching

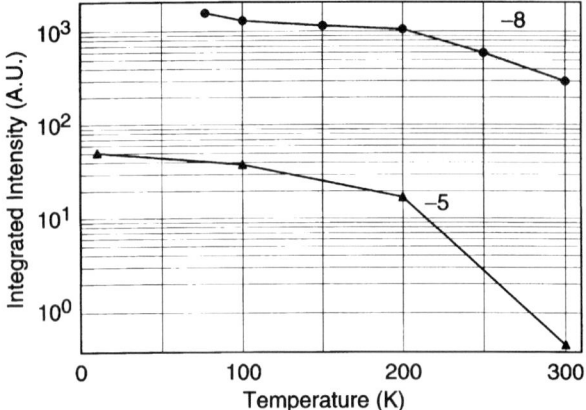

FIG. 11. Temperature dependence of the integrated near-band-edge PL emission for two samples. The sample labeled "−5" shows a quenching of two orders of magnitude while "−8" demonstrates a PL intensity quenching of only a factor of 4 bedroom cryogenic temperatures and room temperature.

of the PL spectrum suggests that the point defect concentrations in the laser active region have been significantly reduced.

III. The ZnSe/GaAs Heterovalent Interface

While some limited work on homoepitaxial growth of ZnSe on ZnSe substrates has been reported at the preparation of this manuscript (Eason et al., 1995), most of the study of epitaxial growth and device applications of ZnSe and its alloys have been carried out using GaAs substrates/buffer layers as the starting epitaxial surface. ZnSe/GaAs is characterized by a chemical valence mismatch at the interface, hence the designation of heterovalent. It was pointed out by Harrison et al. (1978) that such valence mismatch mandates an interfacial reconstruction and intermixing upon the formation of the heterovalent bondings; an atomically flat and abrupt II–V or III–VI transition layer would result in an interfacial dipole field leading to an energetically unstable configuration. So far investigations on the ZnSe/GaAs heterointerfaces have fallen into primarily the following categories: control of interface states, measurement/modification of band offset, vertical transport across the interface, and initial nucleation of ZnSe on GaAs. It is worth noting that the presence of interfacial dipoles, due to the valence mismatch across the II–VI/III–V interfaces, and the associated build-up of localized electric field could also affect the apparent band offset.

1 MBE GROWTH AND ELECTRICAL PROPERTIES OF SEMICONDUCTORS

1. CONTROL OF INTERFACE STATES

In addition to the potential applications as visible light emitters, epitaxial ZnSe on GaAs offers the possibility of passivating GaAs and implementing metal–insulator–semiconductor (MIS) field-effect devices, using the large bandgap ZnSe as the dielectric insulator (Studtmann et al., 1988). In this case the device feasibility depends very much on the control of density of interface states. In principle the employment of the MBE process enables the tailoring of this heterovalent interfacial bonding configuration, from one extreme of As–Zn to the other of Ga–Se bonding, such that both the structural and electrical properties of the heterovalent interface can be investigated. Despite the valence mismatch across the interface, it has been possible to obtain (at least in the case of epilayer/epilayer structures) heterovalent interface state densities comparable to those achieved from typical (Al, Ga)As/GaAs structures (Qian et al., 1989a and 1989b; Qiu et al., 1990).

A series of experiments (Qiu et al., 1990) was performed in which p-type GaAs epilayers were grown on p-type (100) GaAs substrates at 582°C, and then the epilayers was transferred under ultra-high vacuum to a separate growth chamber for the subsequent growth of ZnSe. Prior to the nucleation of ZnSe, the GaAs epilayers were heated in the ZnSe growth chamber to temperatures in the range of 320 to 535°C to modify/reduce the surface As content in stages, resulting in four different GaAs reconstruction patterns, namely c(4 × 4), (2 × 4), (4 × 6), or (4 × 3) Kobayashi, 1988). The (4 × 3) reconstruction, differing from the conventionally reported GaAs surface

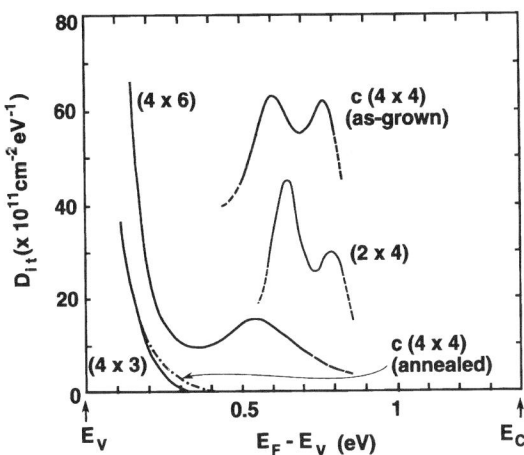

FIG. 12. Plot of interface state density for several different ZnSe/GaAs interface treatments.

reconstruction patterns (Cho, 1976), may have resulted from the "decoration" (Kolodziejski et al., 1986) of the heated GaAs surface by a high vapor pressure element such as Se, or by the reaction of background Se with the heated, Ga-rich surface. In each case, after forming a particular GaAs surface, the substrate temperature was reduced to 320°C for the nucleation of a ZnSe epilayer. Using conventional capacitance-voltage (C-V) measurements (Schroder, 1990), it was determined that the electrical

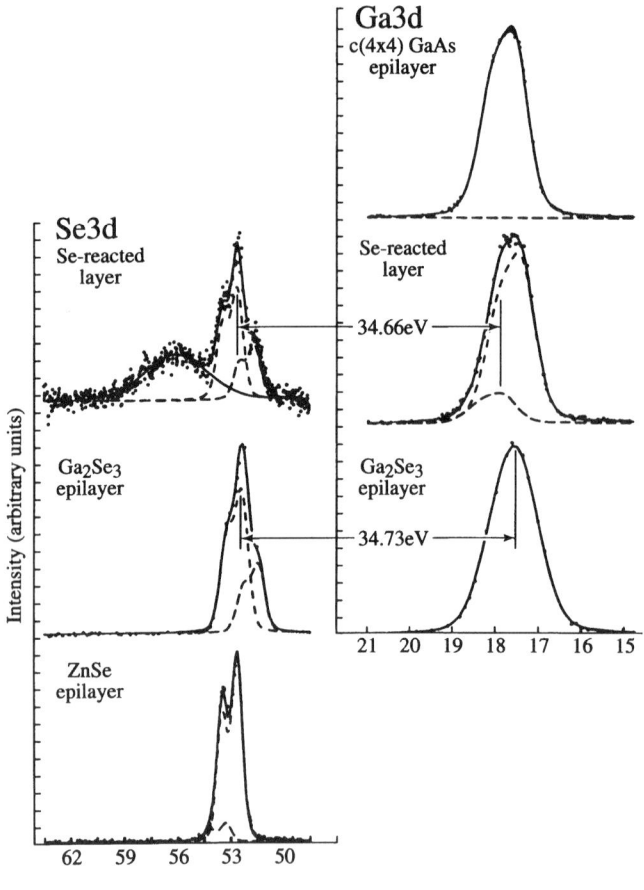

FIG. 13. The Se3d and Ga3d core level photoemission energy spectra curves for (a) the GaAs c(4 × 4) epilayer, (b) the Ga2Se3 epilayer, (c) the Se reacted layer, and (d) the ZnSe epilayer. The features are resolved into several doublets which are attributed to spin-orbit splitting The Se reacted layer includes the As3d plasmon loss feature in addition to the Se3d peaks.

characteristics of the interface improved as the GaAs surface became increasingly As-deficient (Fig. 12). Similar reductions in interface states resulted when the above procedure was modified to react the As-deficient surface with a better controlled flux of Se than was provided by the incidental background pressure of Se in the chamber. When the GaAs was heated to an elevated temperature, a calibrated irradiation of the GaAs epilayer surface was provided from the Se source oven for a predetermined time. The process then became independent of the immediate history of the chamber prior to formation of the low interface state heterovalent interface.

It has been speculated that the ZnSe/GaAs interfaces exhibiting the low interface state densities were associated with the presence of a interfacial layer of Ga_2Se_3 (Qian et al., 1989a, b; Qiu et al., 1990). TEM cross-sectional dark field and high-resolution images of heterojunctions formed on (4 × 3) reconstructed GaAs surfaces indicated the presence of a uniform band interpreted as an interfacial compound having an average thickness of 2 monolayers (~ 6 Å). Image simulations supported the identification of zincblende Ga_2Se_3 as the interfacial layer formed on As-deficient GaAs epilayers (Li et al., 1990). The TEM conclusions were further supported by an investigation of the interfacial bonding using in situ x-ray photoelectron spectroscopy (XPS) (Menke et al., 1991). A comparison (see Fig. 13) of the Se3d core level features from the ZnSe epilayer surface, a Se-reacted As-deficient GaAs surface, and from separately grown "zincblende" (every third Ga site is vacant) Ga_2Se_3 reference epilayers, supported the identification of the interfacial layer as consisting of zincblende Ga_2Se_3.

It is worth mentioning that theoretical predictions (Farrel et al., 1988) suggest that optimum interfacial conditions between II–VI and III–V semiconductors result when a layer of mixed cations is produced at the interface such that charge neutrality is maintained across the interface. It is conceivable that the steps (described above) leading to improved electrical properties of the interface are related to these theoretical predictions.

2. BAND OFFSET

The band offset between ZnSe and GaAs has been reported with a wide range of values (Kowalczyk et al., 1982). A study involving the reduction of interface states by interaction of an As-deficient GaAs surface with a beam of Se at elevated temperatures and employed capacitance-voltage characterization to evaluate the interface state density is described in Section III.1. During the course of these MIS experiments it was found that one could produce an accumulation layer when the GaAs was p-type, but not an electron inversion layer. For n-GaAs an inversion layer could be formed in the ZnSe/GaAs MIS structure, but an accumulation (electron) layer could not. The implication, by comparison with C-V experiments performed on AlGaAs/GaAs heterostructure interfaces, was that the conduction band offset in the ZnSe/GaAs structure is 300 meV or less. Such an alignment is not inconsistent with previous XPS studies (Kowalczyk et al., 1982).

Until recently it was widely held that the band offset could be varied by no more than 100 meV or so. However, there have been proposals for significantly altering the apparent band line-up at heterojunctions by implanting a delta doping layer such as to form a built-in dipole layer at the interface (Capasso and Margaritondo, 1987). More recent reports cite XPS results which show that a considerable variation in the band offset ($\Delta E v$ from 0.6 to 1.2 eV) can result when the anion/cation flux ratio is varied on nucleation of ZnSe on GaAs by MBE (Franciosi et al., 1994; Vanzetti et al., 1995).

3. Vertical Transport across the ZnSe/GaAs Interface

The lack of high-quality conducting II–VI substrates has thus far constrained most of the ZnSe-based diode lasers and LEDs to closely lattice-matched GaAs substrates and/or epilayers. While the structural integrity of the II–VI structures was significantly advanced with the rigid control of alloy composition for achieving lattice compatibility, electrical transport across the n-ZnSe/n-GaAs heterovalent interfaces was presumed to play only a minor role in terms of the overall device I–V characteristics. This conjecture, based on the argument that the conduction band offset between ZnSe and GaAs is relatively small (see above and Kowalczyk et al., 1982), in fact overlooks the detrimental effect of carrier depletion near the heterovalent interface due to the presence of a high density of interface states.

Figure 14a and 14b are the computer simulation output of n-ZnSe/n-GaAs band diagrams with and without the presence of interface traps; the simulation program solves the Poisson and continuity equations using a drift-diffusion current model. (Quantum mechanical tunneling and velocity saturation were not taken into account for the present version.) The doping levels of ZnSe and GaAs were chosen to simulate the real device, a conduction band offset of 300 meV was used between ZnSe and GaAs, and an interface state density of 4×10^{12} cm^{-2} was chosen from our previous studies of ZnSe/GaAs interfaces (Qiu et al., 1990). The extension of the depletion layer into n-ZnSe (in addition to the depletion region due to the

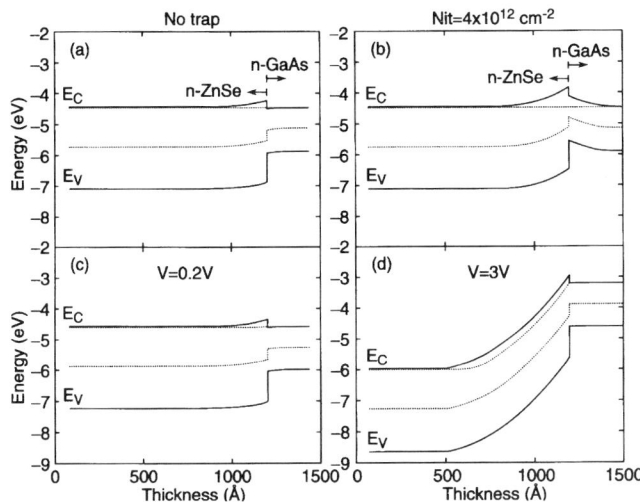

FIG. 14. Band diagrams of a ZnSe/GaAs heterointerface (i) under equilibrium without (a) and with (b) the presence of interface states, (ii) under high level injection without (c) and with (d) the presence of interface states.

band offset in Fig. 14a) is apparent in Figure 14b. The hindrance to carrier transport is exacerbated under high level injection where, due to the sparseness of the carriers in the depletion region, a high electric field is essential to initiate an appreciable drift current component (Fig. 14c and d).

The width of the depletion region is intimately related to the doping concentration, and therefore can be reduced by employing heavy doping near the ZnSe/GaAs interface. With the introduction of a 200 Å layer of n-ZnSe doped to 2×10^{18} cm^{-3} (compared to the normal doping of 5×10^{17} cm^{-3}), we have indeed observed a reduction in the voltage of about 1.7 V at a current density of 300 A/cm^2 from two typical (Zn, Mg)(S, Se) single quantum well lasers. It is worth pointing out that the approach of using heavy doping is inevitably accompanied by a compromise in material quality, which is particularly undesirable at the beginning of the epitaxial growth. To circumvent such a dilemma, an experiment based on our previous study (Qiu et al., 1990) in controlling the stoichiometry of GaAs before the nucleation of ZnSe was performed. Two n-Zn(S, Se) epilayers of 1 μm thick (sample A and B) were grown on thin layers of ZnSe, which were first nucleated on GaAs epilayers. The ZnSe layer of sample A, the reference sample, was nucleated on a c(4 × 4) As-rich GaAs epilayer (a common practice for the nucleation of ZnSe on GaAs). Doping levels of the ZnSe and

Zn(S, Se) were 2×10^{18} cm^{-3} and 5×10^{17} cm^{-3}, respectively. On the other hand, the ZnSe of sample B was nucleated on a Ga-rich (4 × 6) GaAs surface, and a reduced doping level (uniform) of 3×10^{17} cm^{-3} was employed throughout the II–VI layers. In order to somewhat alleviate the tendency for threading dislocation generation, which is a characteristic of the island coalescence stage following the 3D nucleation of ZnSe on a Ga-rich GaAs surface, atomic layer epitaxy was employed for the first few monolayers by alternative deposition of Se and Zn atoms while monitoring the recovery of RHEED intensity. As is described in Section III.1, the nucleation of ZnSe on a Ga-rich GaAs surface could lead to a reduction in the density of interface states by orders of magnitude as compared to nucleation on a As-rich GaAs surface. (A ZnSe layer of 500 Å with doping exceeding 1×10^{19} cm^{-3} was grown as a top contacting layer for both samples.) The n-II-VI/n-GaAs structure studied here can be thought of as one part of an actual laser device. Ti/Au was used to contact n-ZnSe (Miyajima et al., 1992), and a mesa etch was employed (etching down to the GaAs layer) to eliminate current spreading. It was shown that for an injection current density of 300 A/cm^2, the operating voltage of sample B (1.7 V) is less than that of sample A (2.3 V) despite the fact that the doping was significantly lower in sample B.

4. NUCLEATION OF ZnSe AND ITS ALLOYS ON GaAs

Studies of the initial nucleation of ZnSe on GaAs substrates and epilayers indicated that, by modifying the stoichiometry of the starting GaAs surface, one can vary the growth mode from a two-dimensional layer-by-layer growth to a three-dimensional island mode as evident by *in situ* RHEED monitoring (Tamargo et al., 1988; Qiu et al., 1990). The breakthrough in II–VI diode lasers which prompted the need for device quality heterostructures has generated a renewed interest in the field of heterovalent nucleation. It has been shown (Guha et al., 1993; Hua et al., 1994; Kuo et al., 1994a) that the degradation of II–VI blue-green laser diodes originated from the nucleation-induced extended structural defects which intersect the II–VI quantum wells, and became the source for the generation of dark-line defects during laser operation. In terms of the device performance, the effect of defect density on the cw laser lifetime was recently investigated and correlated by Ishibashi (1995).

Despite the inevitable uncertainties introduced due to GaAs substrate surface preparation, both Gaines et al. (1993) and Guha et al. (1993b) were able to derive from their investigations the trend that an initial exposure of the GaAs surface to a Se flux prior to nucleation resulted a 3-D nucleation and a high density of extended defects (stacking faults, etc.), while an irradiation with Zn flux led to a 2-D growth and a reduced extended defect density (Kuo et al., 1994b). It was proposed that the 2-D growth mode can be further facilitated by using the so-called migration enhanced epitaxy (MEE) technique (Gaines et al., 1993). Recently Li and Pashley (1994) employed a scanning tunneling microscope (STM) to examine the initial island formation of 2 to 3 monolayers of ZnSe on a Se-terminated GaAs surface. It was noted that the (2 × 1) Se-terminated GaAs surface is quite different from the (2 × 1) ZnSe surface in terms of the preferential "wetting" of ZnSe layers. An electron counting argument was developed to explain the relatively inert Se–GaAs surface and consequently the difficulty in obtaining 2-D growth on such a surface. It should be also noted that Wu et al. (1993) reported that the exposure of GaAs to a sulfur (S) flux could significantly increase the stacking fault density.

Two issues have to be addressed in attempting to control and suppress the extended defect density. Given the fact that the stacking fault density (for the heterovalent nucleation of ZnSe on a GaAs buffer layer) can be maintained below 10^6 cm^{-2} (or even considerably lower), the "capture" of one stacking fault becomes statistically challenging using cross-sectional TEM, and the estimate of defect density from plan-view TEM can be somewhat limited in accuracy. Other defect-revealing procedures have also been employed including photoluminescence (PL), cathode luminescence (CL), and electroluminescence (EL) microscopies (Haugen et al., 1995) which tend to require the insertion of a quantum well for enhanced radiative transition at room temperature. A bromine–methanol based defect etch was developed by Kamata and Mitsuhashi (1994) as a convenient way to reveal dislocations in ZnSe and ZnSSe; the same formula, however, failed to work on Mg-containing ZnMgSSe quaternary epilayers due to the appearance of a rough surface morphology after etching (Ng et al., 1995).

At a defect density of around 10^6 cm^{-2} or lower, extrinsic factors such as (chemical) wafer cleaning, wafer cleaving, (indium) wafer mounting, and even the presence of contamination or particulates inside of a UHV system could become undesirable sources tending to mask the more "intrinsic" behavior related specifically to the ZnSe/GaAs heterovalent nucleation, and hence serve to mislead the investigation. Figure 15 shows the Nomarski surface morphology photos, taken at various stages during the course of a defect control study in our laboratory, of three etched ZnSSe epilayers. The initial defect density of around 10^6 cm^{-2} (Fig. 15a) was subsequently reduced, through the usage of whole "3" GaAs wafer (cleaving-free) and indium-free mounting, to a level below 10^5 cm^{-2} (Fig. 15b). Further reduction of the nucleation-related defects, to a level as low as mid-to-high 10^3 cm^{-2} (Fig. 15c), was obtained through special nucleation procedures

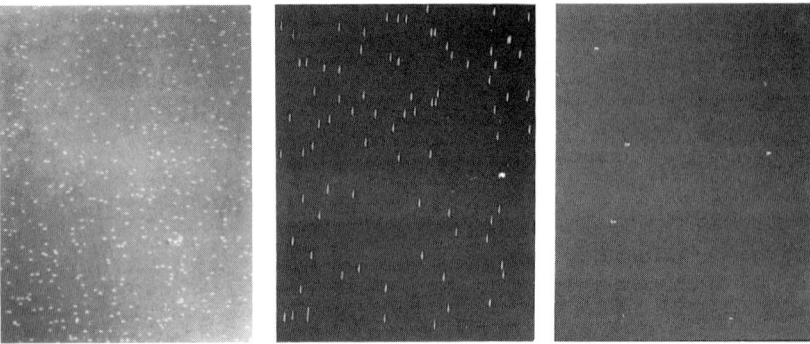

FIG. 15. Nomarski microscopic images showing the progressive reduction of etch pit density from (a) $\sim 1 \times 10^6 \text{cm}^{-2}$, to (b) $\sim 1 \times 10^5 \text{cm}^{-2}$, and to (c) $<1 \times 10^4 \text{cm}^{-2}$.

including Zn flux pretreatment, low-temperature MEE of ZnSe, and the employment of somewhat lower sulfur background at the time of nucleation.

IV. Electrical Contact to p-ZnSe

ZnSe has long been recognized as a compound semiconductor having a deep Γ-point valence-band edge (hole affinity $\sim 6.7\,\text{eV}$). This fact, in conjunction with the attainable p-doping concentration thus far of mid- to high-$10^{17}\,\text{cm}^{-3}$ (by rf nitrogen plasma doping during MBE growth [Qiu et al., 1991; Fan et al., 1992]), makes contacting to p-ZnSe with conventional metalization a challenging (if not impossible) task. Traditionally the investigations of the incorporation of various p-type dopants (As, P, Li, N) relied mainly on contactless optical probing, electron paramagnetic resonance (EPR), or junction techniques (such as capacitance–voltage transients). The quest for a suitable contacting scheme to p-type ZnSe was ignited almost immediately after the first demonstrations of ZnSe-based laser diodes in 1991 when lasing threshold voltages exceeding 25 volts were routinely observed (Haase et al., 1991; Jeon et al., 1991). The excessively high voltage was associated with the breakdown of the reversely-biased Au/p-ZnSe Schottky contact ($\phi_B \sim 1\,\text{eV}$) required for the injection of holes into the p-type II–VI layers to provide and sustain the optical gain. The employment of a p-ZnSe/p-GaAs heterojunction injection scheme (metal contact to n-type II–VI top layers) also failed to alleviate the problem; the large valence band offset ($\sim 1\,\text{eV}$) between ZnSe and GaAs again presents a formidable valence band energy barrier to carrier transport. The reduction of lasing voltages, through the development of a viable contacting scheme, became an important objective toward making the II–VI blue-green laser diodes practical. Furthermore, the availability of an ohmic contact to p-type ZnSe would enable a wide range of electrical characterizations such as current–voltage correlations and Hall-effect measurements from which insight as to the nature of nitrogen incorporation would be gained.

1. Zn(Se, Te) Graded Contact

Prior to the summer of 1992, the most commonly used method for contacting p-ZnSe involved the direct deposition of Au or Pt, though substantial barrier height still remained; no suitable metal having a sufficiently high work function had been found.

In a parallel investigation using the newly available nitrogen rf plasma during the MBE growth of ZnTe (a binary II–VI compound with potential as a green/yellow light emitter), it was found that the same nitrogen source could be used to implement the most effective p-doping yet achieved for the MBE growth of ZnTe (Han et al., 1993). The successful p-doping of ZnTe led to the eventual development of the heterostructure Zn(Se, Te) graded contact. A brief digression to discuss the p-type doping of ZnTe will be introduced in the following section.

a. Nitrogen Doping of ZnTe

Undoped MBE-grown ZnTe is generally lightly p-type and highly resistive. Although ZnTe is, in principle, easily doped p-type, in fact considerable difficulty has been experienced in achieving high-quality p-type ZnTe using MBE. In general it has been difficult to obtain a physical incorporation of the dopant species due to a low sticking coefficient. For example, the p-doping of MBE-grown homoepitaxial layers on ZnTe substrates was performed using antimony (Sb) where Sb flux levels were comparable to that of the host elements; a maximum doping level of $1.0 \times 10^{18} \, \text{cm}^{-3}$ was reported (Kitagawa et al., 1981). Hishida et al. (1989) used elemental phosphorus (P) to obtain p-doping of MBE-grown ZnTe on GaAs substrates, also with the P beam equivalent pressure comparable to that of the zinc and tellurium sources. The P-doping experiments were, however, somewhat less successful, reaching a maximum doping of $4.0 \times 10^{17} \, \text{cm}^{-3}$ with deteriorated crystalline quality as shown by a broad x-ray diffraction rocking curve.

As an extension of the nitrogen plasma doping (of ZnSe) into ZnTe, it was found that a doping level exceeding degeneracy can be easily obtained.

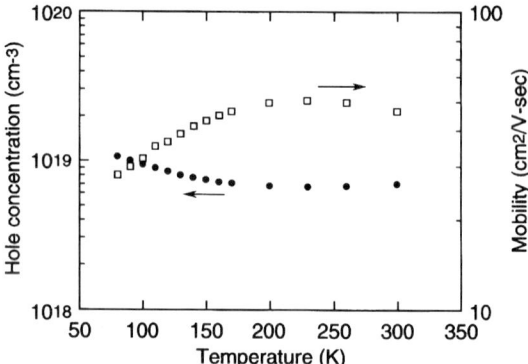

FIG. 16. Free hole concentration and hole mobility as a function of temperature for a ZnTe:N epilayer.

Figure 16 shows the free hole concentration and mobility derived from Hall-effect measurements as a function of temperature. (Van der Pauw samples were prepared by evaporating Au which forms an ohmic contact to the nitrogen-doped ZnTe epilayers.) Given the weak and unusual temperature dependence of carrier concentration in terms of a single acceptor level model, it seems reasonable to assume transport involving an impurity band formed by the overlap of impurity wave functions. At approximately 10^{19} cm^{-3} acceptor density, the average spacing between impurity atoms is indeed of the same order as the size of the Bohr radius for a hydrogen-like acceptor (Baltensperger, 1953). Our study of nitrogen doping at various growth temperatures indicated that, from 275 to 400°C, substrate temperature only has a minor influence on the nitrogen incorporation. It is worth noting that, even under such a degenerate doping level, the structural quality of p-ZnTe was not severely compromised (as compared to undoped samples), as indicated by the FWHM of x-ray rocking curves and ratios between near band-edge and deep level emission from photoluminescence spectrum (Han et al., 1993b).

b. Implementation of the Zn(Se, Te) Contact

The discovery of heavy p-doping of ZnTe, and the observation that metals (Au, for example) can easily form an ohmic contact to p-ZnTe, led to the consideration of the use of heavily p-doped ZnTe as an intermediate layer for contacting p-ZnSe. Figure 17(a) shows the expected energy band lineup at a p-ZnTe/p-ZnSe interface, where it is seen that the valence band offset

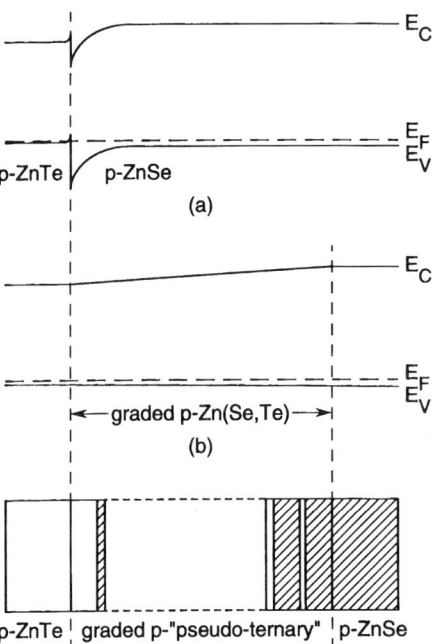

FIG. 17. Schematic diagram of the energy band lineup at the (a) p-ZnTe/p-ZnSe interface and the (b) p-ZnTe/p-Zn(Se,Te) graded layer/p-ZnSe. (c) Schematic drawing of the structure of the pseudograded Zn(Se,Te) contact layer.

($\sim 1\,\text{eV}$) between ZnTe and ZnSe forms a barrier to hole injection. It is understandable why the initial attempt of growing p-ZnTe directly on p-ZnSe still resulted in a Schottky-like current–voltage characteristic. It was envisioned that a possible means for removing the energy spike in the valence band was to introduce a Zn(Se, Te) layer having a graded bandgap, as shown in Figure 17b. A similar contact scheme, but employing a graded (In, Ga)As region, had been employed by Woodall et al. (1981) to form an ohmic contact between a metal–InAs junction and n-GaAs. (It is interesting to note that the lattice constant mismatch, which is approximately 7%, is very similar between the binary compounds at the physical extremes in both the II–VI and III–V examples.) A particular difficulty in realizing the band structure shown in Figure 17b by MBE growth is controlling the Te fraction in the graded alloy region due to the competition between Te and Se species for incorporation on the growth surface (Kolodziejski et al., 1988). To circumvent this problem, we designed and grew a multilayer structure, consisting of 17 cells, each of 20 Å thickness. In each cell the thicknesses of both the ZnTe and ZnSe layers are varied as shown in Figure 17c to approximate a linearly graded region. The first cell next to the p-ZnSe epilayer contained 18 Å of p-ZnSe and 2 Å of p-ZnTe, the next cell 17 Å of p-ZnSe and 3 Å of p-ZnTe, and so on. The fact that the cited thicknesses represented fractional monolayers was viewed as an advantage; a degree of disorder should enhance the simulation of a graded alloy.

The contacts were first evaluated in a configuration designed for Hall measurements, as one objective was to obtain contacts suitable for the evaluation of the free hole concentration in ZnSe:N epitaxial films. In one experiment, three different contacting schemes consisting of Au/p-ZnSe, Au/p-ZnTe/p-ZnSe, and Au/p-ZnTe/p-Zn(Se,Te)/p-ZnSe were evaluated. The p-ZnSe was a nominally 2-μm-thick nitrogen-doped epilayer grown on an undoped semi-insulating GaAs substrate, and had a hole concentration of $3.2 \times 10^{17} \text{cm}^{-3}$. The I–V characteristics, measured between pairs of coplanar contact pads for the three contact configurations, are compared in Figure 18. As seen in Figure 18a, the graded contact appears to be perfectly ohmic, showing a straight line through the origin. Figure 18b shows the characteristics of the contact formed by gold deposited onto an as-grown ZnSe:N epilayer. The I–V characteristic corresponds to two back-to-back Schottky diodes; the observed turn on voltage is the reverse bias breakdown

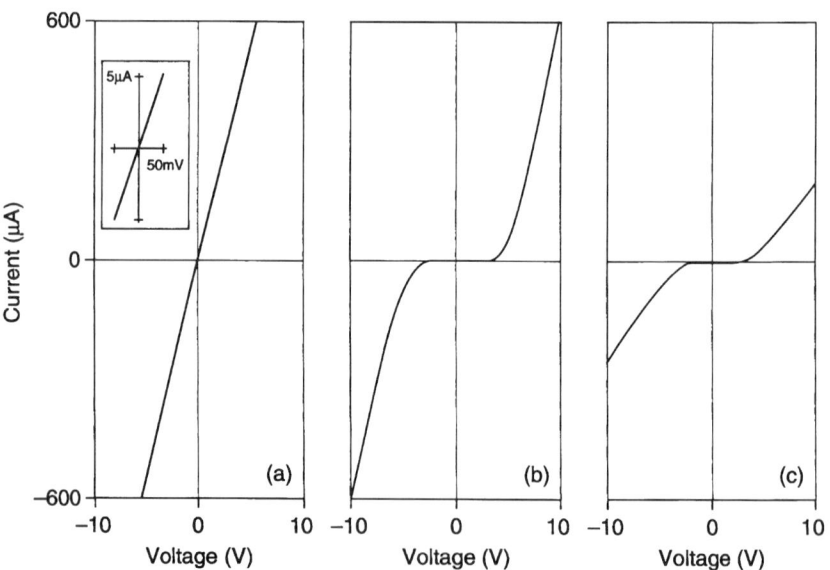

FIG. 18. I–V characteristics of different contact schemes measured by an HP-4145 parameter analyzer at room temperature, (a) Au/Zn(Se,Te) graded layer/p-ZnSe, with the inset showing the I–V characteristic where the voltage range has a 50 mV scale maximum, (b) Au/p-ZnSe, and (c) Au/p-ZnTe/p-ZnSe.

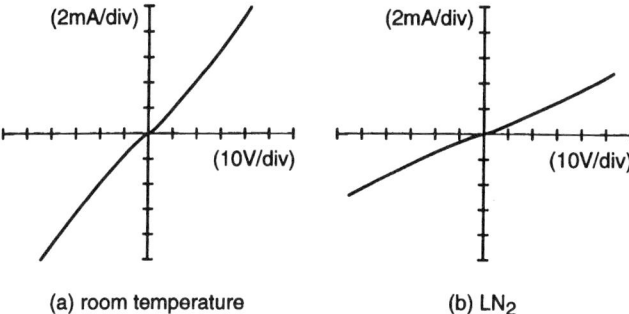

FIG. 19. I–V characteristics at 300 and 77 K from a structure having a Zn(Se,Te) graded layer contact which is identical with Figure 18(a), but with the contact size reduced to a 50-μm dot.

voltage. Figure 18c shows the I–V characteristic when p-ZnTe is used to inject holes into the ZnSe epilayer in the absence of the graded region. The deviation from ohmic behavior for the structure of Figure 18c is attributed to the hole barrier arising from the valence band offset between the two semiconductors.

The I–V characteristics from the graded Zn(Se,Te) contacting scheme were also evaluated at high (compatible with the lasing threshold current) injection levels. Figure 19 shows the I–V characteristics at 300 and 77 K from a structure identical to that of Figure 18a, but with the contact size reduced to 50 μm in diameter, and with current densities of up to 500 A/cm^2; some deviation from linearity is seen. We have observed that a Au/p-ZnTe contact with an area roughly twice that of the 50-μm dot also exhibits a deviation from linearity. Moreover, as shown in Figure 17, the discontinuous (or discrete) nature of the grading, plus the difference in doping levels between p-ZnSe and p-ZnTe layers, are expected to create a spatial Coulombic field and consequently a small effective energy barrier to holes, which possibly contributes to the slight deviation from linearity. (It is worth noting the latter effect can be suppressed by tailoring the gradient profile into a parabolic grading [Hayes *et al.*, 1983; Fan, 1995]). However, of importance to transport studies, it is seen that the graded contact remains pseudo-ohmic even at the lower temperatures.

c. *Evaluation of the Contact*

The specific contact resistance of the Au/p-ZnTe/graded layer/p-ZnSe contact was determined by a standard transmission-line model (TLM) measurement (Han et al., 1994b), where the resistance between paired electrodes of identical size and different spacings is plotted as a function of the spacing. The slope of the TLM line yields material resistivity and the intersect to the ordinate provides the information of specific contact resistance (Schroder, 1990). The validity of TLM measurement greatly depends both on the relative magnitude of the contact and material resistance, as well as the sample configuration; it is necessary that the horizontal component of current flow (between the TLM metal contacts) be essentially one-dimensional. An Ar ion beam mill was employed to isolate the TLM contact mesas down to the semi-insulating GaAs substrates. The Zn(Se, Te) graded layer between the Au electrodes (50 μm × 95 μm) was also removed using the same milling technique. The contact resistance of the metal/p-ZnTe/p-Zn(Se, Te)/p-type ZnSe graded scheme can be viewed as comprising two components in series. One element is the metal/p-type ZnTe contact resistance and the other is that of the p-type Zn(Se, Te) pseudo-graded region. The resistance due to the former part was analyzed by the TLM measurements on a metal/p-type ZnTe sample, which gave a specific contact resistance of 2.4×10^{-6} Ω-cm^2. Figure 20 shows a typical plot (of total resistance versus electrode spacing) from an (unannealed) Au/Pd/ Zn(Se, Te) graded-region/p-ZnSe contact. A specific contact resistance of 4×10^{-4} Ω-cm^2 was derived from the plot using Berger's equation (Schroder, 1990). The reduction of specific contact resistance as compared to a Au/Zn(Se, Te) graded-region/p-ZnSe contact (Fan et al., 1992) was in

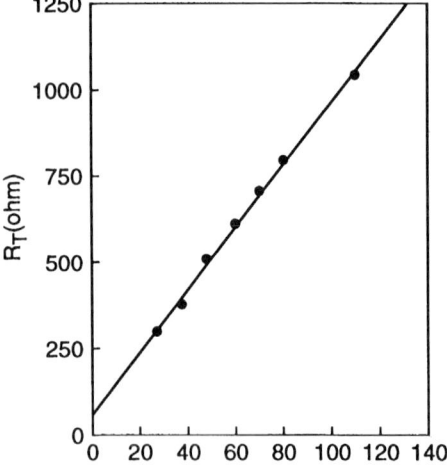

FIG. 20. The plot of measured total resistance vs. the spacing between electrodes for a Au/Pd/p-ZnTe/p-graded Zn(Te, Se)/p-ZnSe contact.

qualitative agreement with the recent study of Pd contacts to p-ZnTe (Ozawa et al., 1993). The material sheet resistivity, obtained from the slope of the plotted line, was 4.6 kΩ/square which is very close to the value (4.4 kΩ/square) derived from a separate Hall measurement. Given the slight deviation in the I–V characteristics from a straight line near the origin as was observed for small contacts under high current level (Fan et al., 1992), the TLM resistance values were taken at a current density of about 250 A/cm^2. The injection level was determined with a knowledge of the "effective contact area," a quantity which reflects the extent of current crowding near the edge of the electrodes and can be derived from the intersection of the data line to the abscissa (Schroder, 1990). The comparison of the measured TLM contact resistances between metal/p-ZnTe and the overall graded contact seems to suggest that the limiting factor of the graded contact scheme lies at present in the pseudo-graded Zn(Se, Te) region.

2. OTHER TECHNIQUES FOR FORMING CONTACTS TO p-ZnSe: HgSe, ZnSe–ZnTe RESONANT TUNNELING CONTACT

In addition to the Zn(Se, Te) graded contact to p-ZnSe, alternative contacting schemes have been demonstrated or proposed with various degree of success. A so-called ZnSe–ZnTe multiple quantum well resonant tunneling contact, which bears some resemblance to the pseudo-graded scheme (Fan et al., 1992), was proposed by Hiei et al. (1993). In spite of the similar geometrical configuration between the two designs, the explanations as to the underlying carrier (hole) transport mechanisms are quite different. The term "resonant tunneling" derived from the speculated enhancement of carrier injection when the quantized levels in each ZnTe well are aligned under forward bias condition. However, the presence of a high degree of lattice disorder/defect (due to a $\sim 7\%$ lattice mismatch) and the lack of electrical or optical confirmation of the existence of quantized levels, places the assertion of resonant tunneling as questionable. In another heterostructure bandgap engineering attempt, Lansari et al. (1992) reported the epitaxial growth of HgSe, a semimetal, together with a $ZnSe_{1-x}Te_x$ transition layer to facilitate hole injection. The use of semimetallic HgSe has the effect of lowering the Schottky barrier. The very high conductivity of HgSe also makes it attractive as a top current-spreading layer for blue and green LEDs.

The aforementioned contacting schemes all involved the use of lattice-mismatched heteroepilayers where the presence of a high density of structural defects due to strain relaxation could (eventually) compromise the device performance. A recent claim of heavy p-type doping using postgrowth diffusion of LiN into MOCVD-grown ZnSe indicated a possibility of achieving an ohmic contact using only metals (Lim et al., 1994).

V. P-Type Doping of ZnSe-based Alloys

It was known since the early 1960s that n-type ZnSe could be obtained by various bulk growth techniques. (Sometimes a postgrowth annealing under a Zn-rich environment is required.) However, the inability to achieve amphoteric conductivity control, in this case p-type conduction, had hampered the growth of the II–VI field for nearly 30 years. Given the absence of p-n junctions, it was not feasible to consider the making of injection light-emitting diodes and lasers in the blue and green spectral range. The traditional difficulty associated with the p-doping of ZnSe was usually explained in terms of so-called "self-compensation" through native defects based on arguments assuming thermodynamic equilibrium, a condition applicable to most of the bulk growth techniques near the melting point. The self-compensation picture was favored due to the conceptually plausible argument that the formation energy of the (compensating) native defects (vacancies and interstitials, for example) is reduced by roughly the amount of bandgap energy for every compensated carrier, therefore greatly increasing the likelihood of their existence. This picture can also provide an explanation as to the traditional difficulty in doping any wide bandgap semiconductor. Despite its apparent popularity, the validity of the self-compensation picture remains quite controvertial (refer to Chapter 4 for a detailed discussion).

The advance of modern epitaxial growth techniques such as MBE and metal-organic chemical vapor deposition (MOCVD), where the growth takes place under highly nonequilibrium conditions, implies the possibility of revisiting the doping of the wide bandgap II–VIs, and in particular ZnSe, with the hope that the dopant incorporation behavior will be dictated by the surface kinetics (desorption, adsorption, and surface diffusion), a process which could be brought under control through variations of growth parameters. We will confine our discussion here to the p-type doping of ZnSe using MBE, as this technique has proved the most successful.

1. BACKGROUND (P, As, O, AND Li DOPING)

It is quite obvious that the prime candidates to serve as a p-type dopant of Zn(IIB)Se(VIB) should come from the group IA and group VB elements

1 MBE GROWTH AND ELECTRICAL PROPERTIES OF SEMICONDUCTORS

in the periodical table. Both lithium (Li)- (Cheng et al., 1989; Ren et al., 1990) and sodium (Na)-doping (Cheng et al., 1989) have been explored. The sodium-doping attempt, in spite of a clear presence of an acceptor-bound exciton peak from PL, was limited by the available source purity and the associated compensation. Lithium doping indeed resulted in p-type conductivity as confirmed by both electrical and optical characterizations. The control of the doping profile, however, was compromised by the high diffusion length (on the order of mm) of lithium atoms. The net acceptor concentrations also saturated near the high -10^{16} cm^{-3} range despite further increase of Li incorporation; the formation of compensating Li insterstitial donors (due to its small atomic size) is likely responsible for the doping limit. For the group VB elements, phosphor (using Zn_3P_2) (Yao and Okada, 1986) and arsenic (elemental As and Zn_3As_2) (Shibi et al., 1990) doping has been reported. (Although the physical incorporation of both dopants was realized, the p-type conductivity was inhibited by compensation, under most cases proportional to the doping levels.) A novel attempt of using oxygen (O) from a ZnO source during MBE growth of ZnSe as an isoelectronic dopant source was reported by Akimoto et al. (1989a and 1989b). In spite of the initial promising evidence from PL emission and I–V, the doping result reported has not been widely reproduced.

2. NITROGEN-DOPING OF p-ZnSe

It was pointed out in the 1980s that nitrogen, when incorporated substitutionally on a Se site, forms a shallow acceptor level (with Ea ~ 110 meV) in ZnSe (Stutius et al., 1982). The subsequent effort mainly involved the search for a viable means to effectively incorporate nitrogen atoms into ZnSe. The use of N_2, NH_3 (Park et al., 1985) and the employment of nitrogen ion source (Mitsuyu et al., 1986) had all been explored with limited success; the increase of nitrogen incorporation often comes at the expense of lattice damage. The breakthrough came in 1990 when Park et al. (1990) and, independently, Ohkawa et al. (1991) employed a particular nitrogen source, namely the active nitrogen species excited from a remote radiofrequency (RF) plasma source manufactured by Oxford Applied Research, U.K., during MBE growth; the result was an unambiguous and reproducible p-type conduction in ZnSe. The controversial subject as to the exact species (nitrogen ions, excited molecules, or atomic nitrogen) responsible for the p-doping of ZnSe was mostly resolved by Vaudo et al. (1993) where the N_2 plasma emission spectrum was analyzed and correlated with the doping concentration. It is generally accepted now that atomic nitrogen is responsible for the much-increased doping efficiency. Subsequently, several variations of nitrogen doping of ZnSe have been demonstrated including the use of an electron–cyclotron resonance (ECR) plasma cell (Okuyama et al., 1992), and recently the employment of thermal cracking of nitric oxide (NO) as an alternative atomic nitrogen source (El-Emawy et al., 1995).

a. Electrical Transport Hall Measurement

Due to the lack of a suitable ohmic contacting scheme to p-ZnSe, characterizations of the dopant incorporation and behavior initially relied mainly on optical measurements (photoluminescence, photoreflectance, etc.), capacitance–voltage profiling techniques, or electromagnetic reflection techniques which could be implemented without the requirement of an ohmic contact.

The demonstration of the Zn(Se, Te) graded bandgap contact to p-ZnSe (Fan et al., 1992) and its ability to maintain the ohmicity down to cryogenic temperatures enable the performance of classical temperature-dependent Hall-effect measurements where transport parameters such as doping, compensation, and activation energy associated with the nitrogen acceptors could be quantitatively derived (Fan et al., 1993). Of special importance is the preparation and configuration of Hall samples, where the four-point "van der Pauw" pattern has been the most widely used so far. It is worth noting that, as was pointed out in the original paper by van der Pauw

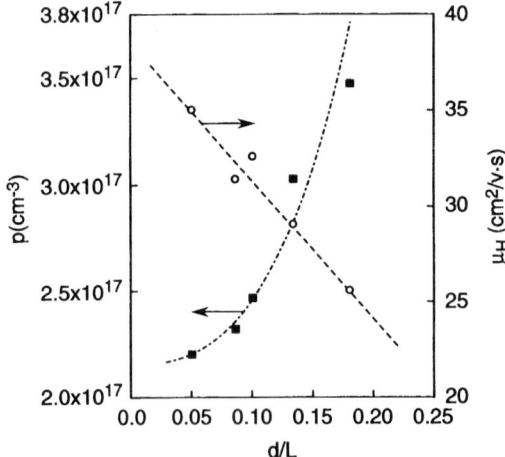

FIG. 21. Dependence of measured apparent hole concentration and mobility on the contact/sample size ratio. The measurements were performed at room temperature.

(1958), error in free hole concentration and mobility is introduced by the finite contact size on the Hall measurement sample. This effect was later shown to be especially pronounced for materials with low mobility (Chwang et al., 1974) such as p-type ZnSe. In order to investigate the effect of the ratio of contact to sample size on the measured transport parameters, a series of Hall samples from the same nitrogen-doped ZnSe epilayer was prepared with the contact/sample size ratio ranging from 5% to 18%, and measurements were performed at room temperature. The measured apparent hole concentration increased by 59%, and the measured room temperature mobility decreased by 26% over that range (Fig. 21). In the author's laboratory the sample size was standardized at typically 7 mm square with 200-μm contacts; the corresponding contact/sample ratio was about 3%. The estimated error in hole concentration for such a ratio was below 7% (Fan et al., 1994).

Ohmic behavior (as shown in Fig. 22) between ZnSe and the Au/Zn(Se, Te) graded contact was maintained over the temperature range (300 to 77 K) of the Hall measurements. Figure 23 shows both the experimental

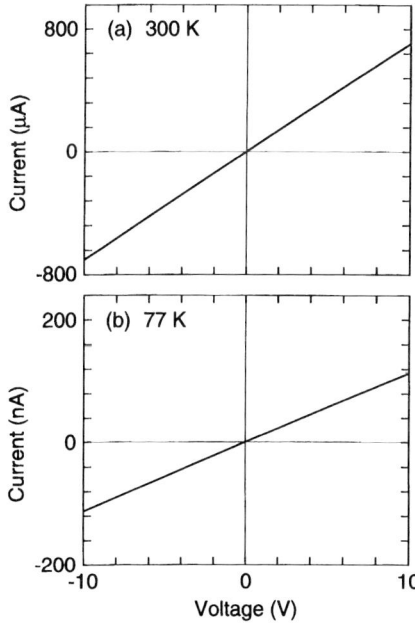

FIG. 22. I–V characteristics of a typical Hall sample at room temperature (a) and 77 K (b) as plotted by a HP 4145B semiconductor parameter analyzer. The sample was 7×7 mm^2 with four contact dots nominally 200 μm in diameter at the corners.

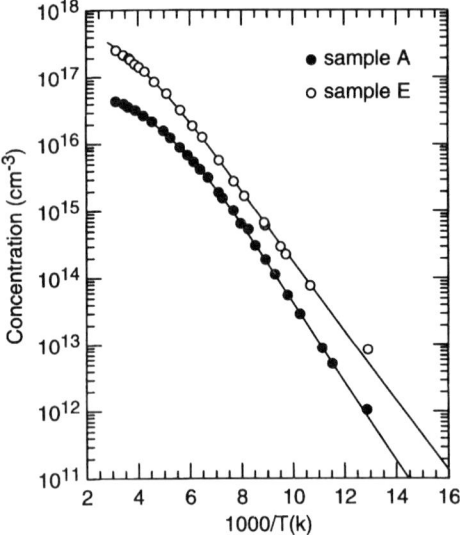

FIG. 23. The temperature dependence of free hole concentrations for samples A and E (Table I). The solid line represents the calculated concentrations for both samples using the curve-fitting parameters listed in Table I.

and (fitted) theoretical temperature dependence of the free hole concentration from two nitrogen-doped ZnSe epilayers. An expression

$$p(p + Nd)/(Na - Nd - p) = Nv/g \exp(-Ea/kT) \qquad (8)$$

which was derived directly from the charge-neutrality requirement assuming a nondegenerate semiconductor with a single acceptor level, was fitted to the experimentally measured temperature-dependent hole concentration using a least-square curve fitting program. Na and Nd are the acceptor and donor concentrations, respectively, p is the free hole concentration, $Nv = 2(2\pi m^*kT/h^2)^{3/2}$ is the effective density of states, and m^* is the valence band density of states effective mass. Ea is the activation energy of the impurity, and g is the degeneracy factor. The deviation from linear dependence in Figure 23 in the low temperature range was observed previously in n-type GaAs, and was attributed to impurity band conduction (Stillman and Wolfe, 1976). The curve fitting was performed using data at temperatures above where the aforementioned deviation occurs. Table I lists the results of curve fitting for several samples. The light and heavy hole effective masses

TABLE I

SUMMARY OF TEMPERATURE DEPENDENT HALL EFFECT MEASUREMENTS ON ZnSe:N FILMS

Sample	p at 293 K ($\times 10^{17}$ cm^{-3})	μ_p (cm^2/Vs)	Ea (meV)	N_a ($\times 10^{17}$ cm^{-3})	N_d ($\times 10^{16}$ cm^{-3})	N_d/N_a
A	0.40	39.3 (293 K) 596.4 (77 K)	106	0.776	0.692	0.09
B	1.26	38.5 (293 K) 422.0 (86 K)	99	3.77	2.36	0.06
C	1.83	33.7 (293 K) 239.0 (77 K)	97	6.98	4.95	0.07
D	2.15	35.6 (293 K) 220.5 (77 K)	95	8.73	6.53	0.07
E	2.20	34.3 (293 K) 147.1 (77 K)	92	9.54	10.10	0.11

were taken at $0.15m_0$ and $0.78m_0$, (Kranzer, 1974) respectively, and g was assumed to be 4 to account for the valence band degeneracy at the Γ point. Over the samples studied, the activation energy Ea was found to be relatively insensitive to the degeneracy factor, changing by only 0.1 to 0.8 meV when g was decreased from 4 to 2. With the choice of g equals to 4, the compensation ratio (Nd/Na) in p-ZnSe varied between 0.06 and 0.11.

b. *Nitrogen Acceptor — Hydrogenic or not?*

After the success of rf nitrogen plasma doping, speculations arose as to the nature of nitrogen acceptors which made ZnSe p-type. Even though the activation energy of nitrogen in ZnSe has been measured indirectly by photoluminescence (Ohkawa et al., 1991) to be approximately 112 meV, in close agreement with the hydrogenic model, Bowers et al. (1994) nevertheless observed, using a HgSe-based contacting scheme to p-ZnSe (Lansari, 1992), a significant deviation of the activation energy from the hydrogenic model. The simple hydrogenic effective-mass model mandates that the substitutional dopant (with one extra electron or hole) behaves like a free hydrogen atom (with very similar eigenstates and energies) when the dielectric (ε) and crystalline (m^*) properties are taken into account.

The activation energy (or in the case of the hydrogen atom, ionization energy) Ea of substitutional impurities, as derived from temperature-de-

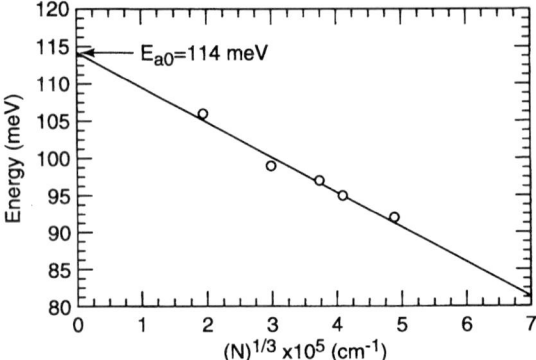

FIG. 24. Variation of nitrogen activation energy Ea in ZnSe with the cubic root of average ionized acceptor concentration.

pendent Hall measurements, is expected to be a function of doping level. The reason for the reduction of Ea with increasing doping level has been attributed to mechanisms such as screened Coulomb interaction (Debye and Conwell, 1954) and the overlap of impurity excited states (Stillman and Wolfe, 1976). The relation, $Ea = Ea_0 - a(N)^{1/3}$, first used by Debye and Conwell (1954), and later shown applicable to II–VI compounds, (Woodbury and Aven, 1973; Crowder and Hammer, 1966; El Akkad, 1987) was adopted here for a phenomenological description of the Ea-doping level dependence in nitrogen-doped ZnSe. Following Debye and Conwell (1954), N is the concentration of ionized acceptors averaged over the temperature range effective for the determination of Ea. Figure 24 shows Ea as a linear function of the cubic root of N having a slope $a = 4.7 \times 10^{-5}$ meV cm, which is close to the value (4.1×10^{-5}) found for shallow donor levels in n-type ZnSe (Woodbury and Aven, 1973). The activation energy Ea_0 at infinite dilution was extrapolated from Figure 23 to be 114 meV; this value is in good agreement with the activation energy (112 meV) previously derived from photoluminescence measurement (Ohkawa et al., 1991) as well as the value (128 meV) predicted by a simple hydrogenic model using a relative dielectric constant of 9.1 (Berlincourt et al., 1963). The linear dependence between Ea and $(N)^{1/3}$ over the range shown in Figure 24, as well as the approximate consistency between the extrapolated Ea at infinite dilution and the theoretical value, are supportive of the hydrogenic impurity model. In Section V.4 we will describe how a deviation from the hydrogenic behavior of nitrogen acceptors was observed when Mg and/or S are added into binary ZnSe to increase the bandgap energy.

3. COMPARATIVE STUDY OF NITROGEN DOPING FOR ZnSe AND ZnTe

As is discussed in Section IV.1a, ZnTe can be doped p-type while n-type doping has not been convincingly demonstrated. When the ZnTe epilayers were grown nominally at the same temperature, growth rate, and with the same nitrogen plasma operating conditions as ZnSe, free hole concentrations exceeding 1×10^{19} cm^{-3} were easily obtained. The striking difference in p-type doping between ZnTe and ZnSe has stimulated several theoretical studies as the answer to achieving a higher p-doping in ZnSe is intertwined with the underlying limiting factor. Two proposals, the so called "self-compensation" (by either native defect [Jansen and Sankey, 1989] or dopant-lattice relaxation [Chadi, 1991]) and the solubility limitations of various dopants (Laks et al., 1993; Van de Walle et al., 1993) have so far received the most attention. The compensation picture, however, does not appear to provide the limiting factor (at least for the growth conditions reported here), as transport (detailed in Section V.2a) revealed a compensation ratio in p-type ZnSe:N epilayers of only about 6% to 11%.

econdary ion mass spectroscopy (SIMS) was used in an effort to uncover the total incorporation of nitrogen atoms for the analysis of the impurity concentrations encountered in the semiconductor samples of the author's study. To determine a N concentration profile in a ZnSe or ZnTe epilayer, nitrogen was first implanted into undoped materials at a given energy and dose; the secondary ion signal was calibrated by arranging for the total amount of nitrogen atoms in the sample to essentially equal the implanted nitrogen dose.

Figure 25 shows nitrogen concentration versus depth profiles for (relatively) heavily doped ZnSe and ZnTe epilayers grown at 245°C. The nitrogen density [N] in ZnSe was found to be 1.2×10^{18} cm^{-3}, a value which is surprisingly consistent with the nitrogen acceptor concentration of Na = 9.5×10^{17} cm^{-3} derived from temperature-dependent Hall effect measurements on this sample. It is therefore concluded that, at the doping levels and particular growth conditions reported here, close to 100% of the nitrogen atoms ([N]) incorporated in ZnSe are substitutional at the Se lattice sites and serve as active acceptors (Na). While larger nitrogen concentrations, in the 10^{19} cm^{-3} range, have been reported from other SIMS studies (Qiu et al., 1991), the net acceptor concentrations were below 10^{18} cm^{-3}. The implication is that in those experiments either the excess nitrogen was not incorporated substitutionally, or there was a significant degree of compensation from donor-like defects.

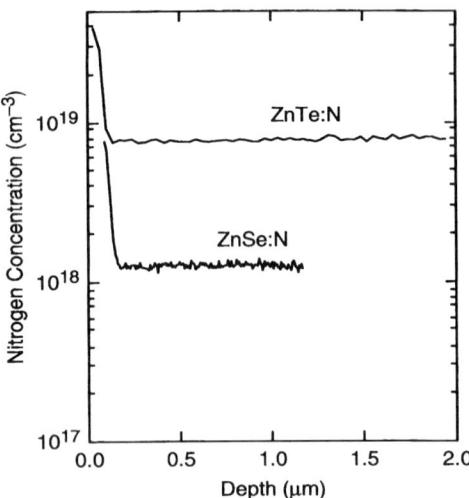

FIG. 25. Profiles of nitrogen concentration versus depth for heavily doped ZnSe and ZnTe epilayers grown under similar conditions.

In the case of ZnTe the nitrogen density [N] was found to be much higher, for example 8×10^{18} cm^{-3} as shown in Figure 25, than in the case of ZnSe even though the growth conditions were nominally the same. Temperature-dependent Hall effect measurements were performed on the heavily doped ZnTe samples, but no freeze-out of free carriers was observed (Han et al., 1993a). Due to the lack of freeze-out the value of Na could not be obtained from curve fitting to the charge neutrality relation as was done in the case of ZnSe. However, the room temperature free hole concentration in the sample of Figure 25 was 7×10^{18} cm^{-3}, which suggests that the nitrogen acceptor concentration Na is at least 7×10^{18} cm^{-3}. We have studied the effects of growth temperature and flux ratio on nitrogen doping in ZnTe and found that, over the temperature range from 245°C to 400°C, the total nitrogen concentration [N] and free hole concentration are relatively insensitive to growth temperature. It is therefore concluded that almost 100% of the nitrogen atoms incorporated into ZnTe are at the Te lattice sites and serve as acceptors.[1]

[1] For both ZnSe and ZnTe the activation energies associated with the nitrogen acceptor (Na) level, as determined from temperature-dependent Hall measurements, appear to correlate very well with calculations for a substantial impurity assuming a hydrogenic effective-mass model (Fan et al., 1992; Fan et al., 1994). The close similarity between nitrogen ([N]) and net acceptor (Na) concentrations yields evidence that all of the nitrogen dopant atoms are incorporated substitutionally at anion sites.

A theoretical model, based on computing the solid solubility of dopant atoms in various host semiconductors, was proposed by Laks *et al.* (1993) to explain the differences in *p*-doping levels between ZnS and ZnTe. The calculations predicted that, under thermodynamic equilibrium conditions, the solubility of Li and N in ZnTe should be much higher than in ZnSe. The solubility argument, if applied to the kinetic situation obtaining for the nonequilibrium growth by MBE, would pose an explanation as to why it is more difficult to make *p*-type ZnSe than *p*-type ZnTe.

4. NITROGEN-DOPED WIDE BANDGAP ALLOYS OF ZnSe: AX CENTERS?

When extending the study of hole transport in ZnSe into other functionally important wider bandgap alloys such as $ZnS_{0.06}Se_{0.94}$ and $Zn_{0.92}Mg_{0.88}S_{0.12}Se_{0.88}$ (through the use of the Zn(Se, Te) graded contact), the difference in the temperature dependence; i.e., the freeze-out behavior, becomes quite dramatic as revealed by the slopes of the lines in Figure 26. The activation energy of nitrogen in the (Zn, Mg)(S, Se) can be extracted from the linear region of the freeze-out curve; a value of 177 meV is obtained. The activation energy is seen to increase by about 80 meV in going from Zn(S, Se) to ZnMgSSe, a change which can not be accounted for by an effective mass model assuming Vegard's law for physical parameters. Also, at low temperatures, a second slope appears in the ZnMgSSe data; the

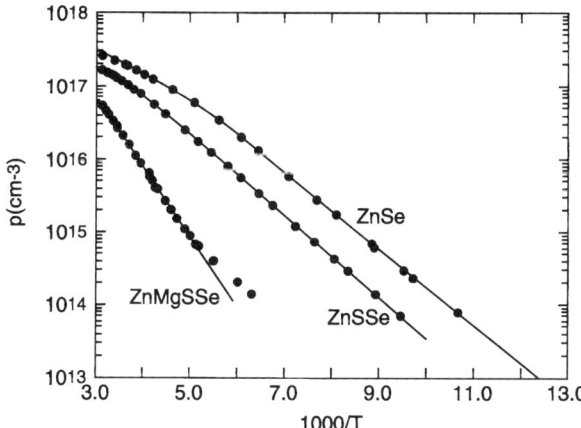

FIG. 26. Measured free hole concentrations of nitrogen-doped ZnSe, ZnSSe, and ZnMgSSe. The difference in the activation energy is reflected by the different slopes of the freeze-out curves.

doping level in the quaternary is too low for this second slope to be related to impurity band conduction (Han et al., 1994b). Given a lack of carrier (hole) saturation up to 320 K, the values of Na and Nd deduced from curve fitting to a simple charge neutrality relation for (Zn, Mg)(S, Se) may be somewhat higher than the real values. (Determination of the compensation ratio Nd/Na is unaffected by the lack of saturation, however.) The trend of rapid increase of nitrogen activation energy (as the bandgap increases with increasing Mg and S fractions), an issue which is certainly of scientific interest, has implications to the technological development of II–VI blue laser diodes. The increasing difficulty of electrical activation of p-type ZnMgSSe:N, which translates to a higher device impedance, was reflected in the high threshold voltage (10.5 V) of the blue laser (460 nm) as compared to the green diode laser (508 nm) which has been operated at around 5 V (Ringle et al., 1994b).

It is known from the study of n-doping of $Al_xGa_{1-x}As$ that an increase of Al fraction (to above 20%) causes both an apparent increase in the activation energy of donors and a second slope to appear at low temperatures in the Hall measurement data (Malloy and Khachaturyan, 1993), a phenomenon associated with the "DX" center behavior. One of the more striking manifestations of a DX center is the presence of persistent photoconductivity (PPC).

Temperature-dependent Hall measurements were repeated with the ZnMgSSe samples, now with the addition of low temperature illumination, to look for a PPC effect (Han et al., 1994c). An Arrhenius plot of the data is shown in Figure 27. The filled circles were taken as the sample was cooled in the dark. After reaching 110 K, the sample was illuminated with a white light for approximately 5 sec; the free hole concentration increased by orders of magnitude. The increase in hole concentration remains even after the light was turned off. (In a separate experiment, resistivity measurements of ZnMgSSe taken at 90 K showed no measurable change in the reduced resistance at 30 min after the illumination was removed.) The open circles in Figure 27a represent data taken as the sample was warmed in the dark after illumination; the initial warm-up process was well-characterized by an exponential rise in the free-hole concentration. Once the temperature exceeded 180 K, the hole concentration decreased and eventually converged to the value observed during the cooling of the sample. The activation energy of nitrogen in ZnMgSSe at low temperatures derived from the Hall data is 93 meV before illumination and 90 meV after illumination. This activation energy is similar to that obtained for ZnSe, and thus is the value expected if the dopant behaves in a hydrogenic manner. (The earlier study of temperature-dependent Hall measurements by Fan et al. [1993] led the authors to speculate that the transport behavior of nitrogen acceptors in

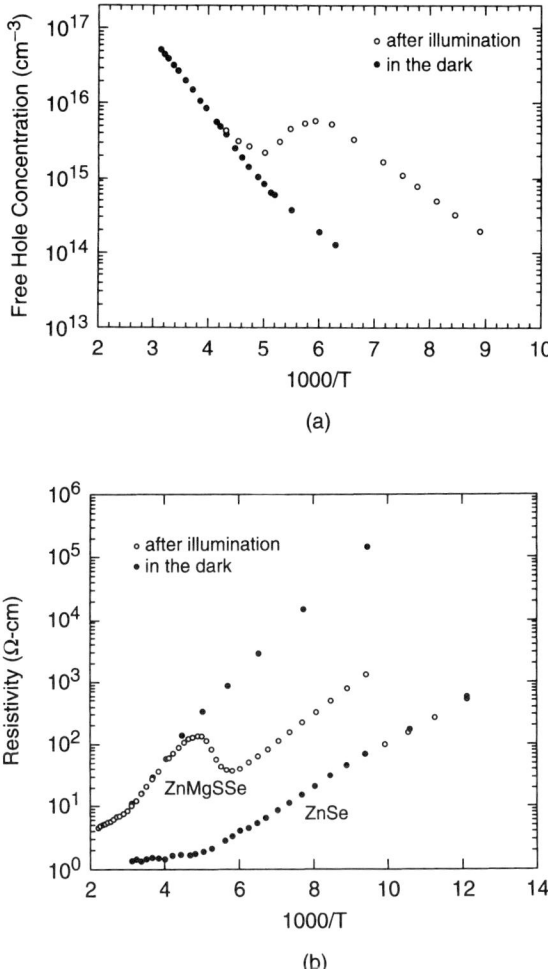

FIG. 27. (a) Free hole concentration versus inverse temperature for a ZnMgSSe epilayer. The solid dots represent data taken in the dark. The second slope in the Arrhenius plot of free hole concentration, which appears below some critical temperature, is interpreted to represent the presence of a shallow hydrogenic acceptor state. The photoionization occurring under illumination for photons above about 1.65 eV results in a metastable persistent photoconductivity supported by free holes in thermodynamic equilibrium with the shallow acceptor state. (b) Resistivity of a rectangular bar versus inverse temperature. The persistent photoconductivity in the quaternary is associated with a drop in resistivity of more than two orders of magnitude. Under similar illumination, the ZnSe epilayer shows about a 10% reduction in resistivity at low temperature.

p-ZnSe supports a hydrogenic effective-mass model.) Temperature-dependent resistivity measurements of ZnSSe also show PPC, but the resistance only changed by a factor of 4 upon illumination; ZnSe only changed by at most 10% when illuminated (Fig. 27b). If the PPC was a macroscopic effect associated with a spatial separation of carriers due to the GaAs/II–VI heterojunction, it is reasonable to expect that it would also have been observed for ZnSe, and further that one would not expect an increase in carrier concentration as the sample was warmed.

While the observation of PPC is a necessary condition for a microscopic lattice relaxation phenonema, it is not sufficient. A more important criterion is a large difference between the thermal and optical activation energies for the defect (Henry and Lang, 1977), implying that the dopant undergoes a strong coupling to the host lattice. Optical cross-section measurements were performed by measuring the photoconductivity transient as a function of wavelength (Fig. 28); the rate of photon capture, as revealed in the rate of increase in conductivity, was interpreted as a measure of optical cross section. Illumination was provided by a tungsten lamp and a particular wavelength was selected using a monochromator. Intensities were normalized using a calibrated silicon photodetector. The optical activation energy derived from this measurement is 1.65 eV, which is indeed much greater than the thermal activation energy (177 meV).

Phenomenological comparisons between the behavior of n-type III–V compounds and that of nitrogen-doped alloys of ZnSe tend to support the

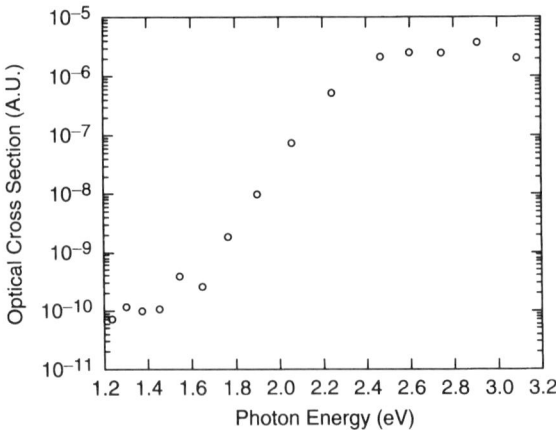

FIG. 28. Optical capture cross section as a function of irradiation wavelengths. The ionization threshold is determined to be about 1.65 eV.

assertion that nitrogen acceptors in *p*-ZnMgSSe possess a DX nature; the proposed relaxed state here should perhaps be called an "A(acceptor)X center" (Han *et al.*, 1994c). The deep localized AX state ($Ea = 177$ meV) and its hydrogenic state ($Ea \sim 90$ meV) are separated by thermal (capture and emission) barriers as a result of phonon coupling. The transition probability (between these two states) therefore can be enhanced or quenched by the ambient temperature. The appearance of the second, less steep freeze-out slope ($Ea \sim 90$ meV) observed in the dark below the critical temperature Tc signifies that the capture of holes into the deep localized levels (AX centers) is inhibited by a thermal barrier, a process similar to the quenching of nonradiative centers at low temperature (Henry and Lang, 1977).

Below Tc, and in the dark, the transition between valence-band holes and the relatively shallow hydrogenic level becomes the only allowed exchange, hence the reduced slope of the freeze-out characteristics. At low temperature, illumination with photon energies greater than the optical ionization energy releases holes from the localized deep states. However, at the low temperatures trapping of holes into the deep AX centers (via multiple phonon absorption/emission) is very unlikely due to the lack of sufficient thermal phonons at low temperatures. The capture barrier thus facilitates a low-temperature, metastable population of holes which are restricted to the valence band–hydrogenic acceptor combination; the result is the observed persistence of photoconductivity. The similarity of the value of Ea for the shallower level of ZnMgSSe:N to that of ZnSe:N provides compelling evidence that both originate from the hydrogenic effective-mass impurity level. The warm-up process below Tc can be modeled as a *p*-ZnMgSSe sample doped with only "shallow" nitrogen acceptors, specifically with Ea of about 90 meV. When the temperature approaches Tc, the thermal capture barrier can be overcome and the trapping into deep AX centers starts to dominate the electrical transport, as seen by the decrease in the free hole concentration from 160 to 200 K. In fact, strikingly similar behavior, in terms of the freeze-out in the dark and the observed PPC, has been reported in *n*-type AlGaAs doped by Te (Dmochowski *et al.*, 1989).

The thermal capture barrier therefore can be measured near this temperature range, since the temperature is low enough such that the emission of AX states back to the hydrogenic form is still not in effect. Upon turning off the illumination, the time rate of decay of the PPC yields the capture cross section for the AX center. The upper temperature range (~ 180 K) was limited by the onset of the emission process (from AX centers), a process which tends to complicate the interpretation of the rate curve; the lower temperature range (below which the decay rate becomes too slow) was set

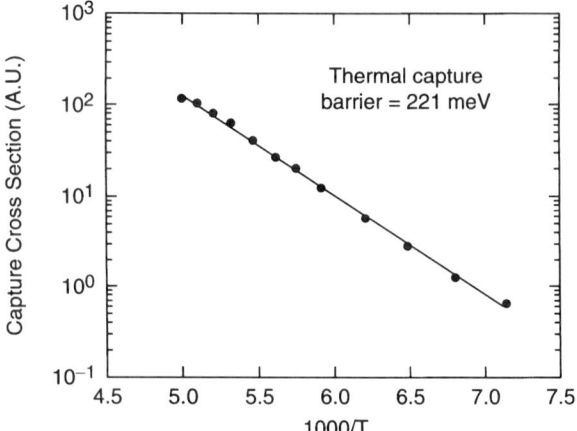

FIG. 29. Arrhenius plot of thermal capture coefficient (from conductance transient measurements at various temperatures). A thermal capture barrier of 221 meV is determined from the slope of this plot.

such that the transient measurement can be performed within reasonable time frame. The slope of an Arrhenius plot of the capture cross section provides a value for the energy (229 meV) needed to surmount the capture barrier (Fig. 29).

Figure 30a plots the measured thermal capture barrier (solid circles) as a function of the bandgap energy for several epilayers. The data of measured E_a are plotted on the same graph. From a configuration–coordinate diagram such as typically employed to explain DX centers, it is straightforward to extrapolate the emission barrier height E_e using the relation $E_e = E_a + E_c$. Figure 30b shows a plot of the derived E_e as a function of the bandgap. A value of about 0.4 eV, relatively independent of the alloy fractions, can be assigned to the (Zn, Mg)(S, Se):N alloy family. It is interesting to compare this result to the measured E_e of DX centers in Si-doped n-type (Al, Ga)As (Malloy and Khachaturyan, 1993). For both n-type III–V and p-type II–VI compounds the emission barrier is quite insensitive to the alloy fractions despite the variations of the capture barrier E_c and the apparent activation energy E_a. The constant emission barrier in (Al, Ga)As has been modeled by a two-step (three levels) transition involving an intermediate satellite (L) conduction band (Theis et al., 1988); the valence band does not have satellite bands. The measured E_e of 0.4 eV is quite small as compared to the thermal barrier expected when the lattice

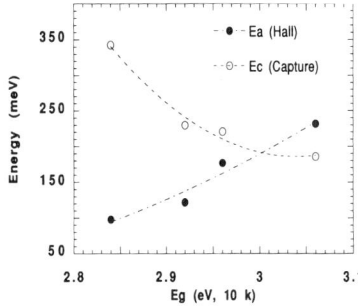 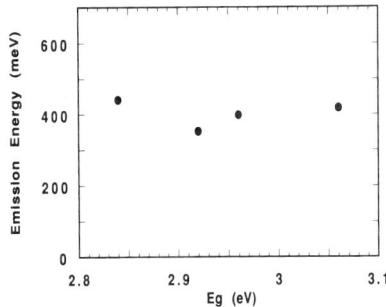

FIG. 30. (a) The measured thermal capture barrier (solid circles) as a function of the quaternary bandgap for several epilayers. Open circles show the measured apparent activation E'_a. (b) The emission barrier height E_e, assuming that the relation $E_e = E'_a + E_c$ holds, as a function of the bandgap.

relaxation bond breaking occurs between host atoms (~1.6 eV [Garcia and Northrup, 1995]). The implications is that the nitrogen acceptor is directly involved in the microscopic lattice relaxation, a mechanism responsible for the observed deepening and compensation of nitrogen doping in p-$(Zn, Mg)(S, Se)$.

We see that the rapid increase of nitrogen activation energy E_a from ZnSe to ZnMgSSe seems closely analogous to that of donors in GaAs–AlGaAs. Thus the "deepening" observed in the II–VI system is likely due to the relative positioning between the Γ-point valence-band edge (sensitive to the alloy composition) and the highly localized deep state (less sensitive to the alloy composition). The addition of Mg and S to ZnSe moves the valence band edge (and the hydrogenic acceptor level) to lower energy, and causes the AX level (now lying within the bandgap) to be energetically more favorable.

VI. Conclusion

The MBE growth technique contributed much to the development of the wide bandgap II–VI. The current low level of extended defects, now into the 10^3cm^{-2}, is respectable for any heterostructure system. The ability to form quantum well structures, to reach useful levels of p and n-doping, the means for forming ohmic contacts to p-ZnSe, and the ability to grow a variety of alloys of ZnSe, led to the present injection laser diodes.

Acknowledgments

This work represents a team effort with a number of graduate students whose names appeared in the cited references. It is especially a pleasure to acknowledge contributions from several collaborators including A. V. Nurmikko (Brown University), L. A. Kolodziejski (MIT), M. Kobayashi (Chiba University, Japan), and N. Otsuka (JAIST, Japan).

References

Akimoto, K., Miyajima, T., and Mori, Y. (1989a). *Phys. Rev. B* **39**, 3138.
Akimoto, K., Miyajima, T., and Mori, Y. (1989b). *Jpn. J. Appl. Phys.* **28**, L531.
Aven, M. (1967). In *II–VI Semiconducting Compounds*, ed. D. G. Thomas. Benjamin, New York. 1232.
Baldereschi, A., and Lipari, N. O. (1973). *Phys. Rev.* **B8**, 2697.
Ballard, S. S., Brown, S. E., and Browder, J. S. (1978). *Appl. Opt.* **17**, 1152.
Baltensperger, W. (1953). *Philos. Mag.* **44**, 1355.
Berlincourt, D., Jaffe, H., and Schlozawa, L. R. (1963). *Phys. Rev.* **129**, 1009.
Blakemore, J. S. (1982). *J. Appl. Phys.* **53**, R123.
Born, M., and Wolf, E. (1965). *Principles of Optics*, 3rd ed. Pergamon, New York.
Bour, D. P., and Shealy, J. R. (1988). *IEEE J. Quantum Electron.* **24**, 1856.
Bowers, K. A., Yu, Z., Gossett, K. J., Cook, J. W. Jr., Schetzina, J. F. (1994). *J. Electron. Mat.* **23**, 251.
Bylsma, R. B., Becker, W. M., Bonsett, T. C., Kolodziejski, L. A., Gunshor, R. L., Yamanishi, M., and Datta, S. (1985). *Appl. Phys. Lett.* **47**, 1039.
Cammack, D. A., Dalby, R. J., Cornelissen, H. J., and Khurgin, J. (1987). *J. Appl. Phys.* **62**, 3071.
Capasso, F., and Margaritondo, G. (1987). *Heterojunction Band Discontinuities*. North-Holland, Amsterdam. 442.
Chadi, D. J., and Chang, K. J. (1989). *Appl. Phys. Lett.* **55**, 575.
Chadi, D. J. (1991). *Appl. Phys. Lett.* **59**, 3589.
Cheng, H., DePuydt, J. M., Potts, J. E., and Haase, M. A. (1989). *J. Cryst. Growth* **95**, 512.
Cho, A. Y., (1976). *J. Appl. Phys.* **47**, 2841.
Chu, C. C., Ng, T. B., Han, J., Hua, G. C., Gunshor, R. L., Ho, E., Warlick, E. L. Kolodziejski, L. A., Nurmikko, A. V., (1996) Appl. Phys. Lett. 69, 602.
Chwang, R., Smith, B. J., and Crowell, C. R. (1974). *Solid State Electron.* **17**, 1217.
Crowder, B. L., and Hammer, W. N. (1966). *Phys. Rev.* **150**, 541.
Debye, P. P., and Conwell, E. M. (1954). *Phys. Rev.* **93**, 693.
Ding, J., Pelekanos, N., Nurmikko, A. V., Luo, H., Samarth, N., and Furdyna, J. K. (1990). *Appl. Phys. Lett.* **57**, 2885.
Dmochowski, J. E., Dobaczewski, L., Langer, J. M., and Jantsch, W. (1989). *Phys. Rev. B* **40**, 9671.
Durbin, S. M., Han, J., O. S., Kobayashi, M., Menke, D. R., Gunshor, R. L., Fu, Q., Pelekanos, N., Nurmikko, A. V., Li, D., Gonsalves, J., and Otsuka, N. (1989). *Appl. Phys. Lett.* **55**, 2087.
Eason, D. B., Ren, J., Yu, Z., Hughes, C., Cook, Jr., J. W., Schetzina, J. F., El-Masry, N. A., Cantwell, G., and Harsch, W. C. (1995). *J. Cryst. Growth* **150**, 718.

El Akkad, F. (1987). *Semicond. Sci. Technol.* **2**, 629.
El-Emawy, A. A., Qiu, Y., Osinsky, A., Littlefield, E., and Temkin, H. (1995). *Appl. Phys. Lett.* **67,** 1238.
Fan, Y., Han, J., He, L., Saraie, J., Gunshor, R. L., Hagerott, M., Jeon, H., Nurmikko, A. V., Hua, G. C., and Otsuka, N. (1992). *Appl. Phys. Lett.* **61**, 3160.
Fan, Y., Han, J., He, L., Saraie, J., Gunshor, R. L., Hagerott, M., and Nurmikko, A. V. (1993). *Appl. Phys. Lett.* **63**, 1812.
Fan, Y., Han, J., He, L., Saraie, J., Gunshor, R. L., Hagerott, M., and Nurmikko, A. V. (1994). *J. Electron. Mat.* **23**, 245.
Fan, Y. (1995). *Appl. Phys. Lett.* **67,** 1739.
Farrel, H. H., Tamargo, M. C., and de Miguel, J. L. (1988). *J. Vac. Sci. Technol.* **B6**, 767.
Fitzpatrick, B. J., Werkhoven, C. J., McGee, III, T. F., Harnack, P. M., Herko, S. P., Bhargava, R. N., and Dean, P. J. (1981). *IEEE Trans. Electron. Devices* **ED-28**, 440.
Franciosi, A., Vanzetti, L., Bonanni, A., Sorba, L., Bratina, G., and Biasiol, G. (1994). *II–VI Blue/Green Laser Diodes.* SPIE Vol. 2346, 100.
Gaines, J. M., Petruzzello, J., and Greenberg, B. (1993). *J. Appl. Phys.* **73**, 2835.
Gaines, J. M., Drenten, R. R., Haberrem, K. W., Marshall, T., Mensz, P., and Petruzzello, J. (1993). *Appl. Phys. Lett.* **62**, 2462.
Garbuzov, D. Z., and Khalfin, V. B. (1993). In *Quantum Well Lasers*, ed. P. S. Zory, Jr., Chapter 6, Academic Press, San Diego.
Garcia, A., and Northrup, J. E. (1995). *Phys. Rev. Lett.* **74**, 1131.
Gasiorowicz, S. (1974). *Quantum Physics*, John Wiley & Sons, New York.
Grillo, D. C., Fan, Y., He, L., Han, J., Gunshor, R. L., Hagerott, M., Jeon, H., Salokatve, A., Nurmikko, A. V., Hua, G. C., and Otsuka, N. (1993). *Appl. Phys. Lett.* **63**, 2723.
Guha, S., DePuydt, J. M., Haase, M. A., Qiu, J., Cheng, H. (1993a). *Appl. Phys. Lett.* **63**, 3107.
Guha, S., Munekata, H., and Chang, L. L. (1993b). *J. Appl. Phys.* **73**, 2294.
Gunshor, R. L., Kolodziejski, L. A., Melloch, M. R., Vaziri, M., Choi, C., and Otsuka, N. (1987). *Appl. Phys. Lett.* **50**, 200.
Gunshor, R. L., Kolodziejski, L. A., Nurmikko, A. V., and Otsuka, N. (1991). *Semiconductors and Semimetals*, Vol. 33, 337.
Haase, M. A., Qiu, J., DePuydt, J. M., and Cheng, H. (1991). *Appl. Phys. Lett.* **59**, 1273.
Haase, M. A., Baude, P. F., Hagedom, M. S., Qiu, J., DePuydt, J. M., Cheng, H., Guha, S., Hofler, G. E., and Wu, B. J. (1993). *Appl. Phys. Lett.* **63**, 2315.
Halliwell, M. A. G., Lyons, M. H., and Hill, M. J. (1984). *J. Cryst. Growth* **68**, 523.
Han, J., Durbin, S. M., Gunshor, R. L., Kobayashi, M., Menke, D. R., Pelekanos, N., Hagerott, M., Nurmikko, A. V., Nakamura, Y., and Otsuka, N. (1991). *J. Cryst. Growth* **111**, 767.
Han, J., Stavrinides, T. S., Kobayashi, M., Gunshor, R. L., Hagerott, M. M., and Nurmikko, A. V. (1993a). *Appl. Phys. Lett.* **62**, 840.
Han, J., Stavrinides, T. S., Kobayashi, M., Gunshor, R. L., Hagerott, M. M., and Nurmikko, A. V. (1993b). *J. Electron. Mat.* **22**, 485.
Han, J., He, L., Grillo, D. C., Fan, Y., Ringle, M. D., Gunshor, R. L., Jeon, H., Salokatve, A., Nurmikko, A. V., Hua, G. C., and Otsuka, N. (1994a). *J. Vac. Sci. Tech.* **B12**, 1254.
Han, J., Fan, Y., Ringle, M. D., He, L., Grillo, D. C., Gunshor, R. L., Hua, G. C., and Otsuka, N. (1994b). *J. Cryst. Growth* **138**, 464.
Han, J., Ringle, M. D., Fan, Y., Gunshor, R. L., and Nurmikko, A. V. (1994c). *Appl. Phys. Lett.* **65**, 3230.
Harrison, W. A., Kraut, E. A., Waldrop, J. R., and Grant, R. W. (1978). *Phys. Rev. B* **18**, 4402.
Haugen, G. M., Guha, S., Cheng, H., DePuydt, J. M., Haase, M. A., Hofler, G. E., Qiu, J., and Wu, B. J. (1995). *Appl. Phys. Lett.* **66**, 358.
Hayes, J. R., Capasso, F., Malik, R. J., Gossard, A. C., and Wiegmann, W. (1983). *Appl. Phys.*

Lett. **43**, 949.
Henry, C. H., and Lang, D. V. (1977). *Phys. Rev. B* **15**, 989.
Hiei, F., Ikeda, M., Ozawa, M., Miyajima, T., Ishibashi, A., and Akimoto, K. (1993). *Electron. Lett.* **29**, 878.
Hishida, Y., Ishii, H., Toda, T., and Niina, T. (1989). *J. Cryst. Growth* **95**, 517.
Hovinen, M., Ding, J., Salokatve, A., Nurmikko, A. V., Hua, G. C., Grillo, D. C., He, L., Han, J., Ringle, M., and Gunshor, R. L. (1995). *J. Appl. Phys.* **77**, 4150.
Hua, G. C., Otsuka, N., Grillo, D. C., Fan, Y., Ringle, M. D., Han, J., Gunshor, R. L., Hovinen, M., and Nurmikko, A. V. (1994). *Appl. Phys. Lett.* **65**, 1331.
Ishibashi, A. (1995). *IEEE J. Selected Topics Quantum Electron.* **1**, 741.
Itoh, S., Okuyama, H., Matsumoto, S., Nakayama, N., Ohata, T., Miyajima, T., Ishibashi, A., and Akimoto, K. (1993). *Electron. Lett.* **29**, 766.
Jansen, R. W., and Sankey, O. F. (1989). *Phys. Rev. B* **39**, 3192.
Jeon, H., Ding, J., Patterson, W., Nurmikko, A. V., Xie, W., Grillo, D. C., Kobayashi, M., and Gunshor, R. L. (1991). *Appl. Phys. Lett.* **59**, 3619.
Kamata, A., and Mitsuhashi, H. (1994). *J. Cryst. Growth* **142**, 31.
Kitagawa, F., Mishima, T., and Takahashi, K. (1981). *J. Electrochem. Soc.* **127**, 937.
Kittel, C. (1976). *Introduction to Solid State Physics*, 5th ed. John Wiley & Sons, New York. 100.
Kobayashi, N. (1988). *Jpn. J. Appl. Phys.* **27**, L1597.
Kolodziejski, L. A., Gunshor, R. L., Otsuka, N., and Choi, C. (1986). *J. Vac. Sci. Technol.* **A4**, 2150.
Kolodziejski, L. A., Gunshor, R. L., Fu, Q., Lee, D., Nurmikko, A. V., Gonsalves, J. M., and Otsuka, N. (1988). *Appl. Phys. Lett.* **52**, 1080.
Kowalczyk, S. P., Kraut, E. A., Waldrop, J. R., and Grant, R. W. (1982). *J. Vac. Sci. & Technol.* **21**, 482.
Kranzer, D. (1974). *Phys. Status Solidi* **A26**, 11.
Kroger, F. A. (1965). *J. Chem. Phys. Solids* **26**, 1717.
Kuo, L. H., Salamanca-Riba, L., DePuydt, J. M., Cheng, H., and Qiu, J. (1994a). *Philosoph. Mag.* **A69**, 301.
Kuo, L. H., Salamanca-Riba, L., Wu, B. J., DePuydt, J. M., Haugen, G. M., Guha, S., and Haase, M. A. (1994b). *Appl. Phys. Lett.* **65**, 1230.
Laks, D. B., Van de Walle, C. G., Neumark, G. F., and Pantelides, S. T. (1993). *Appl. Phys. Lett.* **63**, 1375.
Lansari, Y., Ren, J., Sneed, B., Bowers, K. A., Cook, Jr., J. W., and Schetzina, J. F. (1992). *Appl. Phys. Lett.* **61**, 3160.
Li, D., Gonsalves, J. M., Otsuka, N., Qiu, J., Kobayashi, M., and Gunshor, R. L. (1990). *Appl. Phys. Lett.* **57**, 449.
Li, D. and Pashley, M., (1994). *J. Vac. Sci. Technol.* **B12**, 2547.
Lim, S. W., Hondo, T., Koyama, F., Iga, K., Inoue, K., Yanashima, K., Munekata, H., and Kukimoto, H. (1994). *Appl. Phys. Lett.* **65**, 2437.
Malloy, K. J., and Khachaturyan, K. (1993). *Semiconductors and Semimetals*, ed. R. K. Willardson, A. C. Beer, and E. R. Weber. V.38, 235. Academic Press, Inc.
Mandel, G. (1964a). *Phys. Rev.* **134**, A1073.
Mandel, G., Morehead, F. F., and Wagner, P. R. (1964b). *Phys. Rev.* **136**, A826.
Menke, D. R., Qiu, J., Gunshor, R. L., Kobayashi, M., Li, D., Nakamura, Y., and Otsuka, N. (1991). *J. Vac. Sci. Technol.* **B9**, 2171.
Miotkowski, I., and Ramdas, A. K. (1991). Physics Department, Purdue University, West Lafayette, Indiana.
Mitsuyu, T., Ohkawa, K., and Yamazaki, O. (1986). *Appl. Phys. Lett.* **49**, 1348.

Miyajima, T., Okuyama, H., and Akimoto, A. (1992). *Jpn. J. Appl. Phys.* **31,** L1743.
Nakayama, N., Itoh, S., Nakano, K., Okuyama, H., Ozawa, M., Ishibashi, A., Ikeda, M., and Mori, Y. (1993). *Electronics Lett.* **29,** 1488.
Ng, T., Hua, G. C., Han, J. and Gunshor, R. L. (1995). Unpublished.
Ohkawa, K., Karasawa, T., and Mitsuyu, T. (1991). *Japan. J. Appl. Phys.* **30,** L152.
Okuyama, H., Nakano, K., Miyajima, T., and Akimoto, K. (1991). *Jpn. J. Appl. Phys.* **30,** L1620.
Okuyama, H., Miyajima, T., Morinaga, Y., Hiei, F., Ozawa, M., and Akimoto, K. (1992). *Electr. Lett.* **28,** 1798.
Ozawa, M., Hiei, F., Ishibashi, A., and Akimoto, K. (1993). *Electr. Lett.* **29,** 503.
Park, R. M., Mar, H. A., and Salansky, N. M. (1985). *J. Appl. Phys.* **58,** 1047.
Park, R. M., Troffer, M. B., Rouleau, C. M., DePuydt, J. M., and Haase, M. A. (1990). *Appl. Phys. Lett.* **57,** 2127.
Petroff, P. M., and Kimerling, L. C. (1976). *J. Appl. Phys.* **29,** 461.
Petruzzello, J., Greenberg, B. L., Cammack, D. A., and Dalby, R. (1988). *J. Appl. Phys.* **63,** 2299.
Qian, Q.-D., Qiu, J., Melloch, M. R., Cooper, Jr., J. A., Kolodziejski, L. A., Kobayashi, M., and Gunshor, R. L. (1989a). *Appl. Phys. Lett.* **54,** 1359.
Qian, Q.-D., Qiu, J., Kobayashi, M., Gunshor, R. L., Melloch, M. R., and Cooper, Jr., J. A. (1989b). *J. Vac. Sci. Technol.* **B7,** 793.
Qiu, J., Qian, G.-D., Gunshor, R. L., Kobayashi, M., Menke, D. R., Li, D., and Otsuku, N. (1990). *Appl. Phys. Lett.* **56,** 1272.
Qiu, J., DePuydt, J. M., Cheng, H., and Haase, M. A. (1991). *Appl. Phys. Lett.* **59,** 2992.
Ren, J., Bowers, K. A., Sneed, A., Dreifus, D. L., Cook, Jr., J. W., Schetzina, J. F., and Kolbas, R. M. (1990). *Appl. Phys. Lett.* **57,** 1901.
Ringle, M. D., Grillo, D. C., Han, J., Gunshor, R. L., Hua, G. C., and Nurmikko, A. V. (1994a). *Inst. Phys. Conf. Ser.* No. 141: Chapter 5, 513.
Ringle, M. D., Grillo, D. C., Fan, Y., He, L., Han, J., Gunshor, R. L., Salokatve, A., Jeon, H., Hovinen, M., Nurmikko, A. V., Hua, G. C., and Otsuka, N. (1994b). *Mat. Res. Soc. Symp. Proc.* Vol. 340, 431.
Salokatve, A., Jeon, H., Ding, J., Hovinen, M., Nurmikko, A. V., Grillo, D. C., He, L., Han, J., Fan, Y., Ringle, M., Gunshor, R. L., Hua, G. C., and Otsuka, N. (1993). *Electronics Lett.* **29,** 2192.
Schroder, D. K. (1990). *Semiconductor Material and Device Characterization.* John Wiley & Sons, New York.
Segmuller, A., Noyan, I. C., and Speriosu, V. S. (1989). *Prog. Crystal Growth and Charact.* Vol. 18, 21.
Shibi, S. M., Tamargo, M. C., Shromme, B. J., Schwartz, S. A., Schwartz, C. L., Nahory, R. E., and Martin, R. J. (1990). *J. Vac. Sci. Technol.* **B8,** 187.
SpringThorpe, A. J., and Majeed, A. (1990). *J. Vac. Sci. Technol.* **B8,** 266.
Stillman, G. E., and Wolfe, C. M. (1976). *Thin Solid Films* **31,** 69.
Studtmann, G. D., Gunshor, R. L., Kolodziejski, L. A., Melloch, M. R., Cooper, Jr., J. A., Pierret, R. F., Munich, D. P., Choi, C., and Otsuka, N. (1988). *Appl. Phys. Lett.* **52,** 1249.
Stutius, W. (1982). *Appl. Phys. Lett.* **40,** 246.
Takeuchi, S., Suzuki, K., Maeda, K., and Iwanaga, H. (1984). *Philosoph. Mag.* **A50,** 171.
Tamargo, M. C., de Miguel, J. L., Hwang, D. M., and Farrell, H. H. (1988). *J. Vac. Sci. Technol.* **B6,** 784.
Theis, T. N., Mooney, P. M., and Wright, S. L. (1988). *Phys. Rev. Lett.* **60,** 361.
Van der Pauw, L. J. (1958). *Philips Res. Rep.* **13,** 1.
Van de Walle, C. G., Laks, D. B., Neumark, G. F., and Pantelides, S. T. (1993). *Phys. Rev. B* **47,** 9425.
Vanzetti, L., Bonanni, A., Bratina, G., Sorba, L., Franciosi, A., Lomascolo, M., Greco, D., and

Cingolani, R. (1995). *J. Cryst. Growth* **150,** 765.
Vaudo, R. P., Yu, Z., Cook, J. W. Jr., Schetzina, J. F. (1993). *Optics Lett.* **18,** 1843–1845.
Watts, R. K., Holton, W. C., de Wit, M. (1971). *Phys. Rev.* **B3,** 404.
Woodall, J. M., Freeouf, J. L., Pettit, G. D., Jackon, T., and Kirchner, P. (1981). *J. Vac. Sci. Technol.* **19,** 626.
Woodbury, H. H., and Aven, M. (1973). *Phys. Rev.* **B9,** 5195.
Wright, S. L., Jackson, T. N., and Marks, R. F. (1990). *J. Vac. Sci. Technol.* **B8,** 288.
Wu, B. J., Cheng, H., Guha, S., Haase, M. A., DePuydt, J. M., Meis-Haugen, G., and Qiu, J. (1993). *Appl. Phys. Lett.* **63,** 2935.
Xie, W., Grillo, D. C., Kobayashi, M., Gunshor, R. L., Hua, G. C., Otsuka, N., Jeon, H., Ding, J., and Nurmikko, A. V. (1992). *Appl. Phys. Lett.* **60,** 463.
Yao, T. (1985). In *The Technology and Physics of Molecular Beam Epitaxy*, ed. E. H. C. Park. Plenum Press, New York.
Yao, T., and Okada, Y. (1986). *Jpn. J. Appl. Phys.* **25,** 821.
Yao, T., Okada, Y., Matsui, S., and Ishida, K. (1987). *J. Cryst. Growth* **81,** 518.
Yeh, C., Lu, Z. W., Froyen, S., and Zunger, A. (1992). *Phys. Rev. B* **46,** 10086.

CHAPTER 2

Growth and Characterization of ZnSe-based II–VI Semiconductors by MOVPE

Shizuo Fujita and Shigeo Fujita

DEPARTMENT OF ELECTRONIC SCIENCE AND ENGINEERING
KYOTO UNIVERSITY
KYOTO, JAPAN

I. INTRODUCTION	59
II. BRIEF REVIEW OF PIONEERING WORKS	60
III. RECENT GROWTH TECHNIQUES	61
1. *New Precursors*	61
2. *Photo-assisted Growth*	65
IV. DOPING AND DEVICES	72
1. n-*Type Doping*	72
2. p-*Type Doping*	74
3. *Present Status of Device Applications*	77
REFERENCES	79

I. Introduction

Recent progress of blue-green laser diodes (LDs) and light-emitting diodes (LEDs) with ZnSe-based II–VI semiconductors (Ishibashi, 1995; Eason et al., 1995) has attracted much attention for the development of future optoelectronics. It should be noted, however, that all those devices have been fabricated by molecular beam epitaxy (MBE) with solid sources. Nevertheless, difficulty remains with the precise control of alloy composition, which is now a key to realize highly reliable devices with low dislocation densities in the constituent epilayers. This problem arises from high vapor pressure of source materials.

In metalorganic vapor-phase epitaxy (MOVPE), on the other hand, transport rate of source materials can be precisely controlled by mass flow controllers. Further, MOVPE is eminently suitable for mass production, and it may be possible to achieve selective area growth, which will lead to novel device structures in the future.

However, technological breakthroughs have not occurred with MOVPE for reliable *p*-type doping and high-quality epilayers, which are keys so that

MOVPE can compete with or be superior to MBE in the future as the practical device fabrication technology. In this chapter, the state-of-the-arts of growth and doping of ZnSe-based semiconductors in MOVPE are reviewed, and future prospects of this growth technique for device applications are discussed.

II. Brief Review of Pioneering Works

The MOVPE growth of ZnSe and ZnS began with the series of works by Manasevit (1968) and Manasevit and Simpson (1971). The growth of ZnSe on GaAs, which is the most basic structure in the present studies, was reported for the first time by Stutius (1978) and Blanconnier et al. (1978). Since then, MOVPE of ZnSe-based II–VI semiconductors has been extensively studied, and significant progress has been made toward high-quality epilayers, well-defined multilayered structures, and conductivity control (Kukimoto, 1991).

In the earliest works, source precursors used were dialkyl-zinc such as dimethyl-zinc (DMZn) or diethyl-zinc (DEZn) and hydrogen chalcogenide such as H_2Se or H_2S for groups II and VI elements, respectively. These source combinations allow low temperature growth at around 300°C. However, the serious problem lies in unwanted premature reactions between these precursors, which tend to result in nonuniform thickness and poor surface morphology.

The premature reactions are eliminated by using dialkyl-chalcogen (Mitsuhashi et al., 1985, 1986) such as dimethyl-selenide (DMSe), diethyl-selenide (DESe), or diethyl-sulfide (DES). The growth of uniform and smooth epilayers has been reported with these precursors. However, the growth temperature should be as high as 500°C, which is related to the high decomposition temperature of dialkyl-chalcogen. The high growth temperature enhances formation of native defects such as vacancies of constituent elements, which is inherent to ZnSe-based semiconductors containing high vapor pressure elements. Further, interdiffusion at substrate/epilayer and epilayer/epilayer interfaces become heavier, resulting in impurity incorporation and/or destruction of multilayered structures (Ohmi et al., 1987; Parbrook et al., 1992).

From these points of view, much of the research thereafter has been directed toward new source precursors such as monoalkyl-hydrides and new growth techniques such as photo-assisted growth, which can allow low-temperature growth without unwanted premature reactions. Details of the more recent research will be described in the later sections.

III. Recent Growth Techniques

1. New Precursors

a. Zn Adducts

Using Lewis acid-based adducts of dialkyl-zinc with dialkyl-selenide, such as a DMZn-DMSe adduct, together with H_2Se or H_2S, it is possible to grow ZnSe or ZnS, respectively, at about 300°C (Yasuda et al., 1986). However, the unwanted premature reactions are not completely eliminated with these precursors.

Wright et al. (1989) used the triethyl-amine adduct of dimethyl-zinc, DMZn-$(NEt_3)_2$, and showed the growth of ZnSe with H_2Se at 350°C without any noticeable premature reactions. The undoped layer showed n-type with the electron concentration of 9.8×10^{14} cm^{-3} and the mobility of 476 cm^2/V·s at room temperature. This mobility was close to the highest value reported at that time (550 cm^2/V·s by Yao et al. (1983) with MBE. Further, due to marked difference in volatility between DMZn-$(NEt_3)_2$ and other possible impurities such as methyl-iodine (CH_3I) produced during the synthesis, this adduct may be more readily purified than DMZn (Jones et al., 1991).

b. Se and S Precursors

Recent studies have reported a variety of novel group-VI precursors. Monoalkyl-hydrides, where one of the alkyl groups of dialkyl-chalcogen is replaced by hydrogen, were investigated in the earlier studies. Fujita et al. (1988a) used methyl-mercaptan (MSH) as a S source and the growth temperature of ZnS was reduced by 50°C compared to the growth using DES. For Se source, Hirata et al. (1990) and Nishimura et al. (1993) reported methyl-selenol (MSeH) and tertiarybutyl-selenol (t-BuSeH), respectively. These precursors scarcely caused premature reactions with dialkyl-zinc, and allowed low-temperature growth at about 300–400°C.

Use of allyl-compounds of Se, such as methyl-allyl-selenide (MASe) or diallyl-selenide (DASe), is proposed by Giapis et al. (1989) and Patnaik et al. (1991). The growth temperature of ZnSe with DMZn was about 400°C, but the epilayer showed high concentration (10^{20}–10^{21} cm^{-3}) of carbon impurities. As an alternative, Danek et al. (1994) proposed tertiarybutyl-allyl-selenide (t-BuASe). The growth was carried out with DMZn-$(NEt_3)_2$ as a Zn precursor. The growth temperature was 350°C and carbon concentration in the epilayer was lower than the detectable limit (5×10^{17} cm^{-3}). The low carbon concentration was attributed to the tertiarybutyl group precluding internal rearrangement reactions responsible for carbon incorporation. They also investigated tripropyl-phosphine-selenide (TPPSe) and showed the growth of ZnSe at around 400°C.

Kuhn et al. (1992) proposed ditertiarybutyl-selenide (Dt-BuSe). The growth of ZnSe was carried out at 330°C, and the electron mobility of as-grown n-type layer was as large as 500 cm^2/V·s.

A general strategy for the development of new group-VI precursors has been to weaken the Se–C bonds by replacing methyl and ethyl groups with substituents forming stable radicals. This has allowed reduction in decomposition temperature of the precursor, resulting in low-temperature growth. In addition, the new precursor should have the potential to avoid contamination of the grown epilayers with halogen, carbon, and hydrogen impurities, which result in residual donors, poor crystallinity, and acceptor passivation, respectively.

c. *Mg Precursors*

The ZnMgSSe system lattice-matched to a GaAs substrate is the key material for blue/green light emitter structures. The MOVPE growth of ZnMgSSe has been investigated by Toda et al. (1994) with bismethylcyclopentadienyl-magnesium ((MeCp)$_2$Mg) as a Mg precursor with DMZn, DES, and DMSe. At the growth temperature of 480°C, the solid composition of Mg was well controlled up to 13% by the flow rate of (MeCp)$_2$Mg, as shown in Figure 1. The 4.2 K photoluminescence (PL) of ZnMgSSe layers exhibited no remarkable deep level emissions, but the full widths at half maximum (FWHMs) of the PL band edge peak and of the x-ray rocking curve were slightly large as 12.2 meV and 170–200 arcsec, respectively.

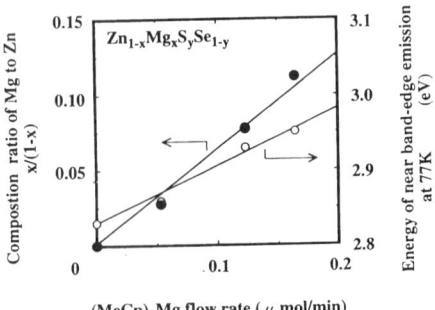

FIG. 1. Solid composition of Mg in ZnMgSSe against the flow rate of (MeCp)$_2$Mg. The growth temperature was 480°C (Toda et al., 1994).

These data suggest remaining problems of composition uniformity and/or crystalline quality. However, the ZnMgSSe-based heterostructure has demonstrated photopumped and pulsed current-injection lasing (Toda *et al.*, 1994, 1995) at room temperature and 77 K, respectively.

d. Purity Problems of Precursors

High purity of precursors is inevitably important for high-quality epilayers. Several years ago, undoped ZnSe epilayers tended to show strong donor-related peaks in PL and *n*-type conductivity, especially grown by photo-assisted technique at low temperature 300–350°C with the source combination of dialkyl-zinc and dialkyl-selenide (Fujita *et al.*, 1987; Yasuda *et al.*, 1989). Kukimoto (1990) suggested that halogen (probably chlorine) impurity in alkyl-zinc was responsible for donors, and that use of high-purity Zn precursors drastically reduced the donor concentrations. Figure 2 shows the comparison of PL spectra of undoped ZnSe grown by conven-

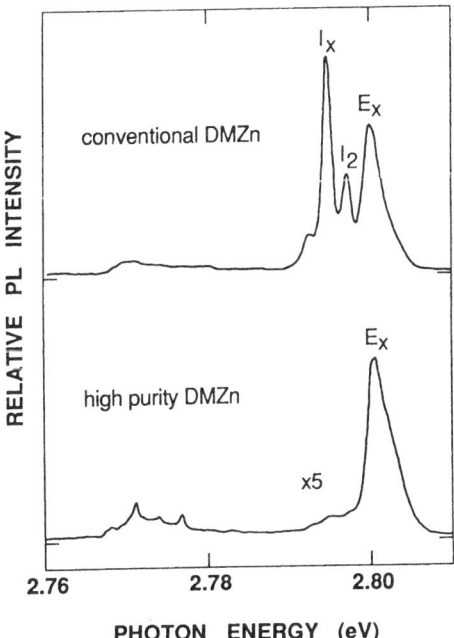

FIG. 2. Comparison of PL spectra at 10 K of undoped ZnSe grown by conventional DMZn and high purity DMZn. Here, the emission line labeled E_x is due to free excitons, and the lines I_2 and I_x are due to excitons bound to donors (Kukimoto, 1990).

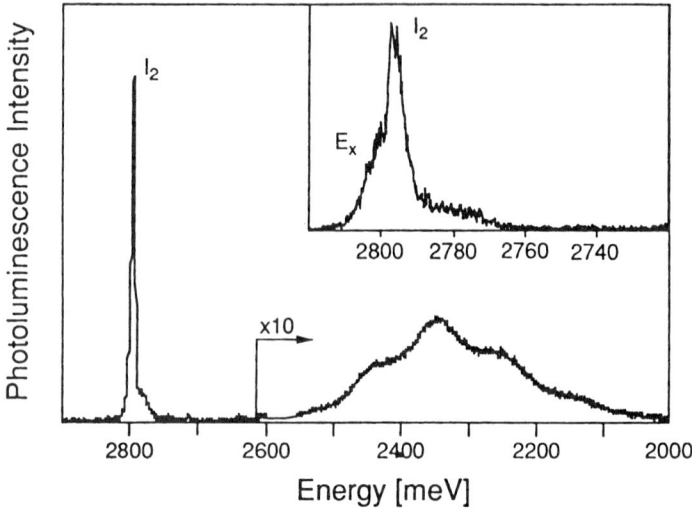

FIG. 3. PL spectrum at 10 K of undoped ZnSe grown by DMZn-(NEt$_3$)$_2$ and t-BuASe combination at 350°C (Danek et al., 1994).

tional DMZn and high purity DMZn. The difficulty in purifying dialkyl-zinc seems to lie in the fact that the boiling point of dialkyl-zinc (e.g., 44°C for DMZn) is close to that of halogen impurities used in synthesis (e.g., 42°C for CH$_3$I).

The purity of dialkyl-zinc commercially available seems to have gradually improved, and recently, the layers grown with conventionally available dialkyl-zinc generally show strong free excitonic emissions in PL. However, there still remain purity problems especially in dialkyl-selenide, which seriously affect the p-type impurity doping (Fujita, Sz. and Fujita, Sg., 1994a).

The purity problem is now severe in new precursors. Although promising low-temperature growth characteristics are achieved by using a novel precursor, the purity of the precursor tends to be too poor, probably because of the lack of experience in purifying the precursor and of the cost required, to achieve the p-type doping. Figure 3 shows the PL of the undoped ZnSe grown by DMZn-(NEt$_3$)$_2$ and t-BuASe combination at 350°C (Danek et al., 1994). The strong I$_2$ line suggests high incorporation of donors. A similar problem was also recognized when using Dt-BuSe (Stanzl et al., 1994). Together with development of new precursors, it seems necessary that makers and users should discuss how the purity of precursors can be improved and whether the new precursor is suitable for high purification.

2. Photo-assisted Growth

Photo-assisted technique has been widely applied for low-temperature growth and deposition of metals, semiconductors, and insulators. The most fundamental and important processes in this technique have generally been considered as direct photodecomposition or photoexcitation of precursors in gas phase (Nishizawa, 1994). Also for the photo-assisted MOVPE of II–VI semiconductors, irradiation of 253.7 and 184.9 nm lines from a low-pressure mercury lamp (Ando et al., 1985) or 193 nm line of an ArF excimer laser (Kawakyu et al., 1986), which are in the absorption bands of source precursors in gas phase, have successfully reduced the growth temperature and resulted in reduction of intrinsic defects and residual impurities.

On the other hand, Fujita et al. (1983, 1987) observed the growth rate enhancement in ZnSe and ZnS by the irradiation of visible light with which direct photodecomposition or photo-excitation of the source precursors in gas phase is hardly expected. This phenomenon has been especially remarkable in the growth of wideband gap II–VI semiconductors, and attributed to the promotion of surface reactions by assistance of electrons and/or holes generated at the growth surface under the photoirradiation (Fujita et al., 1988b).

From the viewpoint of ZnSe-based semiconductor growth, this type of photo-assisted MOVPE growth technique has attracted much attention because of the following features (Fujita, Sz. and Fujita, Sg., 1994b):

1. One can use a conventional irradiation source such as a xenon lamp, a high-pressure mercury lamp, or an argon ion laser.
2. Compared to direct photodecomposition or photoexcitation in gas phase, irradiated photons can effectively associate with surface reactions because of higher absorption coefficient of the growing materials.
3. Gas phase processes sometimes make particles, which affect the surface morphology. Irradiation of the wavelength longer than that absorbed in gas phase is essentially free from this problem.
4. Highly nonequilibrium conditions at the growth surface, due to presence of photo-generated carriers, may influence the thermodynamics for defect generation and reduce the defect density. Details will be discussed later.

a. Growth Characteristics

Photo-assisted MOVPE has brought solutions for high growth temperature with source combinations of dialkyl-zinc and dialkyl-chalcogen, which

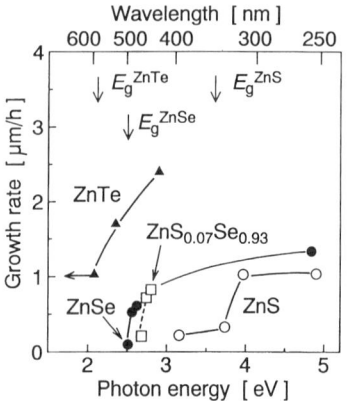

FIG. 4. Wavelength dependence of growth rate of ZnSe, ZnSSe, ZnS, and ZnTe in photo-assisted MOVPE (Fujita, Sz. and Fujita, Sg., 1995a).

scarcely cause premature reactions in gas phase. Using a xenon lamp (Fujita et al., 1987), a high-pressure mercury lamp (Yasuda et al., 1989; Taskar et al., 1993; Akram and Bhat, 1994; Gokhale et al., 1996), or 458–515 nm lines of an argon ion laser (Yoshikawa et al., 1990), the growth temperature of ZnSe can be as low as 300–400°C. The irradiation intensity has been less than 200 mW/cm^2, which is much lower than that used for direct photodecomposition/excitation.

The growth characteristics reported are briefly summarized as follows:

1. The growth rate enhancement is observed in the temperature range of reaction-limited growth (Fujita et al., 1987, 1988b).
2. Photons with energies higher than the bandgap of the growing material are responsible for the growth rate enhancement (Fujita et al., 1988b, 1995a; Yoshikawa et al., 1990). See the wavelength dependence of the growth rate shown in Figure 4.
3. The quantum yield, i.e. (number of adhered molecules by irradiation)/(number of irradiating photons), is as high as the order of 1–10% (Fujita et al., 1988c; Yasuda et al., 1989; Yoshikawa et al., 1990).
4. Hydrogen gas plays an important role in the growth; without hydrogen the growth reactions are hardly enhanced (Yoshikawa et al., 1990; Fujita et al., 1991a).
5. Surface morphology is seriously influenced by the growth conditions, especially by the irradiation intensity and flow rate of dialkyl-zinc (Yasuda et al., 1989; Gokhale et al., 1995).

6. The growth rate is very low at the initial stage of growth on the substrate (Yoshikawa et al., 1990; Fujita et al., 1991b) and is different on different substrates (Okamoto and Yoshikawa, 1991). There also is a report that the growth cannot be achieved directly on GaAs (Yasuda et al., 1989).

The results (2) and (3) suggest association of photo-generated carriers under above-band gap photoirradiation. This is quite similar to the chemical processes in the decomposition of H_2O by irradiation of the surface of TiO_2 dipped in H_2O (Fujishima and Honda, 1972), which was characterized by photocatalytic reactions, i.e., photo-generated electrons and holes cause reduction and oxidation, respectively. Therefore, fundamental processes of the photo-assisted MOVPE seem to be interpreted by the photocatalytic reactions. The result (6) is also supported by photocatalytic reactions because they cannot occur without catalyst, i.e., the growing material. The result (4) suggests that a fundamental chemical reaction is to make volatile hydrocarbon gases by the reaction between alkyl groups in source precursors and H_2.

b. Growth Mechanism

Detailed investigations on chemical processes under photoirradiation have been done by mass analysis (Fujita et al., 1991b, 1991c). The results are as follows:

1. Thermal decomposition of dialkyl-zinc occurs at the substrate temperature (T_s) higher than 300°C.
2. Under photoirradiation, remarkable enhancement of decomposition is observed for dialkyl-zinc at $T_s > 300°C$. On the other hand, it is not detected at $T_s < 300°C$.
3. Photo-enhanced decomposition of dialkyl-zinc is observed under irradiation of ZnSe or ZnS surface with photon energy higher than their bandgaps.
4. Thermal decomposition of dialkyl-chalcogen occurs at T_s higher than 450–500°C.
5. Under photoirradiation, the decomposition characteristics of dialkyl-chalcogen are hardly influenced.
6. In the growth atmosphere, i.e., in the simultaneous flow both of dialkyl-zinc and dialkyl-chalcogen, photo-enhanced decomposition is observed both for dialkyl-zinc and dialkyl-chalcogen.

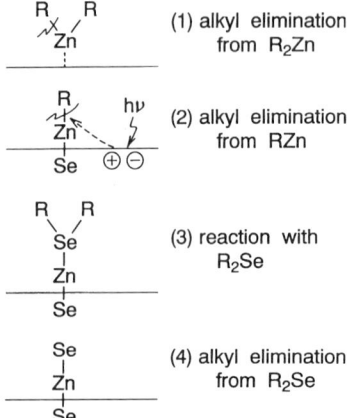

FIG. 5. Plausible model of chemical processes in the growth of ZnSe under photoirradiation (Fujita, Sz. and Fujita, Sg., 1994b).

7. At the initial stage of growth, mass spectra are hardly changed with irradiation. After growing the layers with few hundred Å in thickness, photo-enhanced decomposition of alkyl-zinc is observed.

From these results, a model for the growth processes has been deduced as shown in Figure 5 (Fujita, Sz. and Fujita, Sg., 1994b), i.e., (1) dialkyl-zinc releases one of the alkyls during pyrolysis and chemisorbs at the growth surface, (2) it releases another alkyl being assisted by the photo-generated carriers, (3) resultant active Zn bonds attract dialkyl-chalcogen (dialkyl-selenide in this figure) and weak Zn-Se bonds are formed, (4) dialkyl-chalcogen releases alkyl(s) and makes strong Zn-Se bonds.

As described before, the growth rate becomes markedly small with reducing the partial pressure of H_2 gas in the carrier gas. This implies that one of the most fundamental chemical processes in the growth reactions may be the formation of volatile RH gases (CH_4, C_2H_6, etc.) with the reaction R- + H_2.

On the other hand, another model which attributes the decomposition of dialkyl-chalcogen as a fundamental reaction under photoirradiation has been proposed (Yohikawa and Okamoto, 1991, 1992). This model is based on the following results:

1. The average bonding energy of alkyls is generally much higher in dialkyl-chalcogen than in dialkyl-zinc. Therefore, the decomposition of dialkyl-chalcogen should limit the growth rate at low temperatures.

2. According to one of the band bending models (Yoshikawa et al., 1990; Okamoto and Yoshikawa, 1991), holes accumulate at the growth surface, which may cause oxidation reactions. In dialkyl-selenide, for example, Se is negatively polarized (-0.164 esu) and has the oxidation number of -2; hence it attracts holes and is easily oxidized.
3. For the combination of dialkyl-zinc and H_2Se, the growth rate is enhanced by irradiation at temperatures as low as $<250°C$, while it is hardly enhanced at these temperatures for the combination of dialkyl-zinc and dialkyl-selenide. This difference is due to different Se sources.
4. Photoluminescence for the samples grown under irradiation is essentially identical to that for the samples grown at higher H_2Se flow rate.

Further research with *in situ* monitoring during the growth may disclose more detailed chemical processes in the future.

c. Reduction of Compensating Defects

Photoirradiation during the crystal growth is effective on thermodynamics of defects as well as growth rate and doping efficiency. It is well recognized that the concentration of charged defects is influenced by the carrier concentration (Kröger and Vink, 1956). For example, donor-like defects (positively charged defects) increase with acceptor doping, i.e., with an increase of the hole concentration. This phenomenon is called self-compensation. Carriers can also be created by photoirradiation, and therefore the photo-generated carriers may influence the defect concentrations, as do carriers from impurity atoms.

Consider a donor-like defect X. Neutral X, designated as X^0, can be ionized through the reaction

$$X^0 = X^+ + e^- \qquad (1)$$

where X^+ is the ionized state of X and e^- an electron. Under the mass action law, it is concluded that X^+ concentration decreases with increasing electron concentration in the crystal. Directly extending this conclusion to the nonequilibrium case, one may predict that X^+ concentration decreases with photoirradiation (with an increase of the electron concentration). However, considering the reaction of capturing a hole h^+

$$X^0 + h^+ = X^+ \qquad (2)$$

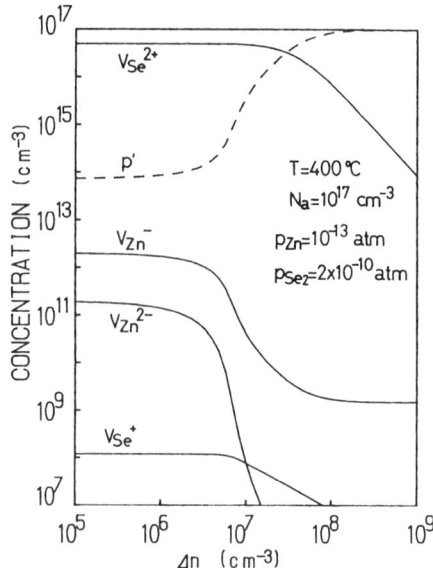

FIG. 6. Calculated defect and hole concentrations in acceptor-doped ZnSe under photoirradiation as a function of carrier concentration generated by photoirradiation (Ichimura et al., 1991).

then X^+ increases with an increase of the hole concentration due to the photoirradiation. One generally cannot predict whether the photoirradiation increases or decreases defects without specific calculation, because the phenomena depend on electronic properties of defects such as capture cross sections and energy levels.

The result of calculation for defect concentrations in acceptor-doped ZnSe is shown in Figure 6 (Ichimura et al., 1991). Here, V_{Se}^{2+} and V_{Se}^+ denote the concentrations of positively ionized Se vacancies, V_{Zn}^{2-} and V_{Zn}^- those of negatively ionized Zn vacancies, and p' the hole concentration. The horizontal axis, Δn, represents the excess carrier concentration from the thermal equilibrium, i.e., the carrier concentration generated by photoirradiation. The growth temperature T and the acceptor concentration N_A were 400°C and 10^{17} cm^{-3}, respectively. It should be noted that in the calculation it was assumed that the acceptor impurity was completely ionized, and thus the hole concentration at room temperature in a practical system, where the ionization energy of acceptors is about 100 meV, should be lower than that shown in Figure 6.

It is clearly seen that the acceptors are strongly compensated without the irradiation, but they are almost free from the compensation at

$\Delta n > 10^7 \, \text{cm}^{-3}$ due to reduction of defect concentrations. In a steady state, Δn is given by $g\tau$, where g is the excitation density and τ the lifetime of excess carriers. For example, if a 365 nm line of a mercury lamp is used, g is roughly estimated as $10^{20} \, \text{cm}^{-3}\text{s}^{-1}$ for an output power of $1 \, \text{mW/cm}^2$, assuming the absorption coefficient of $5 \times 10^4 \, \text{cm}^{-1}$ at 365 nm. Although the lifetime at the growth temperature is not known, the condition $\Delta n > 10^7 \, \text{cm}^{-3}$ can be easily achieved even if the lifetime is as small as 10^{-12} s. The results shown here are encouraging for the reduction of compensating defects and for the *p*-type doping as influence of the photoirradiation. Similar calculation also demonstrated the stabilization of Li impurities (Ichimura *et al.*, 1993).

d. Composition Modulation with Photoirradiation

Photoirradiation significantly influences the solid composition of alloy semiconductors. As an example, in the growth of $Zn_xCd_{1-x}Se$ alloys, correlation between the solid composition x of Zn and the gas phase composition of Zn ($=[\text{DMZn}]/([\text{DMZn}]+[\text{DMCd}])$) is shown in Figure 7 (Fujita *et al.*, 1993a). Here, the source precursors used were dimethyl-zinc (DMZn), dimethyl-cadmium (DMCd), and diethyl-selenide (DESe). Irradiation intensity was $40 \, \text{mW/cm}^2$ and the substrate temperature was 400°C.

The photoirradiation results in a remarkable increase of solid composition x of Zn at a constant gas phase composition, as well as increase of the growth rate r_g. Calculating the effective growth rate of ZnSe as $r_g x$ and that of CdSe as $r_g(1-x)$, it has been derived that the effective growth rate of

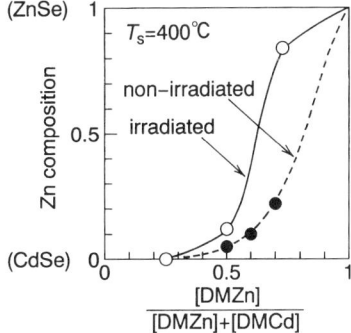

FIG. 7. Correlation between the solid composition x of Zn and the gas phase composition of Zn ($=[\text{DMZn}]/([\text{DMZn}]+[\text{DMCd}])$) in the growth of $Zn_xCd_{1-x}Se$ alloys (Fujita *et al.*, 1993a).

ZnSe is remarkably increased while that of CdSe is unchanged with the irradiation.

The growth characteristics of ZnCdSe alloys shown above are interpreted as follows. At low substrate temperatures, e.g. 400°C, decomposition of DMZn is not enough to make Zn–Se bonds without irradiation. When irradiated, the decomposition of DMZn is enhanced and results in enhancement of effective growth rate of ZnSe. On the other hand, DMCd is thermally well decomposed even at the substrate temperature as low as 400°C; hence there seems to remain no room for photoirradiation to enhance the decomposition of DMCd and to enhance the effective growth rate of CdSe.

With the results above, the irradiation/nonirradiation sequences during the growth of ZnCdSe at a constant gas phase composition can fabricate multilayered structures. The solid composition also depends on the irradiation intensity. Therefore, it is possible to grow various multilayered or graded structures, controlling the photoirradiation conditions with keeping the gas phase composition constant. Attempts to grow $Zn_{0.53}Cd_{0.47}Se/Zn_{0.97}Cd_{0.03}Se$ single quantum well (QW) are reported by Fujita et al. (1993a). Growth of ZnCdS QWs and their characterization by time-resolved photoluminescence have also been shown by Dumont et al. (1995a, 1995b). These phenomena can be recognized as a *soft* technique compared to the conventional *hard* technique for growth control of multilayered structures.

Tokumitsu et al. (1989) have shown the composition modification by ArF excimer laser irradiation in the MOMBE of AlGaAs from the source combination of triisobutyl-aluminum (TiBAl), triethyl-gallium (TEGa), and triethyl-arsenic (TEAs). Since the molar extinction coefficient of TiBAl was three times larger than that of TEGa, they reported that the solid composition of Al was 0.5 with irradiation and 0.4 without irradiation. Contrary to this result, II–VI semiconductors have shown the more drastic variation of composition with and without irradiation under much weaker light intensity. This is probably because of the great difference of photo-induced decomposition characteristics between DMZn and DMCd, and also of high quantum efficiency involved in the photocatalytic surface reactions.

IV. Doping and Devices

1. *n*-TYPE DOPING

N-type doping in MOVPE has been achieved by doping with group III elements such as aluminum (Kamata et al., 1988), or group VII elements

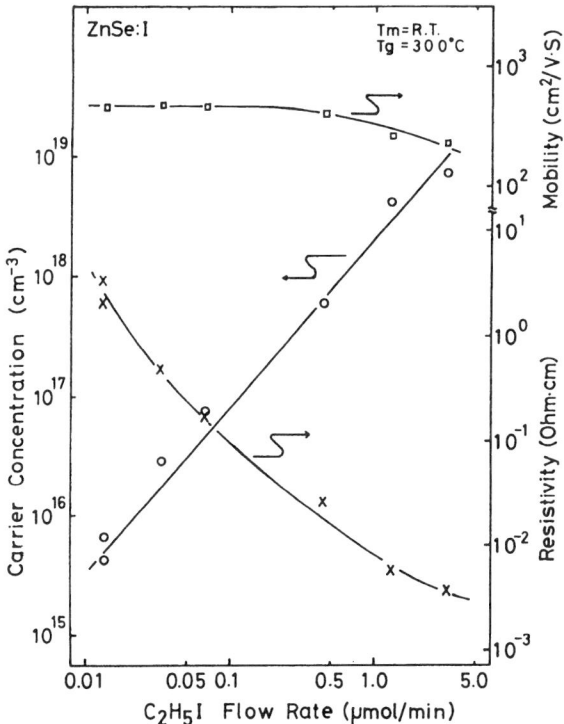

FIG. 8. Relationship between flow rate of C_2H_5I and electron concentration in n-type doping of ZnSe (Yoshikawa et al., 1988a).

such as chlorine (Kamata et al., 1988, 1989) or iodine (Shibata et al., 1988; Yoshikawa et al. 1988a). Among these dopants, iodine seems to be one of the best candidate to realize well-controlled doping over a wide range of electron concentration. An example of the relationship between flow rate of C_2H_5I and electron concentration is shown in Figure 8 (Yoshikawa et al., 1988a). It is recognized that the electron concentration is successfully controlled from 5×10^{15} cm^{-3} to 8×10^{18} cm^{-3} by the flow rate of dopant. However, the appearance of deep level emission in PL at the electron concentration close to 10^{19} cm^{-3} (Shibata et al., 1988) suggests degradation of epilayers due to overdoping, and therefore it seems there is room to improve the quality of doped epilayers with novel doping techniques and/or precursors. In the recent work by Stanzl et al. (1994), n-butyl-chloride (n-BuCl) was used as the dopant source.

2. p-TYPE DOPING

The fact that p-type doping of MBE to the order of 10^{17} cm^{-3} (Park et al., 1990; Ohkawa et al., 1991) was followed in a short time by the first operation of blue/green laser diode (Haase et al., 1991) suggests that reliable and low resistive p-type doping by MOVPE will lead this technology to the successful fabrication of optoelectronic devices. For ZnSe-based semiconductors, many research studies have concentrated on Li-based doping and nitrogen doping, by which the net acceptor concentration of the order of 10^{17} cm^{-3} has already been reported.

a. *Li-Based Doping*

With the co-doping of Li and N, the hole concentration up to 9×10^{17} cm^{-3} and the blue light emitting diode were demonstrated by Yasuda et al. (1988). Several attempts were made afterwards to overcome the low vapor pressure of the dopant used in the above work (Li$_3$N) by other precursors possessing higher vapor pressures such as tertiarybutyl-lithium (t-BuLi) (Yahata et al., 1990), cyclopentadienyl-lithium (CpLi) (Yoshikawa et al., 1988b; Mitsuishi et al., 1990), or dimethyl-amino-lithium ((CH$_3$)$_2$NLi) (Yanashima et al. 1992). These dopants have brought the hole concentrations of 10^{14}–10^{16} cm^{-3}. More recently, thermal diffusion of LiN$_3$ resulted in hole concentrations higher than 9×10^{17} cm^{-3} (Lim et al., 1994). This technique is also useful for making an ohmic contact to p-type layers.

b. *Nitrogen Doping*

Although Li-based doping has achieved p-type conductivity, the possibility of heavy diffusion of Li (Yamada et al., 1989) seems to have obstructed the practical applications to optoelectronic devices constituted with thin multilayered structures. On the other hand, doping of nitrogen, which is now recognized as a promising dopant material in MBE (Park et al., 1990; Ohkawa et al., 1991; Ito et al., 1992), has attracted attention also in MOVPE.

Using NH$_3$ as a dopant source, Ohki et al. (1988) and Suemune et al. (1988) showed the hole concentrations of the order of 10^{14}–10^{15} cm^{-3} in ZnSe and ZnSSe. Taskar et al. (1993) applied rapid thermal annealing of as-grown ZnSe layers doped with N from NH$_3$, and demonstrated the net acceptor concentration up to 3×10^{16} cm^{-3}. The highest hole concentration of 8.8×10^{17} cm^{-3} with the NH$_3$ doping source was reported by Lee *et al.* (1994). They used Se-rich growth conditions in order to reduce donor defects related to Se vacancies.

It is suggested that heavy hydrogen passivation of doped acceptors seriously obstructs the *p*-type conduction (Wolk *et al.*, 1993; Kamata *et al.*, 1993). From this point of view, NH_3 does not seem to be an appropriate dopant precursor because of the presence of three stable N–H bonds in a molecule. In order to overcome the hydrogen passivation problem, one attempt is to use new dopant precursors which are more easily decomposed and possess fewer N–H bonds. Another attempt is to apply postgrowth thermal annealing, which has successfully brought *p*-type conductivity of GaN followed by high efficient light-emitting diodes (Nakamura *et al.*, 1992).

As a precursor which is easily decomposed compared to NH_3, tertiarybutyl-amine (t-$BuNH_2$) has been examined (Fujita *et al.*, 1993b; Fujii *et al.*, 1993), and resulted in *p*-type conductivity with the acceptor concentration of the order of 10^{17} cm^{-3}. The hole concentration of 8.3×10^{17} cm^{-3} is the highest value reported with this dopant (Fujita *et al.*, 1995). Other candidate precursors are phenyl-hydrazine ($PhHN$-NH_2) (Akram and Bhat, 1994), ethyl-azide (EtN_3) (Yamasaki *et al.*, 1993; Kamata, 1994), triallyl-amine (TAN) (Stanzl *et al.*, 1994), diisopropyl-amine (DiPNH) (Toda *et al.*, 1995), and so on. Many of the works mentioned the importance of low-temperature growth either by photo-assisted technique or by new Se precursors for less formation of defects, less interdiffusion of constituent layers, and more incorporation of dopants. On the other hand, it is also pointed out that high-temperature growth may be preferable for less hydrogen passivation with more decomposition of dopant precursors (Fujimoto *et al.*, 1993).

Fujita *et al.* (1994c, 1995b) showed that postgrowth thermal annealing was desirable to realize higher quality *p*-type ZnSe compared to as-grown *p*-type layers. Doping with t-$BuNH_2$ resulted in the as-grown net acceptor concentration $N_A - N_D$ of the order of 10^{17} cm^{-3}. However, the PL was dominated by a broad donor-to-acceptor pair (DAP) emission peak, and the built-in potential of a Au/p-ZnSe Schottky contact obtained from the capacitance-voltage characteristics did not exhibit the ideal value (Fujita *et al.*, 1993b). These poor optical and electrical properties were attributed to compensation of nitrogen acceptors or to degradation of crystallographic properties due to heavy doping. Therefore, attempts were made to reduce the nitrogen concentration in as-grown ZnSe:N layers, by which they may not show *p*-type behavior, but to activate the nitrogen by thermal annealing in N_2 gas atmosphere.

Figure 9 shows the comparison of 4.2 K PL spectra of the *p*-type ZnSe:N layers fabricated by (a) MOVPE; the layer was grown at high flow rate of

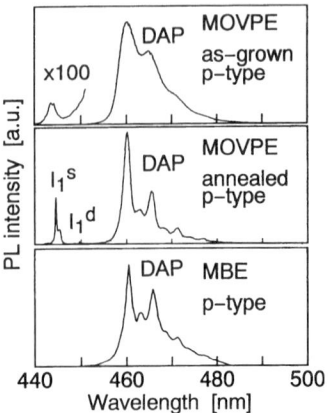

FIG. 9. Comparison of 4.2 K PL spectra of the p-type ZnSe:N layers. (a) MOVPE-grown, as-grown p-type with high flow rate of t-BuNH$_2$ (Fujita et al., 1993b), (b) MOVPE-grown, as-grown high resistivity with low flow rate of t-BuNH$_2$ and showed p-type after annealing in N$_2$ gas, at 500°C, for 30 min (Fujita et al., 1994c, 1995b), and (c) MBE-grown (Hauksson et al., 1992). The net acceptor concentrations of these layers are almost the same at around 1×10^{17} cm^{-3}.

t-BuNH$_2$ and showed as-grown p-type (Fujita et al., 1993b), (b) MOVPE; the layer was grown at low flow rate of t-BuNH$_2$ and showed p-type after annealing at 500°C in N$_2$ gas for 30 min (Fujita et al., 1994c, 1995b), and (c) MBE (Hauksson et al., 1992). The net acceptor concentrations of these layers are almost the same at around 1×10^{17} cm^{-3}. Compared to the as-grown p-type layer, shown in (a), the annealed p-type layer, shown in (b), exhibited clear appearance of deep DAP emission lines at 463 nm and the spectrum shape in DAP region is identical to that of MBE-grown p-type ZnSe:N, shown in (c). It may be concluded that the annealed p-type layers possess the quality similar to that of MBE-grown layers.

Figure 10 shows the variation of capacitance-voltage characteristics of Au/p-ZnSe:N Schottky contacts due to alternate annealing in N$_2$ and in H$_2$. Here, the ZnSe:N layer was (a) annealed in N$_2$ at 500°C for 30 min after the growth, then (b) exposed to H$_2$ at 350°C for 60 min, and (c) reannealed in N$_2$ at 500°C (Ogata et al., 1996). The net acceptor concentration, which was 4×10^{17} cm^{-3} due to annealing in N$_2$, shown by (a), reduced to 2×10^{17} cm^{-3} after exposure in H$_2$, as shown in (b), and then recovered to 4×10^{17} cm^{-3} due to reannealing, as shown in (c). These results suggest that hydrogenation and dehydrogenation occur reversibly, and these processes contribute to passivation and activation of acceptors, respectively.

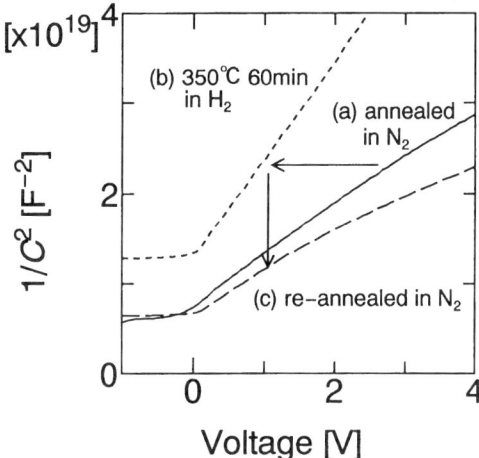

FIG. 10. Variation of capacitance-voltage characteristics of Au/p-ZnSe:N Schottky contacts due to (a) annealing in N_2 at 500°C for 30 min, (b) exposure to H_2 at 350°C for 60 min, and (c) reannealing in N_2 at 500°C (Ogata et al., 1996). The net acceptor concentrations were (a) 4×10^{17} cm^{-3}, (b) 2×10^{17} cm^{-3}, and (c) 4×10^{17} cm^{-3}.

3. PRESENT STATUS OF DEVICE APPLICATIONS

Until now, device-oriented research works by MOVPE have contributed to photo-pumped lasers and LEDs. Nakanishi et al. (1991) showed the photo-pumped lasing of ZnSe/ZnSSe multiple QWs up to 400 K. The threshold at room temperature was 10.5 kW/cm^2, which was the lowest value reported for II-VI semiconductor photo-pumped lasers, although it should be noted that in this experiment the active region was directly excited by a dye laser at 450 nm and the excitation was more effective than that in many of the experiments using the 337 nm or 335 nm lines of N_2 or Nd:YAG lasers, respectively, exciting the cladding layers. For the ZnSe/ZnMgSSe double heterostructure, photo-pumped lasing was reported at room temperature with N_2 laser excitation at the threshold of 70 kW/cm^2 (Toda et al., 1994).

The earliest p-n junction blue LED at room temperature was demonstrated by Yasuda et al. (1988) with ZnSe homojunction where the p-type layer was codoped with Li and N. Fujita et al. (1994d) fabricated the ZnCdSe/ZnSe QW LED, which exhibited the bright electroluminescence from the well (496 nm) at 77 K. Stanzl et al. (1995) showed the blue electroluminescence spectrum at 77 K from ZnSe/ZnSSe QW LED. How-

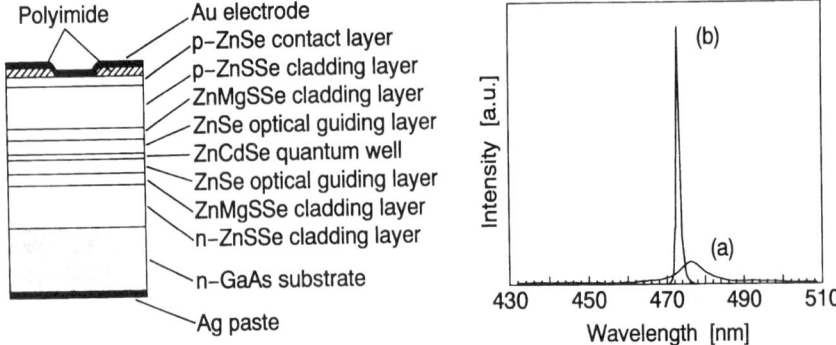

FIG. 11. Sample structure and emission spectra in pulsed current operation at 77 K (a) below and (b) above threshold of the laser diode grown by MOVPE (Toda et al., 1995).

ever, in all of these works, detailed device properties of LEDs have not been investigated.

For current injection lasers, Fujita et al. (1994a) and Yanashima et al. (1994) showed the spectrum narrowing at 77 K in ZnCdSe/Zn(S)Se QW structures. In the ZnCdS/ZnS system, Taguchi et al. (1993) had shown the stimulated emission at 30 K.

Most recently, Toda et al. (1995) reported the successful pulsed laser operation at 77 K in the ZnCdSe/ZnSe/ZnMgSSe separated confinement heterostructure (SCH) shown in Figure 11. Here, the sample was grown under ultraviolet photoirradiation, and the thicknesses of $Zn_{0.8}Cd_{0.2}Se$ quantum well, ZnSe optical guiding layer, $Zn_{0.93}Mg_{0.07}S_{0.1}Se_{0.9}$ inner cladding layer, $ZnS_{0.06}Se_{0.94}$ outer cladding layer, and p-ZnSe contact layer are 6 nm, 120 nm, 60 nm, 500 nm, and 20 nm, respectively. The nitrogen concentration in p-type layers was $1 \times 10^{18} cm^{-3}$, measured by secondary ion mass spectroscopy, but the net acceptor concentration had not been measured. The emission spectra at 77 K, in pulsed operation (1-μs width and 100-Hz frequency), are also shown in Figure 11. The appearance of strongly TE-polarized sharp emission line at 473.3 nm above threshold (0.9 kA/cm^2 at 13.0 V) indicated the laser operation.

Toward device applications, the MOVPE technology has shown continuing progress. It seems that structural quality of QWs grown by MOVPE has already been comparable to that by MBE, and that the most serious problem to be solved lies in reliable and low resistive p-type doping. The breakthrough for this problem will rapidly lead the MOVPE to become the conventional device technology of the future.

References

Akram, S., and Bhat, I. (1994). *J. Cryst. Growth* **138**, 105.
Ando, H., Iniuzuka, H., Konagai, M., and Takahashi, K. (1985). *J. Appl. Phys.* **58**, 8021.
Blanconnier, P., Cerclet, M., Henoc, P., and Jeans-Louis, A. M. (1978). *Thin Solid Films* **55**, 375.
Danek, M., Huh, J. S., Foley, L., and Jensen, K. F. (1994). *J. Cryst. Growth* **145**, 530.
Dumont, H., Fujita, Sz., and Fujita, Sg. (1995a). *Appl. Surf. Sci.* **86**, 442.
Dumont, H., Kawakami, Y., Fujita, Sz., and Fujita, Sg. (1995b), *Jpn. J. Appl. Phys.* **34**, L1336.
Eason, D., Ren, J., Yu, Z., Huges, C., El-Masry, N. A., Cook, J. W., Jr., and Schetzina, J. F. (1995). *J. Cryst. Growth* **150**, 718.
Fujii, Y., Suemune, I., Okamoto, K., Fujimoto, M., and Okamura, K. (1993). *Ext. Abst. 1993 Int. Conf. Solid State Devices and Materials,* Makuhari, p. 65.
Fujimoto, M., Suemune, I., Osaka, H., and Fujii, Y. (1993). *Jpn. J. Appl. Phys.* **32**, L524.
Fujishima, A. and Honda, H. (1972). *Nature* **238**, 37.
Fujita, Sg., Tomomura, Y., and Sasaki, A. (1983). *25th Electronic Materials Conf.,* Burlingon, paper E-7.
Fujita, Sg., Tanabe, A., Sakamoto, T., Isemura, M., and Fujita, Sz. (1987). *Jpn. J. Appl. Phys.* **26**, L2000.
Fujita, Sg., Isemuna, M., Sakamoto, T., and Yoshimura, N. (1988a). *J. Cryst. Growth* **86**, 263.
Fujita, Sz., Tanabe, A., Sakamoto, T., Isemuna, M., and Fujita, Sg. (1988b). *J. Cryst. Growth* **93**, 259.
Fujita, Sz., Takeuchi, F. Y., and Fujita, Sg. (1988c). *Jpn. J. Appl. Phys.* **27**, 2019.
Fujita, Sz., Maruo, S., Ishio, M., Murawala, P. A., and Fujita, Sg. (1991a). *J. Cryst. Growth* **107**, 644.
Fujita, Sz., Hirata, S., and Fujita, Sg. (1991b). *J. Cryst. Growth* **115**, 269.
Fujita, Sz., Hirata, S., and Fujita, Sg. (1991c). *Jpn. J. Appl. Phys.* **30**, L507.
Fujita, Sz., Matsumoto, S., and Fujita, Sg. (1993a). *J. Electron. Mater.* **22**, 521.
Fujita, Sz., Asano, T., Maehara, K., and Fujita, Sg. (1993b). *Jpn. J. Appl. Phys.* **32**, L1153.
Fujita, Sz., and Fujita, Sg. (1994a). *J. Cryst. Growth* **138**, 737.
Fujita, Sz., and Fujita, Sg. (1994b). *Appl. Surf. Sci.* **79/80**, 41.
Fujita, Sz., and Fujita, Sg. (1994c). *J. Cryst. Growth* **145**, 552.
Fujita, Sz., and Fujita, Sg. (1995a). *Appl. Surf. Sci.* **86**, 431.
Fujita, Sz., Tojyo, T., Yoshizawa, T., and Fujita, Sg. (1995b). *J. Electron. Mater.* **24**, 137.
Fujita, Y., Terada, T., and Suzuki, T. (1995). *Jpn. J. Appl. Phys.* **34**, L1034.
Giapis, K. P., Jensen, K. F., Potts, J. E., and Pachuta, S. J. (1989). *Appl. Phys. Lett.* **55**, 463.
Gokhale, M. R., Bao, K. X., Healey, P. D., Jain, F. C., and Ayers, J. E. (1996). *J. Electron. Mater.* **25**, 207.
Haase, M. A., Qiu, J., DePuydt, J. M., and Cheng, H. (1991). *Appl. Phys. Lett.* **59**, 1272.
Hauksson, I. S., Simpson, J., Wang, S. Y., Prior, K. A., and Cavenett, B. C. (1992). *Appl. Phys. Lett.* **61**, 2208.
Hirata, S., Isemura, M., Fujita, Sz., and Fujita, Sg. (1990). *J. Cryst. Growth* **104**, 521.
Ichimura, M., Wada, T., Fujita, Sz., and Fujita, Sg. (1991). *Jpn. J. Appl. Phys.* **30**, 3475.
Ichimura, M., Wada, T., Fujita, Sz., and Fujita, Sg. (1993). *J. Appl. Phys.* **73**, 7225.
Ishibashi, A. (1995). *IEEE J. Selected Topics in Quantum Electronics* **1**, 741.
Ito, S., Ikeda, M., and Akimoto, K. (1992). *Jpn. J. Appl. Phys.* **31**, L1316.
Jones, A. C., Wright, P. J., and Cockayne, B. (1991). *J. Cryst. Growth* **107**, 297.
Kamata, A., Uemoto, T., Okajima, M., Hirahata, K., Kawachi, M., and Beppu, T. (1988). *J. Cryst. Growth* **86**, 285.

Kamata, A., Uemoto, T., Hirahata, K., and Beppu, T. (1989). *J. Appl. Phys.* **65**, 2561.
Kamata, A., Mitsuhashi, H., and Fujita, H. (1993). *Appl. Phys. Lett.* **63**, 3353.
Kamata, A. (1994). *J. Cryst. Growth* **145**, 557.
Kawakyu, Y., Sasaki, S., Hirose, M., and Beppu, T. (1986). *Extended Abst. 1986 Int. Conf. Solid State Devices and Materials, Tokyo*, p. 643.
Kröger, F. A., and Vink, H. J. (1956). In *Solid State Physics* (eds. F. Seitz and D. Turnbull) Vol. 3, p. 307, Academic Press, New York.
Kuhn, W., Naumov, A., Stanzl, H., Bauer, S. Wolf, K., Wagner, H. P., Gebhardt, W., Pohl, U. W., Krost, A., Richter, W., Dümichen, U., and Thiele, K. H. (1992). *J. Cryst. Growth* **123**, 605.
Kukimoto, H. (1990). *J. Cryst. Growth* **101**, 953.
Kukimoto, H. (1991). *J. Cryst. Growth* **107**, 637.
Lee, M. K., Yeh, M. Y., Guo, S. J., and Huand, H. D. (1994). *J. Appl. Phys.* **75**, 7821.
Lim, S. W. Honda, T., Koyama, F., Iga, K., Inoue, K., Yanashima, K., Munekata, H., and Kukimoto, H. (1994). *Appl. Phys. Lett.* **65**, 2437.
Manasevit, H. M. (1968). *Appl. Phys. Lett.* **12**, 1536.
Manasevit, H. M., and Simpson, W. I. (1971). *J. Electrochem. Soc.* **118**, 664.
Mitsuhashi, H., Mitsuishi, I., and Kukimoto, H. (1985). *Jpn. J. Appl. Phys.* **24**, L864.
Mitsuhashi, H., Mitsuishi, I., and Kukimoto, H. (1986). *J. Cryst. Growth* **77**, 219.
Mitsuishi, I., Shibatani, J., Kao, M. H., Yamamoto, M., Yoshino, J., and Kukimoto, H. (1990). *Jpn. J. Appl. Phys.* **29**, L733.
Nakamura, S. Iwasa, N., Senoh, M., and Mukai, T. (1992). *Jpn. J. Appl. Phys.* **31**, 1258.
Nakanishi, K., Suemune, I., Fujii, Y., Kuroda, Y., and Yamanishi, M. (1991). *Appl. Phys. Lett.* **59**, 1401.
Nishimura, K., Nagao, Y., and Sakai, K. (1993). *Jpn. J. Appl. Phys.* **32**, L428.
Nishizawa, J. (1994). *Appl. Surf. Sci.* **79/80**, 1.
Ogata, K., Kawaguchi, D., Kera, T., Fujita, Sz., and Fujita, Sg. (1996). *J. Cryst. Growth* **159**, 312.
Ohkawa, K., Karasawa, T., and Mitsuyu, T. (1991). *Jpn. J. Appl. Phys.* **30**, L152.
Ohki, A., Shibata, N., and Zembutsu, S. (1988). *Jpn. J. Appl. Phys.* **27**, L909.
Ohmi, K., Suemune, I., Kanda, T., Kan, Y., and Yamanishi, M. (1987). *Jpn. J. Appl. Phys.* **26**, L2072.
Okamoto, T. and Yoshikawa, A. (1991). *Jpn. J. Appl. Phys.* **32**, L156.
Parbrook, P. J., Kamata, A., and Uemoto, T. (1992). *Est. Abst. 1992 Int. Conf. Solid State Devices and Materials, Tsukuba*, p. 354.
Park, R. M., Troffer, M. B., Rouleau, C. M., DePuydt, J. M., and Haase, M. A. (1990). *Appl. Phys. Lett.* **57**, 2127.
Patnaik, S., Jensen, K. F., and Giapis, K. P. (1991). *J. Cryst. Growth* **107**, 390.
Shibata, N., Ohki, A., and Zembutsu, S. (1988). *Jpn. J. Appl. Phys.* **27**, L251.
Stanzl, H., Wolf, K., Hahn, B., and Gebhardt, W. (1994). *J. Cryst. Growth* **145**, 918.
Stanzl, H., Resinger, T., Wolf, K., Kastner, M., Hahn, B., and Gebhardt, W. (1995). *Phys. Stat. Sol. (b)* **187**, 303.
Stutius, W. (1978). *Appl. Phys. Lett.* **33**, 656.
Suemune, I., Yamada, K., Masato, H., Kanda, T., Kan, Y., and Yamanishi, M. (1988). *Jpn. J. Appl. Phys.* **27**, L2195.
Taguchi, T., Yamada, Y., Ohno, T., Mullins, J. T., and Masumoto, Y. (1993). *Physica B* **191**, 136.
Taskar, N. R., Khan, B. A., Dorman, D. R., and Shazad, K. (1993). *Appl. Phys. Lett.* **62**, 270.
Toda, A., Asano, T., Funato, K., Nakamura, F., and Mori, Y. (1994). *J. Cryst. Growth* **145**, 537.

Toda, A., Margalith, T., Imanishi, D., Yanashima, K., and Ishibashi, A. (1995). *Electron. Lett.*, **31**, 1921.

Tokumitsu, E., Yamada, T., Konagai, M., and Takahashi, K. (1989). *J. Vac. Sci. & Technol.* **A7**, 706.

Wolk, J. A., Ager, J. W., III, Duxstad, K. J., Haller, E. E., Taskar, N. R., Domen, D. R., and Olego, D. J. (1993). *Appl. Phys. Lett.* **63**, 2756.

Wright, P. J., Parbrook, P. J., Cockayne, B., Jones, A. C., Orrell, E. D., O'Donnell, K. P., and Henderson, B. (1989). *J. Cryst. Growth* **94**, 441.

Yahata, A., Mitsuhashi, H., Hirahata, K., and Beppu, T. (1990). *Jpn. J. Appl. Phys.* **29**, L4.

Yamada, Y., Kidoguchi, I., Taguchi, T., and Hiraki, A. (1989). *Jpn. J. Appl. Phys.* **28**, L837.

Yamasaki, D., Yanashima, K., Watabe, S., Inoue, K., Hara, K., Yoshino, J., and Kukimoto, H. (1993). *Ext. Abst. 54th Autumn Meeting of the Japan Society of Applied Physics*, Sapporo, 29p-ZL-3.

Yanashima, K., Koyanagi, K., Hara, K., Yoshino, J., and Kukimoto, H. (1992). *J. Cryst. Growth* **124**, 616.

Yanashima, K., Yamasaki, D., Watabe, B., Hara, K., Yoshino, J., and Kukimoto, H. (1994). *J. Cryst. Growth* **138**, 755.

Yao, T., Ogura, M., Matsuoka, S., and Morishita, T. (1983). *Appl. Phys. Lett.* **43**, 449.

Yasuda, T., Hara, K., and Kukimoto, H. (1986). *J. Cryst. Growth* **77**, 485.

Yasuda, T., Mitsuishi, I., and Kukimoto, H. (1988). *Appl. Phys. Lett.* **52**, 57.

Yasuda, T., Koyama, Y., Wakitani, J., Yoshino, J., and Kukimoto, H. (1989). *Jpn. J. Appl. Phys.* **28**, L1628.

Yoshikawa, A., Nomura, H., Yamaga, S., and Kasai, H. (1988a). *Jpn. J. Appl. Phys.* **27**, L1948.

Yoshikawa, A., Muto, S., K., Yamaga, S., and Kasai, H. (1988b). *Jpn. J. Appl. Phys.* **27**, L260.

Yoshikawa, A., Okamoto, T., Fujimoto, T., Onoue, K., Yamaga, S., and Kasai, H. (1990). *Jpn. J. Appl. Phys.* **29**, L225.

Yoshikawa, A., and Okamoto, T. (1991). *J. Cryst. Growth* **115**, 274.

Yoshikawa, A., and Okamoto, T. (1992). *J. Cryst. Growth* **117**, 107.

CHAPTER 3

Gaseous Source UHV Epitaxy Technologies for Wide Bandgap II–VI Semiconductors

Easen Ho and Leslie A. Kolodziejski

DEPARTMENT OF ELECTRICAL ENGINEERING AND COMPUTER SCIENCE
RESEARCH LABORATORY OF ELECTRONICS
MASSACHUSETTS INSTITUTE OF TECHNOLOGY
CAMBRIDGE, MASSACHUSETTS

I. INTRODUCTION	83
II. METALORGANIC MOLECULAR BEAM EPITAXY OF ZnSe	85
1. Advantages and Disadvantages of MOMBE	85
2. Experimental Setup	86
3. Growth Rate Limitations Due to Surface Blockage	88
4. Beam-assisted Growth	91
III. GAS SOURCE MOLECULAR BEAM EPITAXY	97
1. Advantages and Disadvantages of GSMBE	97
2. Experimental Details	99
3. n-Type Doping	102
4. p-Type Doping	104
5. Hydrogen Passivation	108
IV. CONCLUDING REMARKS	114
REFERENCES	116

I. Introduction

A number of advanced thin-film epitaxial growth techniques with atomic-layer precision have been developed in the last one to two decades and applied to the deposition of III–V compound semiconductors with much success. These include molecular beam epitaxy (MBE), metalorganic vapor phase epitaxy (MOVPE), chemical beam epitaxy (CBE), metalorganic molecular beam epitaxy (MOMBE), as well as gas source molecular beam epitaxy (GSMBE). Although the merits of some of these techniques, as applied to wide bandgap II–VI materials, will be borne out in more detail in this chapter (as well as elsewhere in this volume), it is useful at this point to understand the ways in which the various techniques are related. Figure

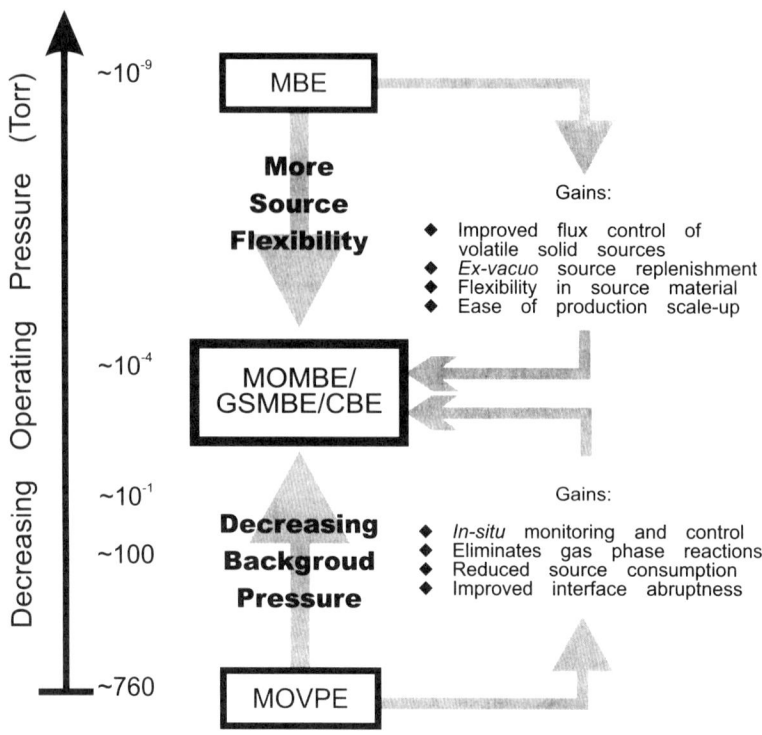

FIG. 1. Comparisons of the physical properties and relative merits of the key nonequilibrium epitaxial growth techniques employed for the fabrication of semiconductor heterostructures.

1 schematically compares these epitaxial growth techniques in terms of their source flexibility, background pressure, and hence their suitability for the application of a variety of useful *in situ* characterization tools. From the figure, it is apparent that the "alternative gaseous source approaches," e.g., MOMBE, GSMBE, and CBE, represent hybrids of the MBE and MOVPE techniques such that the most desirable property of the constituents, i.e., the ultrahigh vacuum (UHV) environment of MBE and the source flexibility of MOVPE, are combined. To date, significant and substantial progress in the field of II–VI device research has been realized by employing conventional MBE using all solid sources. The potential offered by implementing *in situ* feedback and thus excellent compositional control, and source flexibility, as well as the capability for production scaling, constitute the major impetus for our efforts in applying these alternative gaseous source epitaxial methods to the II–VI family of semiconductor compounds.

II. Metalorganic Molecular Beam Epitaxy of ZnSe

1. ADVANTAGES AND DISADVANTAGES OF MOMBE

Of the spectrum of epitaxial growth technologies in existence, the two mainstream techniques of MBE and MOVPE represent two extremes in terms of operating pressure regimes. The UHV environment of MBE allows for easy application of various *in situ* monitoring methods, such as reflection high energy electron diffraction (RHEED), quadrupole mass spectrometry (QMS), and quartz crystal oscillators. Information derived from these *in situ* tools has greatly facilitated the compositional and stoichiometric control of the growth of compound semiconductors, allowing for monolayer-level thickness control. On the other hand, although MOVPE lacks the complement of *in situ* characterization methods found in MBE, clear advantages are realized in terms of source flexibility and ease of flux control and manipulation. MOMBE techniques can therefore be considered a stylized idealization that combines MBE and MOVPE. By using metalorganic gas sources in an UHV-capable environment, MOMBE retains the MOVPE-like source flexibility and precise flux control, while simultaneously allowing for the use of *in situ* techniques that are normally available only in UHV-compatible growth methods like MBE.

Relative to MBE, MOMBE embodies the following important advantages: (1) Use of precision mass flow controllers for both Group II and Group VI precursors allows for precise flux control. (2) *Ex situ* source replenishment effectively results in a semi-infinite supply of the source materials. (3) A rich chemistry, or presence of a variety of precursor molecules, at the growth surface provides additional incorporation pathways such that the reaction can be fine tuned or engineered: e.g., selective area epitaxy.

Advantages of the MOMBE technique as compared to MOVPE, on the other hand, include the following: (1) the UHV background allows for the use of various *in situ* characterization tools such as RHEED and QMS. In addition, other vacuum-compatible sources such as radio frequency (RF) or electron cyclotron resonance (ECR) plasma sources can be utilized. (2) A much smaller flow of precursor (without the necessity of using a large amount of H_2 carrier gas) reduces the safety risks associated with the handling of exhaust gases.

On the down side, however, MOMBE systems tend to be fairly complicated. The full complement of gas and exhaust handling equipment that is typical for MOVPE is still required for MOMBE. In addition, large pumping systems are necessary to achieve both high throughput (to maintain molecular beam conditions during gaseous growth) and low background pressure (to maintain an UHV environment). The battery of *in situ* characterization tools are indispensable in controlling the growth process, and are therefore usually operated in higher pressures regimes than those typically encountered during MBE (which compromises their useful service life).

2. EXPERIMENTAL SETUP

Figure 2a depicts our MOMBE reactor that was used in the growth of ZnSe on GaAs, along with a schematic representation of the metalorganic gas manifold (Fig. 2b). The metalorganic precursors are introduced into the vacuum chamber via gas injectors or gas crackers fed by heated stainless steel tubing. Also shown are additional solid source effusion cells or ovens enabling the conventional MBE mode of growth. The chamber is also outfitted with a RF plasma cell for nitrogen plasma doping and hydrogen plasma cleaning of bulk substrate surfaces. A sapphire viewport inclined 45° relative to the substrate assembly (with the loaded substrate facing down) is used to introduce optical radiation for photo-assisted growth. A water-cooled quartz crystal oscillator fitted with a flexible bellows can be moved under the substrate position to measure the elemental source fluxes.

Diethylzinc (DEZn) and diethylselenium (DESe) were normally used as the metalorganic Group II and Group VI precursor, respectively. These sources have sufficient vapor pressure near room temperature such that a H_2 carrier gas is not required (as is usually the case for MOVPE which demands large flows of precursor gases). In addition, dimethylzinc (DMZn) as well as elemental Zn and Se were used in order to aid our understanding of the complex behavior of the MOMBE growth mechanisms. Due to their thermal stability, DESe and DMZn required thermal decomposition in a pyrolytic boron nitride (PBN) cracker having tantalum baffles. The typical cracking temperatures for DESe and DMZn were 800 and 1050°C, respectively, and were determined using QMS of the gas beam. DEZn was found to pyrolyze readily on the ZnSe surface at normal growth temperatures; the DEZn cracker was set at 50°C in nearly all cases to prevent accumulation in the gas injector.

The majority of the films investigated in this study were grown at a calibrated substrate temperature of 320°C; however, substrate temperatures ranging from 150°C to 475°C have been explored. The substrate temperature was calibrated using a low temperature optical pyrometer to measure the eutectic phase transition (356°C) of 500 Å of Au deposited onto a Ge substrate. Eutectic chips were mounted alongside In-bonded GaAs wafers

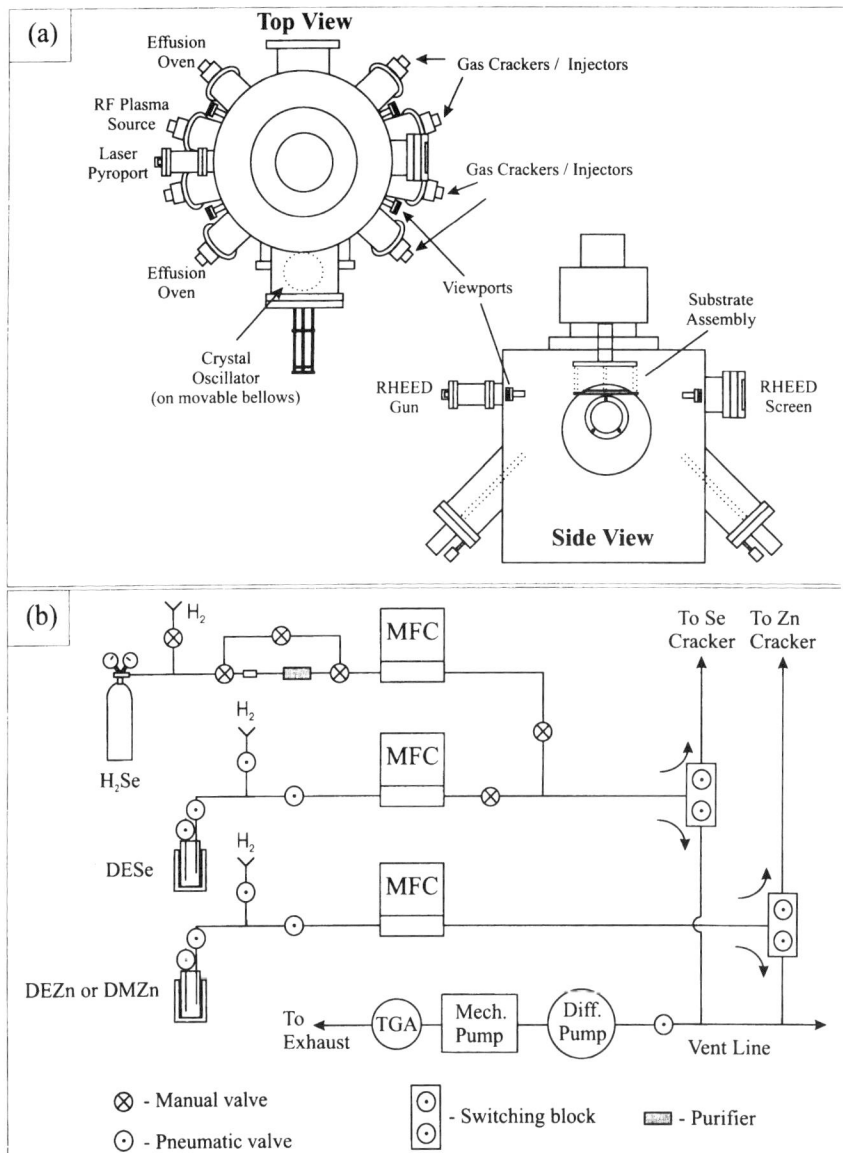

FIG. 2. Schematic diagram of (a) the MOMBE reactor for the growth of ZnSe, and (b) the gas manifolds for delivery of metalorganic sources.

on molybdenum holders to calibrate the thermocouple readings for each experiment.

GaAs wafers of various conduction types were used as the substrates. After a standard degrease with trichloroethane, acetone, and methanol, the GaAs substrates were etched in a 5:1:1 solution of $H_2SO_4:H_2O_2:H_2O$ for 90 sec, followed by oxide formation in deionized water. The etched substrates were then mounted on the molybdenum holder with high purity indium. The native oxide of GaAs was removed by heating to 600°C without an arsenic overpressure while observing the evolving RHEED pattern. Following oxide desorption, the temperature was then immediately lowered to the growth temperature and allowed to stabilize before initiating growth. Typical gas flow rates of 0.5–2.5 SCCM were used for the Group II and Group VI precursors, resulting in chamber pressures of $\sim 1 \times 10^{-4}$ torr. Surface stoichiometry is monitored by observing the RHEED reconstruction pattern during growth, e.g., $c(2 \times 2)$ and (2×1) reconstruction patterns were used to delineate the Zn- and Se-stabilized surfaces, respectively.

3. Growth Rate Limitations Due to Surface Blockage

Figure 3 shows a set of typical RHEED patterns observed during the growth of ZnSe in two azimuthal directions. The presence of narrow streaks and the readily apparent Kikuchi bands suggests the existence of single

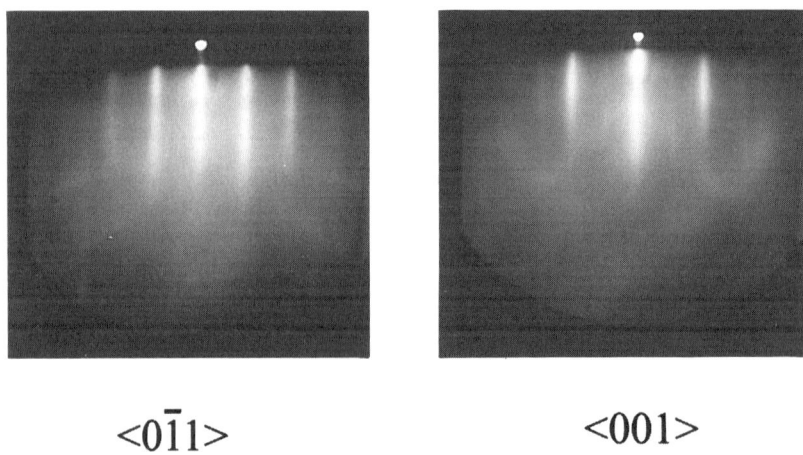

FIG. 3. Typical RHEED patterns observed during the MOMBE growth of ZnSe in the $\langle 0\bar{1}1 \rangle$ and the $\langle 001 \rangle$ azimuthal directions.

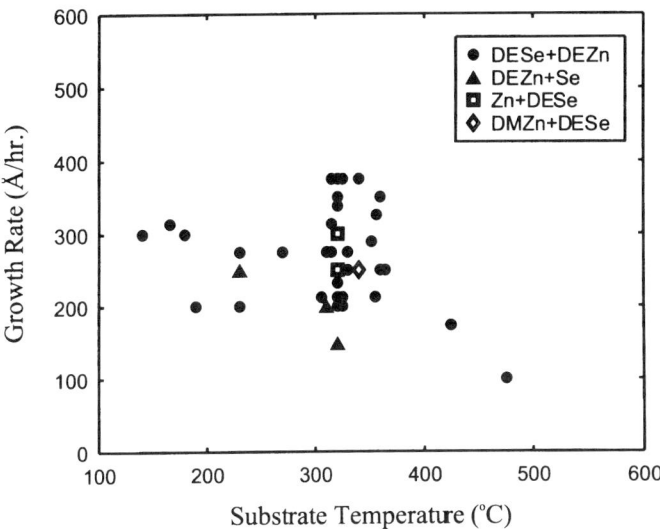

FIG. 4. Growth rate of ZnSe (using at least one ethyl-based metalorganic source) as a function of the calibrated growth temperature for all the experimental conditions and source combinations investigated.

crystalline films having smooth surface morphology. It is worthwhile to note that the typical RHEED patterns observed during MOMBE (as represented by the photographs in Figure 3) are qualitatively similar to those observed during the MBE growth of ZnSe. In this particular case, the presence of the twofold reconstruction in the ⟨001⟩ indicates that growth is occurring under cation-rich conditions.

Utilizing the growth conditions as described above, the MOMBE growth did not proceed as anticipated: in particular, the growth rate was extremely low using DESe and DEZn as sources. Unpublished work (Cunningham, 1992) on the MOMBE of ZnSe indicated a similar problem in achieving reasonable growth rates. Figure 4 shows a plot of the growth rate as a function of the calibrated substrate temperature. Typical growth rates of 100–400 Å/hr were observed regardless of the growth parameters when using the various combination of precursor pairs that are indicated in the figure legend. The growth parameters that were varied included substrate temperature, gas flow rate, VI/II flow rate ratio, intentional addition of hydrogen supplied through a gas injector, and the thermal precracking of each gas by the use of a "cracking" gas injector. For the lowest metalorganic gas flows used, the available amount of zinc and selenium should have yielded a minimum growth rate of 0.2 μm/hr. QMS analysis of the gas beams confirmed that thermal decomposition of DESe and DMZn occurred at 800 and 1050°C, respectively, whereas the DEZn pyrolyzed at temperatures greater than 300°C.

Closer analysis of the DESe + DEZn growth data, in conjunction with growth experiments carried out using various combinations of elemental Zn and Se, as well as DMZn, provided the following clues: (1) the observed low growth rate appeared to be due to a kinetic-limited (as opposed to mass transport–limited) process, and (2) the overall growth rate remained low whenever one ethyl-based metalorganic precursor was used. To ensure that the observed low growth rate was not due to unpyrolyzed DEZn, the DEZn was cracked at 800°C; however, the growth rate still remained very low. Complete avoidance of ethyl-based precursors enabled the growth rate to approach that obtained using a MBE mode of growth using elemental Zn and Se. To further demonstrate that the resultant low growth rate was unique to the use of diethyl-containing metalorganic sources, MOMBE was performed with DMZn (thermally decomposed). In this case, the growth rate achieved using DMZn and elemental Se was typical of that achieved in MBE, and limited only by mass transport to the substrate surface; however, when cracked DESe was substituted for elemental Se and growth proceeded with DMZn, the growth rate was again very low.

Based on these observations, we postulated that the growth rate was limited by the inability of the zinc and selenium precursors to incorporate into the lattice due to surface site blockage by ethyl radicals (Coronado et al., 1992; Ho et al., 1993). These ethyl radicals are believed to be strongly chemisorbed on the ZnSe surface such that incorporation sites for metal atoms are not available. Attempts to increase the desorption rate of the ethyl-based species responsible for surface site blockage, by increasing the substrate temperatures (up to 475°C), were unsuccessful as the resultant growth rate was at the extreme low end (~ 100 Å/hr). We speculate that the desorption rate of the ethyl species was not significantly modified at the high substrate temperatures, but that the desorption of zinc and selenium from the growing surface was significant and dominated the growth. A study of the thermal decomposition of triethylgallium (TEG) on GaAs (100) by Murrell et al. (1990) and others (Banse and Creighton, 1991; Rueter and Vohs, 1992) indirectly supported our hypothesis that a surface species was effectively blocking lattice sites. Murrell and co-workers observed that TEG decomposition resulted in the formation of highly stable, chemisorbed ethyl species that saturate surface sites. It was thus suggested that the decrease in growth rate at low temperatures (between 300 and 400°C) in the chemical beam epitaxy of GaAs using TEG and AsH_3 was due to a reduction in TEG adsorption due to site blockage by the stable ethyl species. We believe that a similar mechanism is at work in the case of the MOMBE growth of ZnSe using ethyl-based metalorganic sources as well. Although the use of diethyl metalorganic sources would be expected to be advantageous for low-temperature growth due to the lower pyrolysis temperatures of DEZn, they appear to be unsuitable for the growth of ZnSe due to the observed surface site blockage which severely curtails the growth rate. However, as will be described in the next section, we found that the use of DEZn and DESe may actually provide some advantages when one is interested in selective area epitaxy by employing photo- or electron-beam-assisted MOMBE growth.

4. BEAM-ASSISTED GROWTH

The use of photon-, as well as electron beam-, illumination was found to alleviate the previously mentioned limitation in the growth rate of ZnSe grown by MOMBE. Under the appropriate photon-illumination conditions during growth (i.e., photo-assisted epitaxy), the growth rate was found to dramatically increase by up to $15 \times$ when compared to the unilluminated growth rates. The effect of photons on the growth was found to be dependent on the power density and energy of the illuminating photons, as well as on the type of precursors selected.

Photo-assisted epitaxy has been reported for the growth of narrow- (Fujii et al., 1988; Fujita et al., 1991; Morris, 1986; Zinck et al., 1988) and wide-bandgap (Bicknell et al., 1986, 1986a; Coronado et al., 1992; Fujita et al., 1988; Gunshor et al., 1990; Harper et al., 1989; Matsumura et al., 1990; Simpson et al., 1992; Yoshikawa et al., 1990) II–VI materials, as well as for a variety of III–V compound semiconductors (Aoyagi et al., 1986; Bedair et al., 1986; DenBaars and Dapkus, 1989; Donnelly and McCauley, 1989; Donnelly et al., 1988; Kukimoto et al., 1986; Nagata et al., 1990; Nishizawa et al., 1986; Sugiura et al., 1990; Tu et al., 1988; Yamada et al., 1992). Several physical mechanisms have been reported or verified experimentally to account for the effect of photon-illumination: (1) selective modification of the desorption rate of adsorbed source precursors; (2) pyrolysis of precursors at the growth front due to a localized increase in the substrate temperature; (3) direct gas-phase photolysis of the source precursors; (4) photocatalysis of the adsorbed molecular species through the creation of free carriers at the growing surface.

Similar effects on the epitaxial growth process have been reported when the surface is illuminated with an electron beam (Ho et al., 1993; Takahashi et al., 1992) as will be described later. In the electron beam–assisted modification of growth, the physical mechanisms that have been postulated include kinetically induced decomposition of adsorbed precursor molecules, or generation of charge carriers that subsequently interact with the surface species. In the photo-assisted epitaxy of II–VI materials, the most common mechanisms include: (1) photolysis of the source materials (as in photo-assisted MOVPE using light sources tuned to specific absorption lines of the precursors); (2) modifications of the relative desorption rates of the precursor species (as in photo-assisted MBE); (3) enhanced photocatalysis of the precursor molecules through the interaction of photo-generated carriers (as in photo-assisted growth of MOVPE and MOMBE).

Photo-assisted growth of MOMBE was performed in our experiments by selectively illuminating a portion of the wafer's surface at a 45° angle to the substrate normal using various optical sources. The laser spot on the wafer was typically an ellipse with major and minor axes of 1.4 and 1.0 cm, respectively. Optical sources that were used include the various emission lines of a 5-watt Ar ion laser (Spectra-Physics, Model 2025), a Ti:sapphire solid-state laser, an organic dye laser containing Coumarin-7 fluorescent dye and pumped by an Ar ion laser, as well as a He–Ne laser. With these sources, a wavelength range of 780 nm to 380 nm is accessible for investigating the wavelength dependence of the photo-assisted phenomenon. The power density was maintained below 200 mW/cm^2 in order to eliminate the possibility of laser-induced local thermal gradients. Layer thicknesses (and thus the photo-enhanced growth rate) was measured both by ellipsometry, as well as by surface profiling of a step formed by selectively etching the ZnSe film.

Figure 5 shows the growth rate enhancement ratio (defined as the ratio of the illuminated growth rate over the unilluminated growth rate) as a function of the energy of the illuminating photons, using diethyl-based as well as dimethyl-based precursors. For these experiments, DEZn was expected to be pyrolyzed at the surface (substrate temperature was 320°C), whereas DMZn molecules were decomposed in the gas injector (1050°C) prior to introduction to the surface. The enhancement ratios for both metalorganics zinc sources were unity (no enhancement) when photons with energies ($\lambda_{incident}$) less than the energy bandgap (or energy of interband defect states) at the growth temperature ($\lambda_{bandgap}$) illuminated the surface. Photo-assisted growth with photon energies greater than the energy bandgap resulted in a nearly constant growth rate enhancement ratio for a given power density. In Figure 5, the approximate location of the energy bandgap at the growth temperature for the films grown with DEZn (substrate temperature was 320°C) and DMZn (substrate temperature was 340°C) is indicated with an arrow (Yoshikawa et al., 1990). These results suggested that photo-generated electron/hole pairs are necessary in order to achieve an increased growth rate. Growth rate enhancement due to gas-phase absorption (Hou et al., 1990) was ruled out as it is expected to occur at wavelengths much shorter (248 nm for DEZn) than the wavelength range

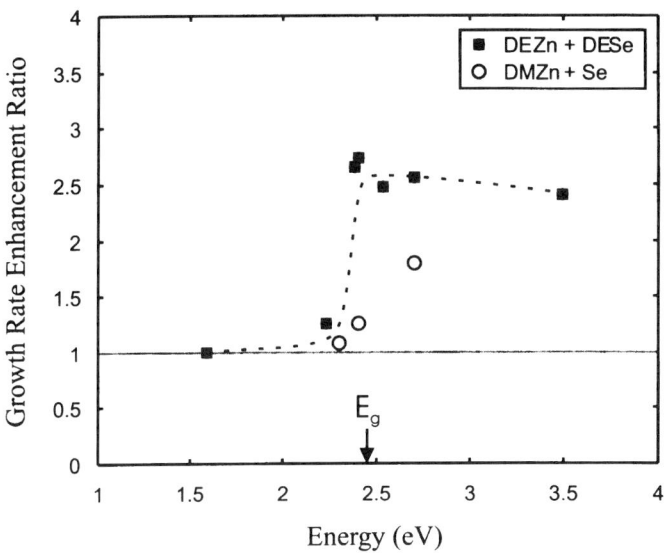

FIG. 5. Growth rate enhancement ratio (defined as the ratio of the illuminated growth rate over the unilluminated growth rate) as a function of the incident photon energy for two combinations of metalorganic sources. The power density of the illumination for all experiments was kept constant at $\sim 180\,\text{mW/cm}^2$.

used in our experiments. We speculate that photon absorption led to the generation of electron/hole pairs within the ZnSe layer, where holes drift to the surface due to band bending. These free carriers were thus able to interact with the adsorbed and unpyrolyzed DEZn molecules leading to the release of the ethyl radicals, which in turn increased the rate of incorporation of Zn. Figure 6 shows a thickness profile of the illuminated region using growth and illumination conditions that lead to growth rate enhancement. The resulting thickness profile (and thus the enhancement ratio) was found to be directly proportional to the TEM_{00} mode of the argon ion laser used. Although we have not actively attempted to optimize our growth conditions to obtain the highest growth rate enhancement, a factor of 15 increase in growth rate with illumination has been observed. Yoshikawa *et al.* (1991) have suggested a similar growth rate enhancement mechanism is observed during photo-assisted MOVPE growth of ZnSe. In their case, the photo-generated carriers in the ZnSe layer participate in the oxidation and reduction reactions with the adsorbed and unpyrolyzed metalorganics to increase the growth rate. The most important difference is that while the photo-assisted MOVPE growth of ZnSe primarily allows the ZnSe to be grown at lower substrate temperatures due to illumination, mass-trans-

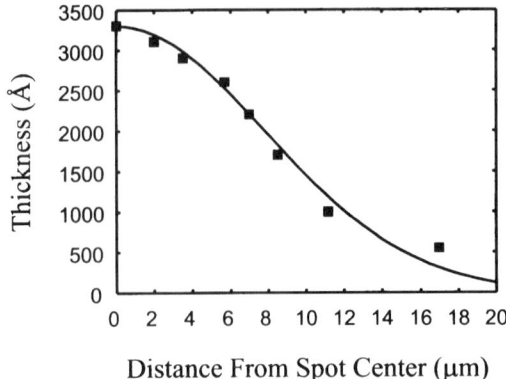

FIG. 6. Lateral thickness profile of a ZnSe region illuminated with photons having energies greater than the energy bandgap at the growth temperature.

port-limited growth of ZnSe was not possible at any substrate temperature without the use of photon-illumination.

In order to determine the precursor species most strongly affected by the photo-generated carriers, elemental Zn and Se were also used in photo-assisted growth experiments along with metalorganic precursors (DEZn, DMZn, and DESe) that were introduced either "cracked" or "uncracked". Table I tabulates the results of these experiments. As can be expected, the use of uncracked DESe with any Zn precursor resulted in no appreciable growth, as DESe is highly stable and remains unpyrolyzed over the range of growth temperatures used here. On the other hand, as long as uncracked DEZn is used with any Se precursor, a significant growth rate enhancement was observed with above-bandgap illumination. When both Zn and Se elemental sources were used, illumination using photon energies sufficient to generate electron/hole pairs resulted in growth rate suppression (i.e., the growth rate decreased with illumination). This phenomenon has been observed in photo-assisted MBE growth of ZnSe to lead to growth rate suppressions (Matsumura et al., 1990) the magnitude of which is consistent with our MOMBE results based on a comparison of the respective power densities used. When DMZn is used as the Zn precursor, a low growth rate is still obtained using DESe; however, normal growth rates were observed when elemental Se was used (i.e., no diethyl precursors involved). The transport-limited mode of growth observed with the combination DMZn + elemental Se indicated that the significant site blockage phenomenon is associated primarily with the ethyl-containing precursors. The photo-assisted growth behavior of DMZn + elemental Se is more complicated due

TABLE I

COMPARISON OF GROWTH RATES OF ZnSe GROWN BY MOMBE WITH PHOTO-IRRADIATION

Source of Zn	Source of Se	MOMBE Growth Rate[a]	Effect of Illumination on Growth Rate
DEZn	DESe	No growth	None
DEZn	Cracked DESe	Low	Enhancement
Cracked DEZn	Cracked DESe	Low	Suppression
Elemental Zn	Cracked DESe	Low	Suppression
DEZn	Elemental Se	Low	Enhancement
Elemental Zn	Elemental Se	High	Suppression
Cracked DMZn	Cracked DESe	Low	Enhancement
Cracked DMZn	Elemental Se	High	Suppression or enhancement

[a]Low: 200–400 Å/hr. High: typical of MBE.

to the necessity for precracking of DMZn: both growth rate enhancement and suppression were observed. In general, lower cracking temperatures led to the observation of growth rate enhancement; while higher DMZn cracking temperatures resulted in growth rate suppression similar to photo-assisted MBE. We are not able to clearly discern the mechanism that is at work involving DMZn because of the insufficient number of experiments that were carried out. Notice that the use of DMZn + Se combination was intended for verification of the site-blocking problem only: use of elemental Se negates the primary aim of the MOMBE technique in replacing the high vapor pressure sources with gaseous ones. Optical characterization of films growth with DMZn + Se further indicated the presence of persistent deep levels that may be caused by undesirable impurities contained in the DMZn sources.

Optical characterization of the ZnSe films grown by MOMBE (both with and without laser assistance) was carried out by low-temperature photoluminescence (PL). Typical 10 K PL spectra of unilluminated regions was dominated by a broad peak centered at ~2.25 eV with weaker excitonic features near the band-edge, suggesting that the film may be highly nonstoichiometric. Regions illuminated by above-bandgap radiation, however, showed a sharp excitonic feature without prominent emission from deep levels. Figure 7 shows the PL collected from a ZnSe layer grown (a) without and (b) with laser illumination. The effect of the laser illumination during growth is dramatically evident. The unilluminated film (1100 Å thick) is dominated by a broad deep level; the illuminated region (3500 Å thick)

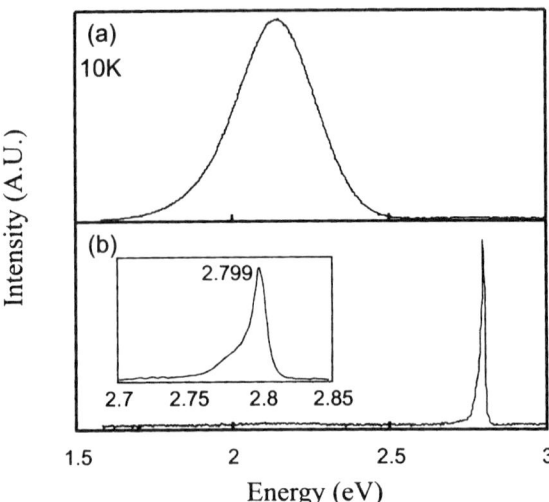

FIG. 7. Photoluminescence of (a) ZnSe grown using DESe + DEZn without photo-irradiation; and (b) regions illuminated with above-bandgap photons along with an inset emphasizing the near-band-edge portion of the spectrum.

exhibits an intense donor-bound exciton peak at 2.799 eV with negligible deep level emission. The strong effect of the laser on the PL properties of ZnSe can be consistently explained by the increased availability of Zn for incorporation as induced by photon illumination. The amount of useful Zn precursor made available by photon-illumination effectively allows one to "tune" the surface stoichiometry. Under conditions where the illuminated regions exhibit desirable PL spectra, one would expect an approximately 1:1 surface stoichiometry. Therefore, it is reasonable to expect that the nonilluminated regions are highly Zn-deficient and defective as observed by photoluminescence.

We further observed that a similar growth rate enhancement/suppression was possible when a high-energy (typically 10 kV) electron beam was incident on the substrate during the MOMBE growth of ZnSe. The RHEED electron gun provided a convenient source of electrons that are focused or defocused as desired. Since RHEED was usually utilized throughout the duration of a growth experiment, electron beam–irradiated areas indicated variations in thickness. The magnitude of electron beam–assisted growth rate modification was found to be dependent on the flux of the incident electron beam, as well as the amount of unpyrolyzed DEZn species present at the surface. The electron beam–assisted growth behaved qualitatively identical to photo-assisted growth when $\lambda_{incident} > \lambda_{bandgap}$: i.e., growth conditions that led to photo-assisted growth rate enhancement (or suppression) also led to electron beam–assisted growth rate enhancement (or suppression). Takahashi *et al.* (1992) have reported growth rate enhancement in the MOVPE of GaAs using trimethylgallium (TMG) and arsine with a focused high energy (40 kV) electron beam. In this case, the authors believe the enhancement mechanism to be kinetically induced decomposition of adsorbed TMG by the incident electron beam; that speculation is supported by the presence of very high levels of carbon incorporation in their films. In the work of Takahashi *et al.* (1992), however, the growth of the GaAs was performed at temperatures considered to be very low and in some cases at room temperature. Secondary ion mass spectroscopy (SIMS) measurements of our ZnSe films, however, failed to find an elevated level of carbon regardless of the electron beam illumination conditions. A slightly elevated carbon concentration localized about the ZnSe/GaAs interface can be seen in some samples and is believed to be a residue from the substrate preparation procedure.

Based on the similarity between photo- and electron beam–assisted growth behaviors, a likely mechanism for the observed electron beam–induced effect was the generation of electron/holes as electrons decelerate in the ZnSe layer. These electron/hole pairs then interact with the adsorbed (and unpyrolyzed) DEZn and DMZn in mechanisms similar to photogenerated carriers. The fact that blue emission has been observed by electron beam illumination during conventional MBE growth further supported this hypothesis. Characterization of the irradiated regions of the ZnSe layers has not been thoroughly performed due to the typically small (1–2 mm) regions that are obtained. Both growth rate enhancement mechanisms described here, e.g., photo- and electron beam–assisted growth, can be applied in the selective area epitaxy of ZnSe. Optical imaging or shadow mask methods can be used with photo-irradiation to define the pattern *in situ*. Similar to electron beam lithography, selective area epitaxy, consisting of very fine lines of ZnSe material, can potentially be achieved by rastering the incident electron beam.

III. Gas Source Molecular Beam Epitaxy

1. ADVANTAGES AND DISADVANTAGES OF GSMBE

Gas source molecular beam epitaxy (GSMBE) is an alternative gaseous source epitaxy growth method that replaces the high vapor pressure elements in the molecular beam epitaxy (MBE) method with hydride sources that are amenable to regulation using a precision mass flow controller. For examples, AsH_3, PH_3, and SeH_2 are used in place of elemental As, P, and Se, respectively, thus greatly reducing the associated problem of flux control. Lower vapor pressure elements such as Ga, In, and Zn are used as effusion sources. GSMBE retains most of the UHV advantages of MBE while avoiding the use of metalorganic precursors which circumvents the issue of carbon incorporation. Furthermore, hydride sources are available with very high purity (ppm impurity levels can be obtained) because of their chemical simplicity, thus avoiding possible source purity issues associated with the use of metalorganic sources. This simplicity in source materials also translates directly into a much simpler array of growth chemistry as compared to MOVPE, and results in the control and prediction of the epitaxial growth process as a more tractable problem. The use of a gas cracker for the hydride sources, however, introduces an additional degree of freedom in controlling the form of the precursors provided to the growing surface. For example, in the GSMBE growth of InP, the dimer molecules produced by cracking the PH_3 have a higher sticking coefficient than the tetramer molecules normally produced by an effusion cell. Several tradeoffs, however, are involved when using the GSMBE method. First and foremost is the use of highly toxic hydrides that necessitate the use of sophisticated gas monitoring equipment and extreme care in the daily operation and in exhaust handling. Since the thermal dissociation of the hydrides generates a large amount of the by-product H_2, the pumping requirements are essentially that of MOMBE. The frequency of required source replenishment is now dictated by the elemental sources and is no longer completely *ex situ*. Overall, the GSMBE method represents a best-case compromise that strives to achieve a balance between (1) the necessity of using a gaseous source to control the high vapor pressure elemental constituents, and (2) the desire to keep source materials and the reactions involved as simple as possible.

Another unique aspect of GSMBE is the copious and unavoidable generation of H_2 and subhydride species (such as As–H and Se–H) that are generated. Whether these H_2 and subhydride species are beneficial or deleterious depends on the material system involved. In the growth of III–V materials, the presence of hydrogen radicals has been postulated to aid in the removal of residual carbon inadvertently introduced into the vacuum system (Okada *et al.*, 1995; Sato, 1995). However, electrical passivation of both intentional and unintentionally introduced dopants has been frequently observed in the growth of Si and III–V compound materials. In these cases, annealing at modest temperatures is usually sufficient to reverse the hydrogen passivation effects. As will be shown later in this section, however, hydrogen plays a central role in the growth and doping of ZnSe by GSMBE, particularly when nitrogen doping is involved.

2. EXPERIMENTAL DETAILS

Ultrahigh purity (6N) Zn and Se, as well as H_2Se (5N) gas, were employed as precursors for the GSMBE growth of ZnSe. The H_2Se gas was thermally decomposed in an EPI hydride gas cracker at temperatures ranging from 700 to 1100°C with the typical cracking temperature usually set at $\sim 1000°C$. The cracking tube is composed of a pyrolytic boron nitride (PBN) tube capped with a tantalum end baffle having an exit aperture ~ 0.7 cm in diameter. The PBN cracking tube assembly also contained a significant amount of internal Ta baffling.

ZnSe epilayers were usually grown on 0.5–0.75 μm epitaxial GaAs buffer layers grown on (100) GaAs epi-ready substrates. The buffer layers were grown in our ultrahigh vacuum interconnected GSMBE reactor dedicated to III–V materials, and transferred into the II–VI reactor *in situ* without surface passivation. [As part of our III–V/II–VI heterostructure effort (which will not be described here), various other ternary and quaternary compounds of (In,Ga,Al)P and (In,Ga)As have also been investigated as epitaxial buffer layers. These buffer layers can be flexibly designed to provide lattice-matched pseudosubstrates to the lattice parameters of ZnSe, GaAs, as well as to a wide range of (Zn,Mg)(S,Se) quaternary layers.] The growth temperature for the experiments was calibrated using Au/Ge eutectics in a manner similar to those employed for the previously described MOMBE growth experiments. A relatively wide range of growth temperatures (between 250 and 370°C) have been explored using GSMBE.

The *n*-type dopant source was solid $ZnCl_2$ contained in an effusion oven whose temperature typically ranged between 150 and 300°C. Either a cryogenically cooled (Model CARS25) or a water-cooled (Model MPD21) radio frequency (RF) plasma source from Oxford Applied Research was used as the nitrogen dopant source for acceptors. RF powers ranging between 200 and 500 W have been used. The plasma emission intensity was monitored using a photodiode coupled to the interior of the quartz plasma chamber via an optical fiber. The exhaust temperature was regulated between -30 and $-50°C$ in the cryogenically cooled plasma source. Various aperture configurations were used and resulted in nitrogen flows that produced chamber pressures between 5×10^{-7} and 2×10^{-5} torr under the high brightness plasma mode of operation (Qiu *et al.*, 1991). The surface stoichiometry was normally maintained near one-to-one for the Zn and Se fluxes as monitored by RHEED during nitrogen doping. Film thicknesses were usually monitored *in situ* using pyrometer temperature oscillations caused by optical interference between the growing ZnSe layer and the underlying GaAs buffer/substrate layers (Ringle *et al.*, 1994) Selective etching of partially masked ZnSe films were also employed to ascertain the actual film thicknesses.

Routine postgrowth characterization of the as-grown films included Nomarski microscopy, photoluminescence (PL), secondary ion mass spectrometry (SIMS), as well as capacitance–voltage (C-V) measurements and Hall effect measurements (for n-type ZnSe). SIMS analyses were performed on Au-coated samples at Charles Evans & Associates using Cs^+ as the primary ion. The typical primary ion raster size was $150 \times 150\,\mu m^2$, with the secondary ions collected from a circular image area of $\sim 30\,\mu m$ in diameter. Electron-beam flooding was employed in all measurements in order to minimize sample charging. Impurity concentrations were calibrated using implanted nitrogen and deuterium profiles in ZnSe epilayers of known dosage. Two device geometries have been employed for C-V measurements depending on the conductivity type of the GaAs substrate and buffer layers. Lateral double-Schottky diodes were used for conductive ZnSe films grown on semi-insulating GaAs substrates. For ZnSe:N grown on p-type GaAs (with p-type GaAs:Be buffer layers), $500\,\mu m$-diameter Schottky contacts (Cr/Au) were used for measurement at 10 kHz. Typical ZnSe:N and ZnSe:Cl film thicknesses for electrical measurements were $\geq 2\,\mu m$.

PL spectra were measured by exciting the samples with the 325 nm line of a focused He–Cd laser, providing a power density of approximately $300\,mW/cm^2$. The samples' luminescence was analyzed by a 0.5-m spectrometer and a photomultiplier tube. As an example of the optical properties exhibited by GSMBE-grown ZnSe, Figure 8 shows the PL spectrum obtained from an undoped layer. For the entire range of growth temperatures investigated, the low temperature (10 K) PL spectrum of undoped layers was typically dominated by an intense donor-bound exciton feature having an energy of 2.798 eV. As can be seen in Figure 8, the spectrum is dominated by the donor-bound (I_2) transition and the free exciton (E_x) transition having nearly equal intensity for a 1.1 μm ZnSe layer grown on an epitaxial buffer layer of (In,Ga)P (Lu et al., 1994). The Y_0 transition at 2.602 eV and the I_v^0 transition at 2.775 eV have been attributed to extended defects and are typically observed in high-purity ZnSe epilayers (Myhajlenko et al., 1984; Saraie et al., 1989; Satoh and Igaki, 1983; Shahzad et al., 1990). (The transition at 1.938 eV is due to emission originating from the (In,Ga)P buffer layer.) Based on the PL peak energy and secondary ion mass spectrometry (SIMS) analysis of the films, we speculate that the unintentional donor impurity was chlorine. The deep level defect-related luminescence band, which is broadly centered about 2.25 eV, is typically at least $100-1000\times$ weaker in intensity than the near-band-edge features. To elucidate the presence of the defect-related luminescence, photoluminescence was also obtained at 77 K as shown in Figure 9 for films grown at various substrate temperatures. The donor-bound exciton remained as the dominant feature. A small amount of luminescence originating from deep levels was

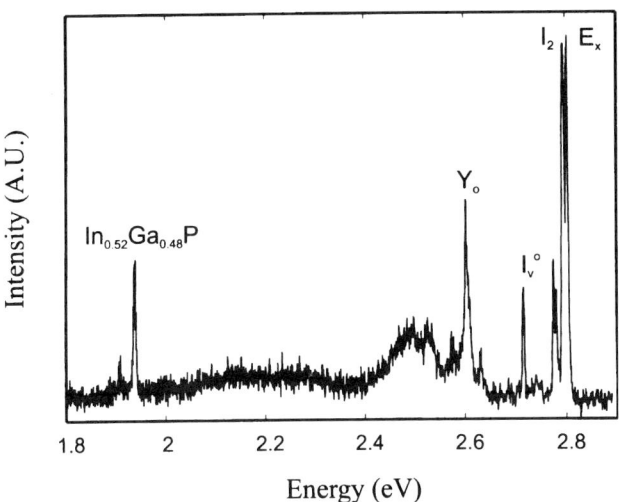

FIG. 8. Photoluminescence of undoped ZnSe (~1.1 μm) grown on $In_{0.52}Ga_{0.48}P$ (partially relaxed, 4.3 μm) epitaxial buffer layers grown on GaAs. The feature at 1.938 eV is attributed to the luminescence from the $In_{0.52}Ga_{0.48}P$ buffer layer. Features at 2.803, 2.795, 2.776, and 2.603 are identified as the E_x, I_2, I_v^0, and Y_0, respectively.

observed only at the growth temperature extremes examined in our experiments. As another indication of the high quality of the ZnSe layers, the integrated PL intensity of the near-band-edge feature was found to decrease to only one-sixtieth of its 10 K value as the sample temperature was increased to room temperature; room temperature luminescence was easily observed by the naked eye.

In order to establish the background carrier type and concentration in unintentionally doped ZnSe, Hall effect measurements were carried out at room temperature. The Van der Pauw geometry was used for the measurement, with annealed In contacts employed as the ohmic contacts. All unintentionally doped films were found to be lightly n-type; the free electron concentrations ranged from the mid 10^{15} cm^{-3} to the low 10^{17} cm^{-3} and appeared to be influenced by the surface stoichiometry. Slightly Zn-rich conditions (as opposed to Se-rich conditions) have been previously (Coronado et al., 1994) found to give rise to photoluminescence (PL) spectra exhibiting a stronger free-exciton feature, and to a lower unintentionally doped free electron concentration as determined using Hall effect measurements.

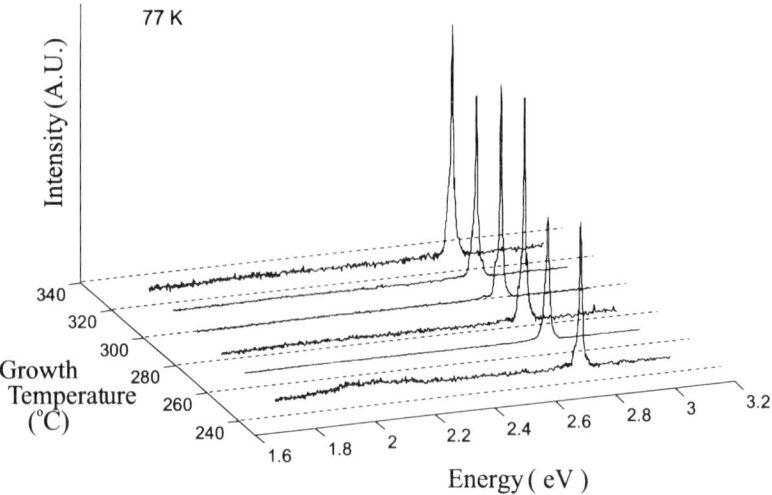

FIG. 9. The 77 K photoluminescence intensity as a function of energy for ZnSe epilayers grown at various substrate temperatures. The ZnSe film at a growth temperature of 284°C was grown on an (In,Ga)P buffer layer. The spectra are dominated by donor-bound exciton (I_2) features at 2.790 eV.

3. *n*-TYPE DOPING

Chlorine has been fairly well investigated in the context of *n*-type doping of ZnSe using the conventional MBE technique and was found to be an effective shallow *n*-type dopant for ZnSe, as well as for wider bandgap (Zn,Mg)(S,Se) materials (Ferreira et al., 1995; Ohkawa et al., 1987; Zhu et al., 1993). In our experiments, we used an effusion source of solid anhydrous $ZnCl_2$ (5 N) to achieve straightforward and effective *n*-type doping of ZnSe by GSMBE. The chlorine concentration (hereafter denoted [Cl]) and growth rate were determined as a function of the $ZnCl_2$ effusion cell temperature, while all other growth parameters were held constant. As shown in the top portion of Figure 10, the chlorine concentration was found to increase with $ZnCl_2$ oven temperature in a well-behaved manner. In Figure 10, the [Cl] was primarily measured by means of SIMS for [Cl] values greater than the detection limit of Cl in the SIMS apparatus ($\sim 10^{17}$ cm^{-3}), whereas the free electron concentrations obtained from Hall effect measurements were used for lower chlorine concentrations. At chlorine concentrations from the low to middle 10^{18} cm^{-3} range (determined by SIMS), the free electron concentrations, as measured by Hall effect measurements, as well as $[N_A - N_D]$ values obtained from C-V measurements all

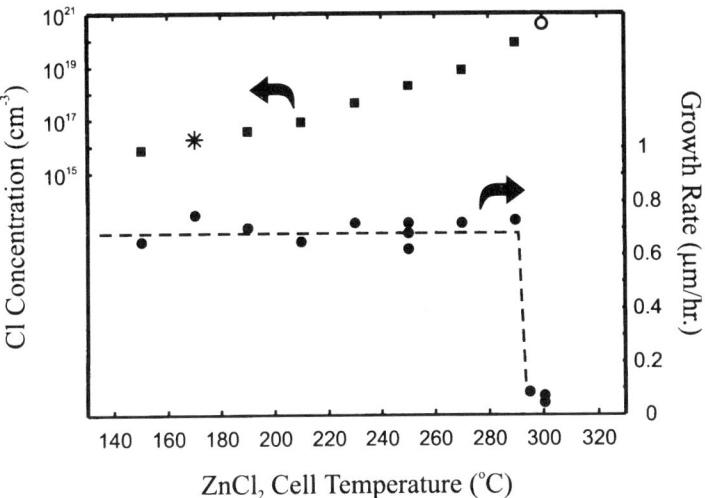

FIG. 10. Chlorine concentration [Cl] (left vertical axis) and the growth rate (right vertical axis) as a function of the ZnCl$_2$ cell temperature. The [Cl] depicted by the solid squares was determined by SIMS, while the starred point was measured using Hall effect measurements. The open circle is an extrapolated value due to the slow growth rate obtained at a cell temperature of 300°C.

agree within the accuracy of the measurement techniques involved. For ZnCl$_2$ oven temperatures from 150 to 290°C, the [Cl] varied from 9×10^{15} to 8×10^{19} cm^{-3}, while the growth rate remained essentially constant as indicated from the bottom portion of Figure 10. At the heavily doped extreme, an abrupt change in the growth rate was observed at an effusion cell temperature of 300°C. The mechanism for this phenomenon is presently unknown, but the sharp transition suggests the possible formation of a separate chemical phase involving the large [Cl] at the surface. These chlorine levels are comparable to the highest levels reported in n-type doped films grown by conventional MBE or by MBE employing a selective planar doping technique (Zhu et al., 1993).

The PL spectra of several ZnSe:Cl films are shown in Figure 11 as a function of increasing [Cl]. For doping levels resulting in [Cl] as high as 4×10^{18} cm^{-3}, the spectra were dominated by a single intense donor-bound excitonic transition. Negligible defect-related deep level emission was observed (at 100× magnification), which is indicative of the high crystalline quality. The intensity of the donor-bound exciton feature increased as the [Cl] was increased, exhibited a maximum in intensity for a [Cl] of 1×10^{17} cm^{-3}, and then decreased as the doping level was increased

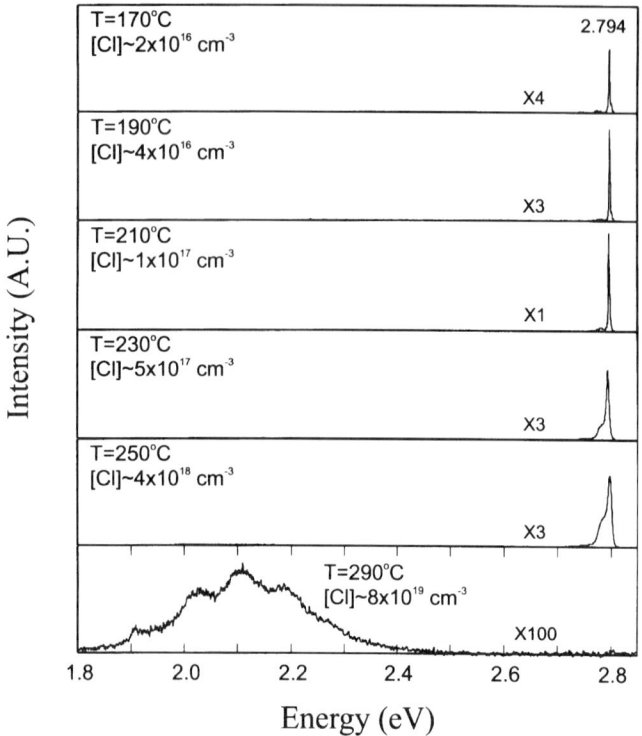

FIG. 11. 10 K photoluminescence of Cl-doped ZnSe films having a progressively higher doping concentration from the top to the bottom subplots. The $ZnCl_2$ cell temperature, relative magnifications used to obtain the spectrum, as well as the [Cl] determined by SIMS are indicated in each case.

further. At a [Cl] of 8×10^{19} cm^{-3}, the donor-bound exciton feature was no longer present, and the spectrum was dominated by a broad band of deep level emission, approximately extending from 1.8 to 2.4 eV. The abrupt change in the PL spectrum is due to defect generation and compensation, and is similar to results reported for high doping levels achieved by MBE (Ohkawa et al., 1987).

4. p-TYPE DOPING

For many decades, the goal of p-type ZnSe proved unattainable for the many researchers in pursuit of II–VI blue light emitter materials using conventional bulk growth technologies. The advent of sophisticated nonequilibrium growth techniques, such as MBE and MOVPE, provided unprecedented atomic-level control and source flexibility in the growth and doping of II–VI compounds, but consistent and practical p-type conductivity of ZnSe nevertheless remained elusive. Although n-type conductivity of ZnSe was achieved relatively easily, the strong tendency for "self-compensation" had prevented the achievement of p-type conversion in ZnSe. In general, acceptor self-compensation refers to the process where increased inclusion of acceptors led to highly compensated (and resistive) resultant films instead of p-type electrically active material. Early theoretical efforts originally argued that native donor defect formation, such as vacancies and interstitials, may electrically counteract the incorporated acceptors (Mandel, 1964). It is now believed that the compensation is caused by a large lattice relaxation around the acceptor site as originally proposed by Chadi and Chang (1989) based on first-principles energy calculations. Chadi and Chang also concluded that nitrogen would form sufficiently shallow acceptor states in ZnSe; other theorists later confirmed their findings (Kwak *et al.*, 1993). But early attempts to introduce nitrogen had not been consistently successful: dopant precursors in the form of N_2 and NH_3 were found to have low sticking coefficients and were thus difficult to incorporate (Park *et al.*, 1985). It was not until the introduction of a plasma source of nitrogen during MBE, reported independently by Ohkawa *et al.* (1991) and Park *et al.* (1990), that consistent and stable p-type conversion of ZnSe became possible. Both experimental as well as recent theoretical evidence, however, indicates that self-compensation, in the form of compensating donors that are generated in response to high levels of nitrogen, limits the maximum achievable hole concentration to be $\leqslant 10^{18}$ cm^{-3} for ZnSe (Fan *et al.*, 1994; Ohkawa *et al.*, 1993; Qiu *et al.*, 1991; Yang *et al.*, 1992), and still lower for wider bandgap quaternaries such as (Zn,Mg)(S,Se) (Han *et al.*, 1994; Ikeda *et al.*, 1995). Since the breakthrough in p-type doping, the field of II–VI research has exploded, leading to both increased understanding of the doping and compensation process in ZnSe, as well as impressive achievements in blue-green laser devices, as documented in this particular volume of *Semiconductors and Semimetals*.

Based on the successes of nitrogen doping in MBE, a plasma source of nitrogen, identical to the ones used in the original MBE nitrogen doping experiments, was used for the growth of ZnSe:N by GSMBE. Figure 12 shows the progression of the PL spectra as a function of the chamber equivalent pressure of nitrogen flow for lightly doped samples; in this figure, the 10 K spectra are shown for samples grown under identical growth conditions, however, the nitrogen gas flow was systematically increased to enhance the incorporation of acceptors into the lattice. As is seen in Figure 12, the characteristic PL features of the ZnSe:N films are seen to be strongly

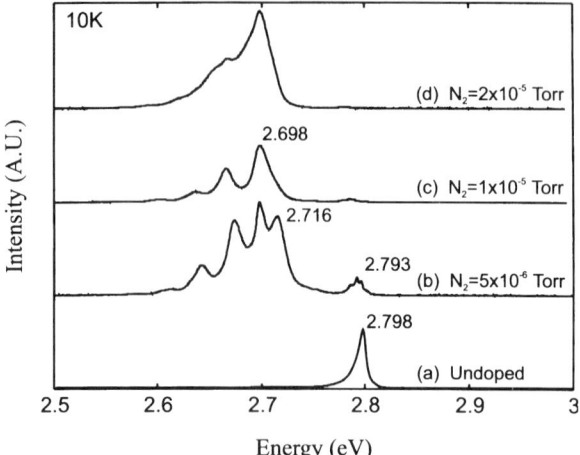

FIG. 12. 10 K photoluminescence intensity as a function of energy demonstrating the evolution of the spectrum as the nitrogen gas equivalent pressure is increased from that obtained from a reference undoped sample (a). In b–d the nitrogen flow was increased to vary the incorporation of the nitrogen acceptor species.

dependent on the degree of nitrogen incorporation. The growth conditions for this series of films were as follows: a 270°C calibrated substrate temperature, a slightly Se-rich surface stoichiometry, and a constant RF power of 100 watts using the water-cooled plasma source. The PL progression as a function of increasing nitrogen flow was characteristic of increasing nitrogen concentration, a trend similar to that reported by Ohkawa et al. (1991) for MBE-grown ZnSe:N. For the sample grown with the lowest nitrogen flow, shown in Figure 12b, the near-band-edge transitions were dominated by the neutral N acceptor peak at 2.793 eV. (PL obtained from an undoped sample is shown for reference in Figure 12a.) The free electron-to-acceptor transition (FA) was present at 2.716 eV to suggest a nitrogen acceptor binding energy of 109 meV, assuming a 10 K bandgap energy of 2.825 eV (Dean et al., 1981). The donor-to-acceptor-pair (DAP) transition peak at 2.698 eV and the associated phonon replicas were also detected. Further increases in the nitrogen flow resulted in a spectrum dominated by the FA and DAP transitions (and their LO phonon replicas), as shown in Figure 12c. At the highest nitrogen gas flows investigated (Figure 12d), the FA transition disappeared, while the DAP and phonon replicas merged into a single broad feature.

By increasing the nitrogen flow, using various combinations of aperture configurations as well as higher applied RF power, significant concentra-

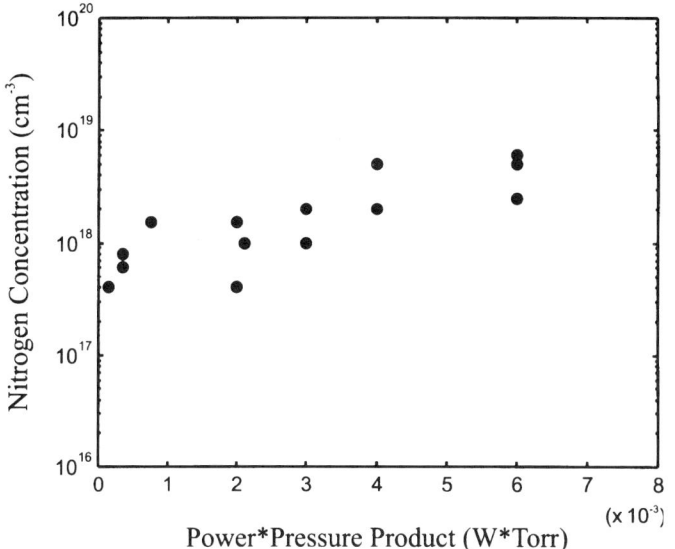

FIG. 13. Incorporated nitrogen concentration, [N], as a function of the nitrogen doping parameter, RF power * gas pressure. Note that the plot does not delineate layers produced using various stoichiometries and temperature conditions.

tions of nitrogen (as measured by SIMS) were found to be incorporated during the GSMBE growth of ZnSe. Figure 13 shows the measured nitrogen concentration [N] as a function of nitrogen doping parameter defined as the product of the RF power and the chamber equivalent pressure of nitrogen flow. The [N] is seen to increase proportionally with the doping parameter as either the RF power or the nitrogen flow is increased while maintaining the high brightness mode of operation. At higher doping levels ([N] $\geq 10^{18}$ cm^{-3}), the PL spectrum becomes progressively broader and lower in energy; this is consistent with the current speculation that donor defects of increasing depth from the valence band-edge appear as the ZnSe becomes more heavily doped.

Electrical measurements of ZnSe:N grown by GSMBE, however, indicated that most of the films were highly resistive regardless of the actual incorporated nitrogen concentration or the appearance of PL spectra that was suggestive of p-type conversion. For example, PL spectra, as shown in Figure 12c and d, would be associated with ZnSe:N films grown using conventional MBE in our chamber that exhibit $[N_A - N_D] > 10^{17}$ cm^{-3}. As will be described in detail in the next section, the culprit of this observed electrical passivation of the incorporated nitrogen acceptors is now known to be the hydrogen molecules that are copiously present in the GSMBE growth environment.

5. HYDROGEN PASSIVATION

Hydrogen passivation of both donors and acceptors has been extensively reported for III–V compounds grown by MOVPE as well as MOMBE (Chevallier et al., 1985; Cho et al., 1993; Johnson et al., 1986; Kozuch et al., 1993; Rahbi et al., 1993). There also exists a substantial body of papers dealing with hydrogen in the Si material system (Pearton et al., 1991). Recent results for ZnSe grown by MOVPE (Kamata et al., 1993; Wolk et al., 1993), as well as experiments where ZnSe was intentionally hydrogenated with deuterium (Pong et al., 1992), suggest that hydrogen is electrically and optically passivating the nitrogen acceptors in ZnSe:N. Unlike hydrogen in III–V and Si material systems, its counterpart in nitrogen-doped ZnSe is differentiated by the relative difficulty in reversing the passivation behavior by thermal treatments. In most cases, postgrowth thermal treatment at or above the normal growth temperatures is required to modify the behavior of the incorporated hydrogen.

Because of the large amount of hydrogen that is present in the GSMBE growth environment, most as-grown ZnSe:N films were found to be highly resistive due to significant hydrogen passivation. In order to investigate the relationship between the hydrogen and the nitrogen concentration that is incorporated, a variety of structures were grown that contained intentional variations in nitrogen concentration.

Figure 14 shows a SIMS profile of a ZnSe:N layer grown on p-type GaAs by GSMBE. The particular structure depicted contains several undoped ZnSe regions (that enable the determination of the background levels of [H] and [N]) in an otherwise uniformly nitrogen-doped ZnSe film having $[N] \sim 10^{18}$ cm^{-3}. Several striking features are apparent in the figure: (1) the extraordinary tracking behavior of [H] with the variation in [N]; (2) both [N] and [H] decrease to their respective background levels in all of the undoped regions; and (3) in the ZnSe:N region, the ratio [N]/[H] was nearly unity. Similarly tracking behavior between [H] and [N] has been observed for ZnSe:N films grown under a wide variety of GSMBE growth and doping conditions (Fisher et al., 1995; Ho et al., 1994, 1995). These facts suggest that the observed incorporation of hydrogen is directly related to the existence of nitrogen. Nearly all of the as-grown films are highly resistive as determined by the C-V method, further suggesting that the nitrogen acceptors have been effectively passivated by hydrogen. These results are in agreement with spectroscopic data obtained for nitrogen-doped ZnSe grown

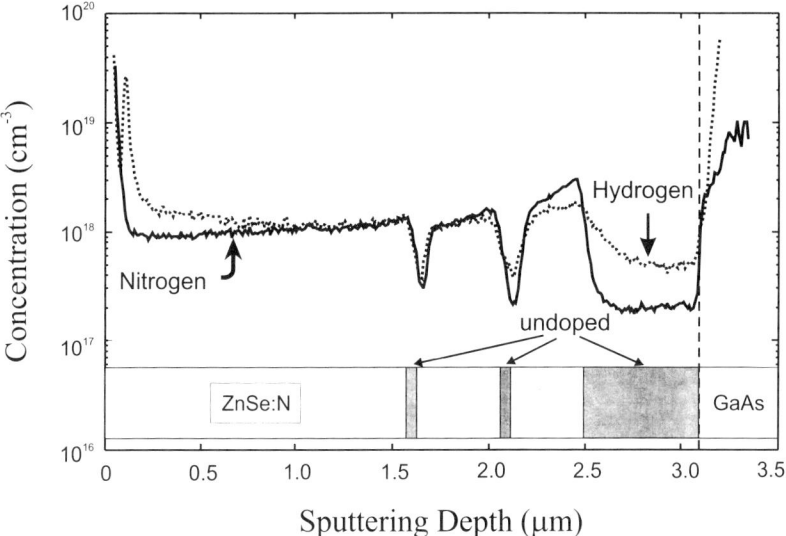

FIG. 14. SIMS depth profile obtained from a uniformly doped ZnSe:N epilayer grown by GSMBE on an undoped ZnSe buffer layer. The undoped ZnSe layers, as indicated in the schematic, provide references for the background concentrations of [N] and [H]. The synergistic correlation between [N] and [H] is clearly seen. The dramatic increase in the signal at the surface and at the ZnSe/GaAs interface are due to surface effects and an artifact of the data normalization procedure, respectively.

by MOVPE, where direct evidence of the N–H bonds has been reported (Kamata et al., 1993; Wolk et al., 1993).

For ZnSe:N films grown by conventional MBE under similar corresponding growth conditions (surface stoichiometry, substrate temperature, and incorporated [N]) in the same chamber, the hydrogen concentration remained below the detection limit of our SIMS instruments, suggesting that incorporation of hydrogen from the UHV background is negligible. Figure 15 shows a similar SIMS profile of a structure containing regions of chlorine-doped ZnSe grown by GSMBE for comparison. Note that the undoped region near the ZnSe/GaAs interface serves as the reference for the background levels of [H] and [Cl]. As the [Cl] is intentionally varied, the [H] remained near the background level throughout the structure, increasing to a large value near the surface due to surface contamination. In stark contrast with ZnSe:N films grown by GSMBE, the absence of the coherent tracking behavior between the dopant [Cl] and [H] also resulted in a negligible amount of electrical passivation. These two observations strongly suggest that the observed hydrogen that is incorporated is directly associated with the presence of the acceptor nitrogen species.

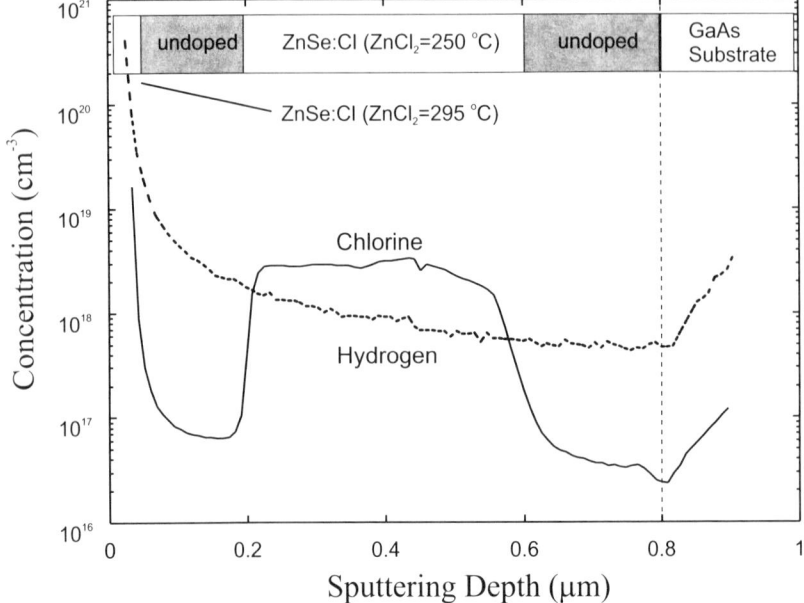

FIG. 15. SIMS depth profile of a structure containing two Cl-doped regions separated by undoped regions (as references for the background concentrations of [N] and [H]). The variation in [H] is seen to be independent of the intentional variation in [Cl]. The $ZnCl_2 = 295°C$ layer was grown for the same amount of time as the $ZnCl_2 = 250°C$ layer. The significantly reduced layer dimension for the higher cell temperature reflects the greatly diminished growth rate at high chlorine doping levels.

Our earlier thermal decomposition or "cracking" studies using *in situ* QMS indicated that molecular hydrogen is the major by-product of the H_2Se decomposition reaction. Figure 16 shows the QMS signal intensity (left vertical axis, in uncalibrated units of detector current) as well as the normalized growth rate (right vertical axis, per SCCM of H_2Se scaled to unity for the highest growth rate obtained at 1000°C) as a function of the cracking temperature of the H_2Se cracker cell. The growth rate increases dramatically above 600°C, indicating the onset of H_2Se decomposition. The hydrogen signal is also seen to increase above 600°C with increasing cracker temperature and is accompanied by a corresponding decrease in the H_2Se signal. The arrow at the upper left corner of Figure 16 indicate the level of QMS signal when 3 SCCM of H_2 is injected into the chamber. Thus it can be reasoned that during the typical GSMBE growth of ZnSe, the amount of H_2, generated as a decomposition by-product of H_2Se, is similar to that measured by the intentional introduction of ~ 1–2 SCCM of hydrogen gas.

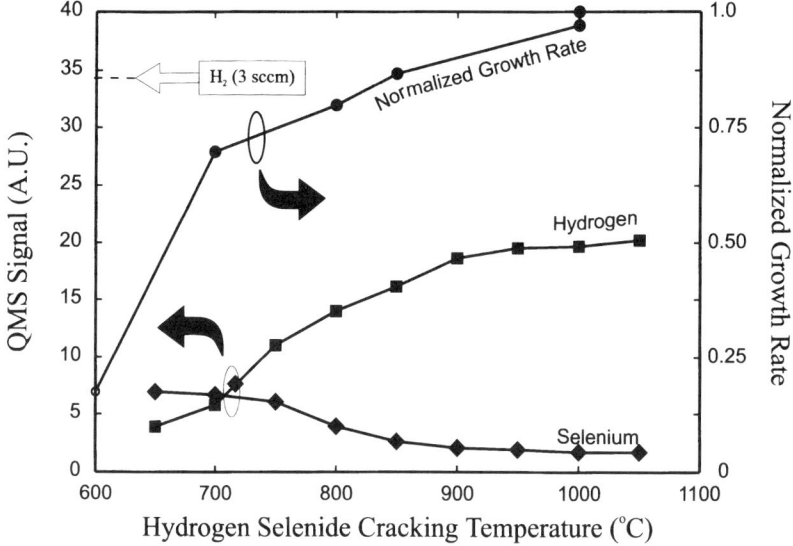

FIG. 16. Comparison of the H$_2$ and elemental selenium signals (left vertical axis) as seen by a QMS that is generated as a by-product of the H$_2$Se cracking; also shown is the signal obtained by 3 SCCM of intentionally introduced H$_2$ gas. The right vertical axis depicts the relative growth rates (normalized to neutralize the effect of varying group II fluxes) as a function of the H$_2$Se cracker cell.

An important question that remains to be answered concerns the passivation that is observed during the GSMBE growth of ZnSe doped with a nitrogen plasma source, i.e. is the dominant source of electrical passivation derived from molecular hydrogen that is produced by cracking, or from hydrogen radicals still attached to the Se precursor? In order to quantify the role of molecular hydrogen in the passivation mechanism of nitrogen acceptors, experiments were carried out where various amounts of hydrogen were introduced into the chamber during conventional MBE growth. Figure 17 shows a SIMS profile of a structure containing intentionally hydrogenated regions (lined regions in the structure schematic) where different amounts of molecular hydrogen were introduced; the layer also contains undoped regions (shaded regions) as well as a doped region that was not exposed to a flux of hydrogen. The main features to note in Figure 17 are as follows: (1) similar to GSMBE-grown ZnSe:N, [H] is seen to closely track [N], with both impurities decreasing to near their background levels in all of the undoped regions; (2) a clear increase in [H] coincident with the addition of 1 SCCM H$_2$ (from the schematically indicated "ZnSe:N" region to the "ZnSe:N + 1.0 SCCM H$_2$" region) indicates

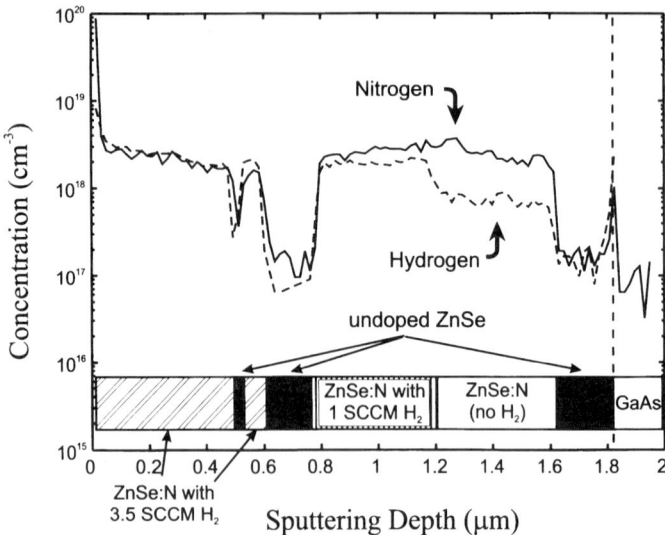

FIG. 17. SIMS profile of a ZnSe:N layer grown by MBE on an undoped ZnSe buffer layer (on GaAs). As the depth into the ZnSe layer increases, various amounts of H_2 were introduced to the growth front, with two undoped regions as indicated. The undoped regions serve as markers for the background [H] and [N] concentrations. Variations in the H_2 flow are as indicated in the figure.

that nitrogen-induced hydrogen incorporation has occurred; and (3) the [N]/[H] ratio in the "ZnSe:N + 1.0 SCCM" region appears to be greater than the [N]/[H] in the "ZnSe + 3.5 SCCM H_2" region which is consistent with the assumption that the observed increase in [H] is directly related to the amount of injected H_2.*

A comparison of C-V measurements performed on unhydrogenated ZnSe:N films (reference films without H_2) and intentionally hydrogenated films (ZnSe:N + various amounts of H_2) lends further support to our hypothesis that molecular hydrogen acts to passivate the nitrogen acceptors in MBE using a nitrogen plasma cell. Several series of nitrogen-doped films were grown by MBE, both with and without intentional hydrogen, using identical growth and doping conditions; C-V measurements were subsequently performed on all samples. With a target $[N_A - N_D]$ of $\sim 2 \times 10^{17}$ cm^{-3} for the unhydrogenated reference samples (typical MBE growth), it

---

*The apparent increase in [H] at the interface of "undoped ZnSe" and "ZnSe:N" regions is due to residual levels of hydrogen in the shared hydrogen/nitrogen gas manifold that feeds the RF plasma cell. Therefore some background level of passivation exists for this particular experiment even though H_2 is not intentionally introduced via the cracker.

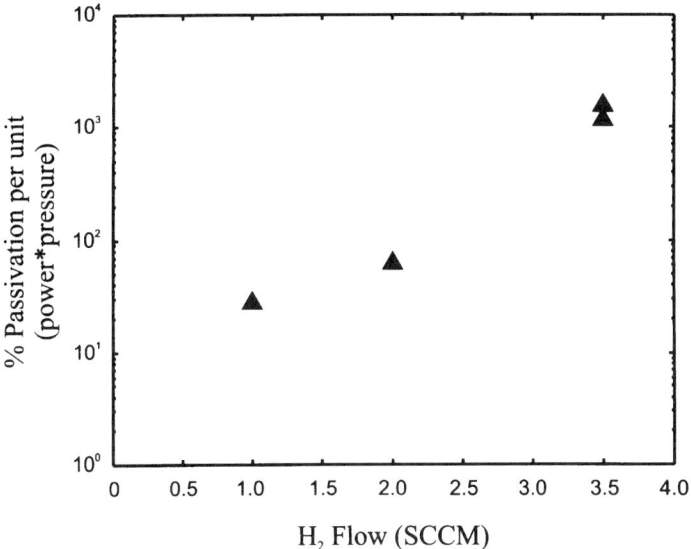

FIG. 18. The percent of acceptor passivation (normalized by the doping parameter previously defined as the product of "power * pressure") as a function of the intentionally injected H_2 flow rate in SCCM. Note that it is the relative amount of passivation that is of interest here and not the absolute magnitudes.

was generally found that a significant decrease in $[N_A - N_D]$ occurred for films that were grown with injected hydrogen. As an example of the magnitudes of passivation that were observed, 25% and 60% passivation were measured for flow rates of 1 and 2 SCCM of H_2, respectively, as compared to the reference (unhydrogenated) MBE-grown samples (where passivation is taken to be the 0% reference).[†] The precise amount of passivation was found to depend on the growth, doping conditions, as well as the hydrogen flow rates that were used. Figure 18 shows a "normalized percent passivation" (normalized by the nitrogen doping parameter defined earlier) as a function of the hydrogen flow rate used during the hydrogenation experiments. The positive slope in Figure 18 highlights a result that we would reasonably expect: that more hydrogen present at the surface for a given amount of active nitrogen species gives rise to a larger degree of observed passivation. Notice that the result of Figure 18 can also be stated in a corollary manner: that the amount of electrical passivation depends on

[†]This particular set of hydrogenation runs were grown at 280°C, under slightly Se-rich conditions, and with the following RF source parameters: 400 W forward power and 2.2×10^{-5} torr chamber background N_2 pressure.

the amount of active nitrogen present for a given flow of hydrogen. These results and postulates advanced above are consistent with recent data on thermal annealing of MOVPE-grown ZnSe:N films obtained by Ogata et al. (1995), where annealing ZnSe:N epilayers (possessing an initial net acceptor concentration of 2.0×10^{17} cm^{-3}) in an H_2 ambient at 350°C was found to provide significant electrical passivation of the nitrogen acceptors. More detailed studies are required to quantitatively understand the passivating relationship between the active nitrogen species and the H_2 present at the growing surface.

As shown in the previous H_2Se cracking experiments, 2 SCCM of injected H_2 provides a background hydrogen signal (as detected using QMS) that is approximately equal to that normally encountered during normal GSMBE growth. Since the degree of electrical passivation of GSMBE-grown ZnSe:N samples is found to be much more significant than that measured for the MBE + H_2 counterparts exposed to similar H_2 background levels, we speculate that an equally important mechanism for nitrogen acceptor passivation is related to incompletely cracked Se–H fragments. It is expected that hydrogen passivation during GSMBE growth can be minimized by cracking the H_2Se precursor as completely as possible; however, our results seem to indicate that some passivation is inevitable due to the presence of H_2 under normal growth and stoichiometry conditions. This reasoning might also play a part in reconciling our present results with other seemingly contradictory results (Imaizumi et al., 1994, 1994a), where hydrogen passivation has been reported to be absent in GSMBE growth of ZnSe using the same precursors. We believe that the hydrogenation behavior described here has significant implications for the GSMBE, as well as MOVPE, growth of ZnSe:N where multiple pathways for hydrogen incorporation may exist. Further experiments are currently underway to study the effect of growth and surface stoichiometry conditions on hydrogen incorporation under both MBE and GSMBE growth conditions. It is our belief that further and fundamental understanding of the surface kinetic reactions that occur with hydrogen during GSMBE and MOVPE will ultimately dictate the usefulness of GSMBE and MOVPE for achieving practical and reproducible p-type conductivity in wide bandgap II–Vs with nitrogen acceptors.

IV. Concluding Remarks

The use of alternative nonequilibrium growth approaches, as applied to wide bandgap II–VI binaries, ternaries, and quaternaries, is clearly highly desirable, particularly to facilitate flux control of high vapor pressure species. The anticipated benefits are expected to parallel the advantages recognized and enjoyed by the III–V community for the growth of (In,Ga)(As,P) employing GSMBE and MOMBE. A few similarities include multiple high vapor pressure constituent species (such as Zn, Se, S, Te and As, P) and the need to utilize quaternaries for lattice-matching and for bandgap engineering of sophisticated electronic and photonic devices. Significant contrasts in the growth of the II–VIs and the III–Vs, however, include substantial differences in the growth temperatures ($\sim 300°C$ for ZnSe compared to $\sim 600°C$ for GaAs), and the fact that the II–VI elements do not have unity sticking coefficients, as does Ga during the growth of GaAs. The similarities and contrasts between the growth of II–VIs and III–Vs impose new constraints in the application of different growth techniques for each material family. Regardless of material family, however, the application of new techniques requires a great deal of research and investigation to discover the appropriate operating parameters required to optimize device performance. It is at this stage of research that we are investigating advantages, and potential barriers, to utilizing gaseous source UHV epitaxy technologies for the growth of wide bandgap II–VI materials. The issue of source purity and hydrogen-containing organic chains, combined with the low substrate temperatures necessary for the II–VIs, suggest that novel metalorganic precursors (Rajavel et al., 1994) are important directions to purue. In the case of GSMBE, hydrogen has been demonstrated to play an important role in passivating nitrogen acceptors in ZnSe. Additional research is necessary to determine the surface kinetic processes that inhibit incorporation of hydrogen, eliminating the need to resort to postgrowth annealing treatments. In fact, interesting results (Imaizumi et al., 1994, 1994a) reporting p-type conductivity and thus the fabrication of optical devices are very encouraging, and suggest that hydrogen incorporation during growth can in fact be minimized. An area that has received very little attention involves the thermal decomposition behavior of II–VI precursors, both metalorganic- and hydride-based, as well as the surface reactions of these molecules in an ultrahigh vacuum environment. It is clear that much more research is necessary before we can determine the suitability of these new growth approaches for wide bandgap II–VI materials, and thus a very rich and interesting field awaits development.

Acknowledgments

We are sincerely grateful for the numerous contributions and dedication of our co-workers at MIT: Dr. G. S. Petrich, Dr. C. A. Coronado, K.-Y. Lim,

J. L. House, K. Lu, P. A. Fisher, and E. L. Warlick. We have benefited greatly from the many contributions of our visiting scientists and collaborators: Drs. C. Huber, H. Nanto, H. Kanie, and N. M. Johnson. We are also very indebted to the groups of Professor R. L. Gunshor at Purdue University and Professor A. V. Nurmikko at Brown University for the unquestioning exchange of samples, supplies, hardware, and for openly sharing their ideas and knowledge. Our research has been sponsored by the following programs: the Advanced Research Projects Agency/Office of Naval Research University Research Initiative (Grant No. 284-25041), the National Science Foundation (Grant No. DMR-9202957), the Joint Services Electronics Program/Army Research Office (Contract No. DAAL-03-92-C-0001), the AASERT Program (Contract No. DAAH-04-93-G-0175), and the National Center for Integrated Photonic Technology (Contract No. 542-381).

REFERENCES

Aoyagi, Y., Kanazawa, M., Doi, A., Iwai, S., and Namba, S. (1986). Characteristics of laser metalorganic vapor-phase epitaxy in GaAs. *J. Appl. Phys.* **60**, 3131.

Banse, B. A., and Creighton, J. R. (1991). The adsorption of triethylgallium on GaAs (100). *Surface Sci.* **257**, 221.

Bedair, S. M., Whisnant, J. K., Karam, N. H., Griffis, D., El-Masry, N. A., and Stadelmaier, H. H. (1986). Laser selective deposition of III-V compounds on GaAs and Si substrates. *J. Cryst. Growth* **77**, 229.

Bicknell, R. N., Giles, N. C., and Schetzina, J. F. (1986). Growth of high mobility n-type CdTe by photoassisted molecular beam epitaxy. *Appl. Phys. Lett.* **49**, 1095.

Bicknell, R. N., Giles, N. C., and Schetzina, J. F. (1986a). p-Type CdTe epilayers grown by photoassisted molecular beam epitaxy. *Appl. Phys. Lett.* **49**, 1735.

Chadi, D. J., and Chang, K. J. (1989). Self-compensation through a large lattice relaxation in p-type ZnSe. *Appl. Phys. Lett.* **55**, 575.

Chevallier, J., Dautremont-Smith, W. C., Tu, C. W., and Pearton, S. J. (1985). Donor neutralization in GaAs(Si) by atomic hydrogen. *Appl. Phys. Lett.* **47**, 108.

Cho, H. Y., Choi, W. C., and Min, S. (1993). Positively charged states of a hydrogen atom in p-type InP. *Appl. Phys. Lett.* **63**, 1558.

Coronado, C. A., Ho, E., Kolodziejski, L. A., and Huber, C. A. (1992). Photo-assisted molecular beam epitaxy of ZnSe. *Appl. Phys. Lett.* **61**, 534.

Coronado, C. A., Ho, E., Fisher, P. A., House, J. L., Lu, K., Petrich, G. S., and Kolodziejski, L. A. (1994). Gas source molecular beam epitaxy of ZnSe and ZnSe:N. *J. Electron. Mater.* **23**, 269.

Cunningham, J. E., AT&T Bell Laboratories, private communication,

Dean, P. J., Herbert, D. C., Werkhoven, C. J., Fitzpatrick, B. J., and Bhargava, R. N. (1981). Donor bound-exciton excited states in zinc selenide. *Phys. Rev. B* **23**, 4888.

DenBaars, S. P., and Dapkus, P. D. (1989). Atomic layer epitaxy of compound semiconductors with metalorganic precursors. *J. Cryst. Growth* **98**, 195.

Donnelly, V. M., and McCaulley, J. A. (1989). Selected area growth of GaAs by laser-induced pyrolysis of adsorbed triethylgallium. *Appl. Phys. Lett.* **54**, 2458.

Donnelly, V. M., Tu, C. W., Beggy, J. C., McCrary, V. R., Lamont, M. G., Harris, T. D., Baiocchi, F. A., and Farrow, R. F. C. (1988). Laser-assisted metalorganic molecular beam epitaxy of GaAs. *Appl. Phys. Lett.* **52**, 1065.

Fan, Y., Han, J., He, L., Gunshor, R. L., Brandt, M. S., Walker, J., Johnson, N. M., and Nurmikko, A. V. (1994). Observations on the limits to p-type doping in ZnSe. *Appl. Phys. Lett.* **65**, 1001.

Ferreira, S. O., Sitter, H., and Faschinger, W. (1995). Molecular beam epitaxial doping of ZnMgSe using $ZnCl_2$. *Appl. Phys. Lett.* **66**, 1518.

Fisher, P. A., Ho, E., House, J. L., Petrich, G. S., Kolodziejski, L. A., Walker, J., and Johnson, N. M. (1995). P-type and n-type doping of ZnSe: effects of hydrogen incorporation. *J. Cryst. Growth* **150**, 729.

Fujii, S., Fujita, Y., and Iuchi, T. (1988). Photo-assisted MOCVD of CdTe using an excimer laser. *J. Cryst. Growth* **93**, 750.

Fujita, Sz., Tanabe, A., Sakamoto, T., Isemura, M., and Fujita, Sg. (1988). Investigations of photo-association mechanism for growth rate enhancement in photo-assisted OMVPE of ZnSe and ZnS. *J. Cryst. Growth* **93**, 259.

Fujita, Y., Terada, T., Fujii, S., and Iuchi, T. (1991). Reaction measurements and growth of HgTe/CdTe by photo-assisted MOCVD using an excimer laser. *J. Cryst. Growth* **107**, 621.

Gunshor, R. L., Kolodziejski, L. A., Nurmikko, A. V., and Otsuka, N. (1990). Strained-layer superlattices: materials science and technology. In *Semiconductors and Semimetals*, Vol. 33 (ed. T. P. Pearsall). Academic Press, Boston. 337–400.

Han, J., Ringle, M. D., Fan, Y., Gunshor, R. L., and Nurmikko, A. V. (1994). X center behavior for holes implied from observation of metastable acceptor states. *Appl. Phys. Lett.* **65**, 3230.

Harper, R. L., Hwang, S., Giles, N. C., Schetzina, J. F., Dreifus, D. L., and Myers, T. H. (1989). Arsenic-doped CdTe epilayers grown by photoassociated molecular beam epitaxy. *Appl. Phys. Lett.* **54**, 170.

Ho, E., Coronado, C. A., and Kolodziejski, L. A. (1993). Elimination of surface site blockage due to alkyl radicals in the MOMBE of ZnSe. *J. Electron. Mater.* (special issue on wide bandgap II-VI materials) **22**, 473.

Ho, E., Fisher, P. A., House, J. L., Petrich, G. S., and Kolodziejski, L. A. (1994). The doping of ZnSe using gas source molecular beam epitaxy. *SPIE Proceedings* **2346**, 61.

Ho, E., Fisher, P. A., House, J. L., Petrich, G. S., Kolodziejski, L. A., Walker, J., and Johnson, N. M. (1995). Hydrogen passivation in nitrogen and chlorine-doped ZnSe films grown by gas source molecular beam epitaxy. *Appl. Phys. Lett.* **66**, 1062.

Hou, H., Zhang, Z., Ray, U., and Vernon, M. (1990). A crossed laser molecular-beam study of the photodissociation dynamics of $Zn(C_2H_5)_2$ and $(Zn(C_2H_5)_2)_2$ at 248 and 193 nm. *J. Chem. Phys.* **92**, 1728.

Ikeda, M., Ishibashi, A., and Mori, Y. (1995). Molecular beam epitaxial growth of ZnMgSSe and its application to blue and green laser diodes. *J. Vac. Sci. Technol.* A **13**, 683.

Imaizumi, M., Endoh, Y., Ohstuka, K., Isu, T., and Nunoshita, M. (1994). Active-nitrogen doped p-type ZnSe grown by gas source molecular beam epitaxy. *Jpn. J. Appl. Phys.* **32**, L1725.

Imaizumi, M., Endoh, Y., Ohstuka, K., Suita, M., Isu, T., and Nunoshita, M. (1994a). Blue light emitting laser diodes based on ZnSe/ZnCdSe structures grown by gas source molecular beam epitaxy. *Jpn. J. Appl. Phys.* **33**, L13.

Johnson, N. M., Burnham, R. D., Street, R. A., and Thornton, R. L. (1986). Hydrogen passivation of shallow-acceptor impurities in p-type GaAs. *Phys. Rev. B* **33**, 1102.

Kamata, A., Mitsuhashi, H., and Fujita, H. (1993). Origin of the low doping efficiency of nitrogen acceptors in ZnSe grown by metalorganic chemical vapor deposition. *Appl. Phys. Lett.* **63**, 3353.

Kozuch, D. M., Stavola, M., Pearton, S. J., Abernathy, C. R., and Hobson, W. S. (1993). Passivation of carbon-doped GaAs layers by hydrogen introduced by annealing and growth ambients. *J. Appl. Phys.* **73**, 3716.

Kukimoto, H., Ban, Y., Komatsu, H., Takechi, M., and Ishizaki, M. (1986). Selective area control of material properties in laser-assisted MOVPE of GaAs and AlGaAs. *J. Cryst. Growth* **77**, 223.

Kwak, K. W., King-Smith, R. D., and Vanderbilt, D. (1993). Pseudopotential total-energy calculations of column-V acceptors in ZnSe. *Physica B* **185**, 154.

Lu, K., House, J. L., Fisher, P. A., Coronado, C. A., Ho, E., Petrich, G. S., and Kolodziejski, L. A. (1994). (In,Ga)P buffer layers for ZnSe-based visible emitters. *J. Cryst. Growth* **138**, 1.

Mandel, G. (1964). Self-compensation limited conductivity in binary semiconductors, I. Theory. *Phys. Rev.* **134**, A1073.

Matsumura, N., Fukada, T., and Saraie, J. (1990). Laser irradiation during MBE growth of ZnS_xSe_{1-x}: a new growth parameter. *J. Cryst. Growth* **101**, 61.

Morris, B. J. (1986). Photochemical organometallic vapor phase epitaxy of mercury cadmium telluride. *Appl. Phys. Lett.* **48**, 867.

Murrell, A. J., Wee, A. T. S., Fairbrother, D. H., Singh, N. K., Foord, J. S., Davies, G. J., and Andrews, D. A. (1990). Surface chemical processes in metal organic molecular-beam epitaxy; Ga deposition from triethylgallium on GaAs (100). *J. Appl. Phys.* **68**, 4053.

Myhaljlenko, S., Batstone, J. L., Hutchinson, H. J., and Steeds, J. W. (1984). Luminescence studies of individual dislocations in II–VI (ZnSe) and III–V (InP) semiconductors. *J. Phys. C* **17**, 6477.

Nagata, K., Iimura, Y., Aoyagi, Y., and Namba, S. (1990). Laser assisted chemical beam epitaxy. *J. Cryst. Growth* **105**, 52.

Nishizawa, J., Abe, H., Kurabayashi, T., and Sakurai, N. (1986). Photostimulated molecular layer epitaxy. *J. Vac. Sci. Technol. A* **4**, 706.

Ogata, K., Kawaguchi, D., Kera, T., Fujita, Sz., and Fujita, Sg. (1995). Paper presented at the Seventh International Conference on II–VI Compounds and Devices, August 13–18, 1995, Edinburgh, Scotland, UK.

Ohkawa, K., Karasawa, T., and Mitsuyu, T. (1991). Characteristics of *p*-type ZnSe layers grown by molecular beam epitaxy with radical doping. *Jpn. J. Appl. Phys.* **30**, L152.

Ohkawa, K., Mitsuyu, T., and Yamazaki, O. (1987). Characteristics of Cl-doped ZnSe layers grown by molecular beam epitaxy. *J. Appl. Phys.* **63**, 3216.

Ohkawa, K., Tsujimura, A., Hayashi, S., Yoshii, S., and Mitsuyu, T. (1993). ZnSe-based laser diodes and *p*-type doping of ZnSe. *Physica B* **185**, 112.

Okada, Y., Sugaya, T., Shigeru, O., Fujita, T., and Kawabe, M. (1995). Atomic hydrogen–assisted GaAs molecular beam epitaxy. *Jpn. J. Appl. Phys.* **34**, L238.

Park, R. M., Mar, H. A., and Salanski, N. M. (1985). Photoluminescence properties of nitrogen-doped ZnSe grown by molecular beam epitaxy. *J. Appl. Phys.* **58**, 1047.

Park, R. M., Troffer, M. B., Rouleau, C. M., Depuydt, J. M., and Haase, M. A. (1990). P-type ZnSe by nitrogen beam doping during molecular beam epitaxial growth. *Appl. Phys. Lett.* **57**, 2127.

Pearton, S. J., Corbett, J. W., and Stavola, M. (1991). *Hydrogen in Crystalline Semiconductors*. Springer-Verlag, Berlin.

Pong, C., Johnson, N. M., Street, R. A., Walker, J., Feigelson, R. S., and De Mattei, R. C. (1992). Hydrogenation of wide-band-gap II–VI semiconductors. *Appl. Phys. Lett.* **61**, 3026.

Qiu, J., DePuydt, J. M., Cheng, H., and Haase, M. A. (1991). Heavily doped *p*-ZnSe:N grown by molecular beam epitaxy. *Appl. Phys. Lett.* **59**, 2992.

Rahbi, R., Pajot, B., Chevallier, J., Marbeuf, A., Logan, R. C., and Gavand, M. (1993). Hydrogen diffusion and acceptor passivation in *p*-type GaAs. *J. Appl. Phys.* **73**, 1723.

Rajavel, D., Zinck, J. J., and Jensen, J. E. (1994). Metalorganic molecular beam epitaxial growth kinetics and doping studies of (001) ZnSe. *J. Cryst. Growth* **138**, 19.

Ringle, M., Grillo, D. C., Han, J., Gunshor, R. L., Hua, G. C., and Nurmikko, A. V. (1994). Compositional control of (Zn,Mg)(S,Se) epilayers grown by MBE for II VI blue green laser diodes. *Inst. Phys. Conf. Ser.* No. 141: Chapter 5, 513.

Rueter, M. A., and Vohs, J. M. (1992). Adsorption and reaction of diethylzinc on GaAs (100). *J. Vac. Sci. Technol. B* **10**, 2163.

Saraie, J., Matsumura, N., Tsubokura, M., Miyagawa, K., and Nakamura, N. (1989:). Y-Line emission and lattice relaxation in MBE-ZnSe and -ZnSSe on GaAs. *Jpn. J. Appl. Phys.* **28**, L108.

Sato, M. (1995). Effect of plasma-generated hydrogen radicals on the growth of GaAs using trimethylgallium. *Jpn. J. Appl. Phys.* **34**, L93.

Satoh, S., and Igaki, K. (1983). Photoluminescence and electrical properties of undoped and Cl-doped ZnSe. *Jpn. J. Appl. Phys.* **22**, 68.

Shahzad, K., Petruzzello, J., Olego, D. J., Cammack, D. A., and Gaines, J. M. (1990). Distortion of excitonic emission bands due to self-absorption in ZnSe epilayers. *Appl. Phys. Lett.* **56**, 180.

Simpson, J., Adams, S. J. A., Wallace, J. M., Prior, K. A., and Cavenett, B. C. (1992). Photoassisted molecular beam epitaxial growth of ZnSe under high UV irradiances. *Semicond. Sci. Technol.* **7**, 460.

Sugiura, H., Yamada, T., and Iga, R. (1990). Mechanism of GaAs selective growth in Ar^+ laser-assisted metalorganic molecular beam epitaxy. *Jpn. J. Appl. Phys.* **29**, L1.

Takahashi, T., Arakawa, Y., Nishioka, M., and Ikoma, T. (1992). Selective growth of GaAs wire structures by electron beam induced metalorganic chemical vapor deposition. *Appl. Phys. Lett.* **60**, 68.

Tu, C. W., Donnelly, V. M., Beggy, J. C., Baiocchi, F. A., McCrary, V. R., Harris, T. D., and Lamont, M. G. (1988). Laser-modified molecular beam epitaxial growth of (Al)GaAs on GaAs and $(Ca,Sr)F_2$/GaAs substrates. *Appl. Phys. Lett.* **52**, 966.

Wolk, J. A., Ager, J. W., Duxstad, K. J., Haller, E. E., Taskar, N. R., Dorman, D. R., and Olego, D. J. (1993). Local vibrational mode spectroscopy of nitrogen–hydrogen complex in ZnSe. *Appl. Phys. Lett.* **63**, 2756.

Yamada, T., Iga, R., and Sugiura, H. (1992). Double-wavelength laser array with InGaAsP/InGaAsP multiple quantum well grown by Ar ion laser–assisted metalorganic molecular beam epitaxy. *Appl. Phys. Lett.* **61**, 2449.

Yang, Z., Bowers, K. A., Ren, J., Lansari, Y., Cook, J. W. Jr., and Schetzina, J. F. (1992). Electrical properties of p-type ZnSe:N thin films. *Appl. Phys. Lett.* **61**, 2671.

Yoshikawa, A., Okamoto, T., and Fujimoto, T. (1991). Effects of Ar ion laser irradiation on MOVPE of ZnSe using DMZn and DMSe as reactants. *J. Cryst. Growth* **107**, 653.

Yoshikawa, A., Okamoto, T., Fujimoto, T., Onoue, K., Yamaga, S., and Kasai, H. (1990). Ar ion laser-assisted MOVPE of ZnSe using DMZn and DMSe as reactants. *Jpn. J. Appl. Phys.* **29**, L225.

Zhu, Z., Yao, T., and Mori, H. (1993). Selective doping of N-type ZnSe layers with chlorine grown by molecular beam epitaxy. *J. Electron. Mater.* **22**, 663.

Zinck, J. J., Brewer, P. D., Jensen, J. E., Olson, G. L., and Tutt, L. W. (1988). Excimer laser–assisted metalorganic vapor phase epitaxy of CdTe on GaAs. *Appl. Phys. Lett.* **52**, 1434.

CHAPTER 4

Doping of Wide-band-gap II–VI Compounds—Theory

Chris G. Van de Walle

XEROX PALO ALTO RESEARCH CENTER
PALO ALTO, CALIFORNIA

I.	INTRODUCTION	122
II.	MECHANISMS THAT LIMIT DOPING	122
	1. Self-compensation by Native Defects	122
	2. Compensation by Other Configurations of the Impurity	124
	3. Formation of Complexes	125
	4. Solubility Limits	125
	5. Compensation by Foreign Impurities	126
III.	FORMALISM FOR CALCULATING DOPING LEVELS	126
	1. Formation Energies	126
	2. Chemical Potentials	128
	3. General Expression for Formation Energy	130
	4. Thermodynamic Equilibrium	131
	5. Charge Neutrality—Self-consistent Solutions	132
	6. First-principles Calculations	133
IV.	NATIVE DEFECTS	135
	1. First-principles Investigations of Native Defects	135
	2. The Zn Interstitial	136
	3. Antisite Defects	137
	4. The Se Vacancy	137
	5. Self-activated Centers	137
	6. Broken-bond Defects	138
	7. Critical Examination of Native Defect Compensation as a Generic Compensation Mechanism	138
V.	p-TYPE DOPING	140
	1. Lithium in ZnSe	140
	2. Sodium in ZnSe	147
	3. Phosphorus and Arsenic in ZnSe	147
	4. Nitrogen in ZnSe and ZnTe	148
	5. Oxygen in ZnSe	151
VI.	n-TYPE DOPING	151
	1. Al and Ga in ZnSe and ZnTe	151
	2. Cl in ZnSe and ZnTe	153

VII. COMPARISON BETWEEN THEORY AND EXPERIMENT 153
  1. Compensation due to Native Defects and Native-defect Complexes 153
  2. Discussion of Doping Saturation Effects 156
  3. Nucleation of Misfit Dislocations 157
  4. Comparison between Different II–IV Materials 157
  5. Solubility-limiting Phases 158
  6. Effect of N Incorporation on the Lattice Constant 158
VIII. CONCLUSIONS AND FUTURE DIRECTIONS 159
  REFERENCES 160

I. Introduction

The recently developed ability to controllably dope ZnSe and other II–VI compounds has had a major impact on the realization of light-emitting devices. A long history of failures to accomplish p-type doping in ZnSe (and similar problems in other wide-band-gap semiconductors) had led to the conventional wisdom that the doping limitations were inherent to the material — the most notable proposed explanation being self-compensation by native defects. Fortunately, the perseverance of the research community has enabled breakthroughs in materials growth and dopant sources, revealing that the doping limitations *can* be overcome.

In spite of these successes, doping of the wide-band-gap semiconductors still suffers from considerable shortcomings. It is therefore important to develop a sound theoretical understanding of the mechanisms that determine doping levels in these materials. Part II of this chapter contains an overview of these mechanisms, providing a general description aimed at fostering physical insights. In Part III, I will describe the theoretical formalism that underlies the calculation of defects, impurities, and doping levels in a semiconductor. This part also contains information about first-principles calculations which can produce key values required in the formalism. Part IV contains results for native defects. Results for p- and n-type doping in various II–VI compounds are discussed in Parts V and VI. These parts also contain various comparisons between theory and experiment; additional comparisons with experiment are described in Part VII. Part VIII, finally, summarizes the chapter, and examines the areas which could benefit from further investigations.

II. Mechanisms that Limit Doping

1. SELF-COMPENSATION BY NATIVE DEFECTS

Native defects (also referred to as intrinsic defects) commonly occur in semiconductors. The three basic types of point defects are vacancies (an

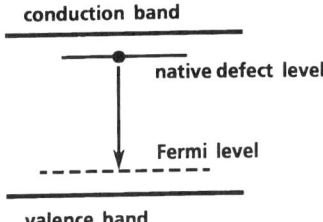

FIG. 1. Schematic illustration of the native-defect compensation process: a donor-like native defect has a level near the conduction band, which in the neutral charge state is occupied with one electron. In p-type material (E_F near the valence band), this electron will drop down to the Fermi level. The energy gained in this process reduces the net formation energy of the native defect.

atom missing from a lattice site), self-interstitials (an additional atom in the lattice), and antisites (in compound semiconductors, when, e.g., a cation is sitting on an anion site). For each of these types, the defect can occur either on the cation site or on the anion site.

Because formation of the native defect often involves breaking or rearranging of bonds, deep levels in the semiconductor band gap are typically introduced. The occupation of the levels determines the charge state of the defect. Depending on the location of the levels in the gap, and the charge states they can assume, native defects can have donor-character or acceptor-character, or even be amphoteric.

For the sake of argument, let us assume that a donor-like native defect can be formed, which in the neutral charge state would have an electron residing in a level near the conduction band (Fig. 1). If we try to dope the material p-type, we are driving the Fermi level down to a position near the top of the valence band. The donor, of course, will become positively charged, by transferring its electron from the gap level to the Fermi level — in the process compensating the electrical activity of an acceptor. The amount of energy gained by this charge transfer can be of the order of the band gap, i.e., a large value in a wide-band-gap semiconductor. The *net* formation energy of the native defect is equal to the energy required to create the *neutral* defect, minus the energy gained by transferring the electron. The above argument indicates that in a wide-band-gap semiconductor the net formation energy of compensating native defects could become very low. This would lead to the formation of a large number of native defects. These donor-type defects of course compensate the acceptors that were introduced to make the material p-type. If this mechanism is indeed active, it would make it impossible to obtain p-type conductivity.

This mechanism was proposed early on to explain the doping difficulties in wide-band-gap semiconductors (Mandel, 1964; Ray and Kröger, 1978).

Unfortunately, there was never any explicit verification of the quantitative aspects of the mechanism. Nonetheless, the hypothesis became widely accepted, to the extent that the possibility of achieving good *p*-type conductivity in ZnSe was often rejected out of hand.

In this chapter, we will show that compensation by native defects can indeed be an issue in ZnSe, just as it is in other semiconductors (not exclusively wide-band-gap). However, this compensation mechanism by no means acts as a show-stopper. Explicit results for ZnSe will be discussed in Part IV.

2. COMPENSATION BY OTHER CONFIGURATIONS OF THE IMPURITY

Suppose an impurity is incorporated into the crystal with the goal of introducing a shallow acceptor level. To obtain this type of electrical activity, it is necessary for the impurity to be located on a substitutional lattice site. For instance, Li in ZnSe behaves as a shallow acceptor if Li is located on a substitutional Zn site. In some cases, the impurity can occupy other locations in the lattice, leading to a different type of electrical activity. In our example of Li in ZnSe, the Li atom can also occupy an interstitial site—and in that location it behaves as a donor. If the growth conditions are such that a large number of Li atoms are incorporated on interstitial sites, as opposed to substitutional sites, significant compensation will occur, as suggested by Neumark (1980). Exactly what the conditions are for this to happen will be discussed in more detail in Section V.1.

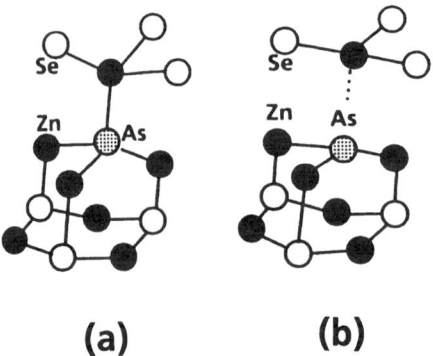

FIG. 2. (a) Structure of an As acceptor in ZnSe in the fourfold-coordinated configuration, stable in the negative charge state. (b) Structure of a state in which As has relaxed off the substitutional site, and an As–Zn bond has been broken (from Chadi and Chang, 1989b).

Another type of compensation occurs when a substitutional impurity undergoes a large lattice relaxation. This type of compensation mechanism received a great deal of attention in the context of the so-called DX centers in Si-doped AlGaAs. Extensive theoretical (Chadi and Chang, 1988b, 1989a) and experimental (for a review, see Mooney, 1992) work has revealed that these centers occur when a Si atom, which behaves as a donor when it sits on a substitutional Ga site, undergoes a large lattice relaxation. The Si atom moves away from the substitutional site into the interstitial region, thereby changing its charge state and compensating the electrical activity of the donors. Chadi has proposed that similar mechanisms could be active in II–VI materials. In 1989, Chadi and Chang (1989b) proposed that P and As acceptors on the Se site in ZnSe would undergo a large lattice relaxation and become positively charged (see Fig. 2). Chadi (1994a) has also suggested that DX-like centers can occur in n-type ZnTe and MgTe. These results will be discussed in Parts V and VI.

3. Formation of Complexes

In Section II.1 we discussed the formation of native defects acting as compensating centers; it was implicit in that discussion that the native defects would occur as isolated point defects. One should also consider the possibility of complex formation between dopant impurities and native defects. Indeed, if the binding energy between a native defect and a dopant is large enough, the formation energy of the resulting complex may well be lower than that of the individual defect. Examples of such complexes will be discussed in Part V.

4. Solubility Limits

In order to dope the material, impurities that act as dopants need to be introduced into the semiconductor. In p-type material, the hole concentration is obviously limited by the concentration of acceptor impurities incorporated in the lattice. If this concentration is limited, it can lead to a limitation on the achievable doping level.

The limited incorporation of dopants can be due to a solubility limit. In thermodynamic equilibrium, at a given temperature, the concentration of substitutional impurities indeed has an upper limit. We emphasize that we use the term *solubility limit* to refer to the maximum concentrations of impurities located on the lattice sites where they exhibit the desired electrical activity. Once this solubility limit is exceeded, the concentration of impurities (as measured, e.g., by secondary-ion mass spectroscopy [SIMS]) may still increase, but the impurity will be incorporated in different lattice positions, or in the form of precipitates; in these other configurations, the impurity does not contribute to the electrical activity. Only the electrically active form of the impurity should be counted as contributing to a maximum concentration given by the solubility limit. Examples of solubility limits leading to limitations on doping levels are given in Part V.

5. COMPENSATION BY FOREIGN IMPURITIES

This compensation mechanism may seem too trivial to mention: if one is trying to dope the semiconductor p-type, one should be very careful not to unintentionally incorporate other impurities which act as donors. With the advent of modern growth techniques such as molecular beam epitaxy (MBE) and metal-organic chemical vapor deposition (MOCVD) and the advances in ultra-clean environments and purity of source materials, there may have been a tendency to underestimate the potential for contamination.

One impurity that probably plays a major role in limiting doping of II–VI's is hydrogen. Hydrogen tends to be the major background impurity in MBE growth chambers. More importantly, hydrogen is omnipresent during MOCVD growth, being used as a carrier gas, as well as being a component of many of the source gases. Hydrogen is well known to be electrically active in many semiconductors, and to compensate or passivate deep as well as shallow impurities (Pankove and Johnson, 1991). The lack of success in obtaining well-conducting p-type ZnSe with the MOCVD growth technique is probably attributable to the incorporation of hydrogen during the growth. N-H complexes have indeed been observed in MOCVD-grown ZnSe (Wolk et al., 1993).

III. Formalism for Calculating Doping Levels

1. FORMATION ENERGIES

The equilibrium concentration, C, of a defect or impurity in a semiconductor is determined by its formation energy, E_{form}:

$$C = N_{sites} \exp\left[-\frac{E_{form}}{kT}\right] \qquad (1)$$

where N_{sites} is the appropriate site concentration; e.g., for a substitutional impurity on the Zn site in ZnSe, N_{sites} is the number of Zn sites, 2.2×10^{22} cm^{-3}. The above definition applies under conditions of thermodynamic equilibrium. The assumption of thermodynamic equilibrium is discussed in Section III.4.

The energy appearing in Eq (1) is in principle a Gibbs free energy, which, in addition to the *formation energy* includes a contribution from the *formation entropy*. In Section III.6 we will discuss how formation energies can be obtained from state-of-the-art first-principles calculations. Such calculations typically produce zero-temperature formation energies. At finite temperatures, the formation energy would also contain a term due to vibrational energy. The entropy term has a vibrational component as well, which in principle can be evaluated by studying the vibrational spectrum of the system. An additional contribution is due to configurational entropy, associated with the various potential ways in which the defect can be incorporated in the lattice, and with the occupation of the defect levels. The configurational entropy can usually be evaluated rather easily, and can be regarded as part of the prefactor. The vibrational contributions to the free energy are more difficult to calculate. Fortunately, these terms tend to be rather small (Laks *et al.*, 1991), and they also tend to cancel when comparing relative free energies (Qian *et al.*, 1988).

We will focus on the definition of the *formation energy*, which is the major contribution to the free energy appearing in the exponent of the expression of Eq (1). Before giving a general definition, in Section III.3, we illustrate the concept of a formation energy with the example of a Li atom on a substitutional Zn site in ZnSe. The formation energy of a Li atom on a substitutional Zn site in ZnSe is defined as follows:

$$E_{form}(Li_{Zn}^-) = \mathscr{E}(Li_{Zn}^-) - \mu_{Li} + \mu_{Zn} - E_F. \tag{2}$$

$\mathscr{E}(Li_{Zn}^-)$ is the energy of a system containing the Li_{Zn}^- impurity; we will discuss in Sections III.3 and III.6 how this quantity can be obtained from first-principles calculations. The other terms in Eq (2) are chemical potentials. μ_{Li} is the chemical potential of Li. This term enters because the formation energy is the difference between the energy of Li as an impurity, and its energy in a reference state. The reference corresponds to a reservoir of Li atoms, whose energy (at $T = 0$) by definition is the chemical potential. This chemical potential depends on the abundance of Li under the relevant growth conditions (Kröger, 1964). The Zn chemical potential, μ_{Zn}, appears in Eq (2) because, in order to make room for the substitutional impurity, a Zn atom has to be removed to its reservoir. Chemical potentials are discussed in more detail in the next section.

In the example of Eq (2), the Li acceptor has been put in a negative charge state (which occurs when the acceptor has donated a hole to the valence band). The last term in Eq (2) represents the energy of the reservoir supplying the electron responsible for the negative charge on the impurity; this energy corresponds to the Fermi level, E_F, which is, thermodynamically, the energy of the electron reservoir.

2. CHEMICAL POTENTIALS

a. Chemical Potential vs. Partial Pressures

As mentioned above, the chemical potentials represent the energies of the reservoirs with which the various atomic species can be exchanged. In older work on defects in semiconductors, defect concentrations have often been expressed as a function of partial pressures. For an element in thermal equilibrium with the gas phase, the chemical potential can be related to the partial pressure of the gas (Kröger, 1964); for an ideal gas with partial pressure p one has $\mu = \mu^0 - kT \ln p$. We find it preferable to work with chemical potentials for the following reasons:

1. Chemical potentials are thermodynamically defined as energy values, which can be directly related to the energies that we calculate from first principles.
2. As discussed in Section III.4, the assumption of thermodynamic equilibrium is likely to be satisfied within the growing solid, allowing the use of expressions such as Eq (1); however, it is uncertain to what extent equilibrium is established between the solid and a surrounding gas under experimental conditions such as MBE. Knowledge of the chemical potential in the gas may therefore not necessarily reflect the relevant chemical potential for the solid.
3. Even if thermodynamic equilibrium with the gas is assumed, the relationship between chemical potential and gas pressure is not well known since the gas sources used in MBE or MOCVD do not obey simple ideal gas laws.

These arguments indicate that a quantitative determination of chemical potentials in terms of experimentally accessible quantities is difficult; however, we will see (in Section III.2.c) that the chemical potentials are subject to rigorous bounds that can be directly related to experimental conditions.

b. Chemical Potentials of the Host Atoms

Regarding the chemical potential of the host atom (in our example, μ_{Zn}), it is very important to realize that it should be treated as a variable. This is different from the situation in an *elemental* semiconductor such as Si; there, the chemical potential of the host atoms, in equilibrium with bulk Si, is simply the energy of one Si atom in the crystal, which can simply be obtained by calculating the energy of a unit cell of Si bulk and dividing by the number of Si atoms in the unit cell. In the case of a *compound* semiconductor, such as ZnSe, this approach would produce the energy of a two-atom unit of the ZnSe crystal. Thus, in a compound semiconductor only the *sum* of the chemical potentials of the constituents is fixed, and equal (at $T = 0$) to the energy of a two-atom unit of the material:

$$\mu_{Zn} + \mu_{Se} = \mu_{ZnSe}. \tag{3}$$

This additional degree of freedom needs to be taken into account when presenting results for formation energies of defects. Equation (3) fixes μ_{Se} once μ_{Zn} is chosen; alternatively, μ_{Se} could be chosen as the free variable, leading to a fixed μ_{Zn}.

c. Bounds on the Chemical Potentials

While the chemical potentials are variable parameters, they are subject to specific boundary conditions. To establish the range of the chemical potentials, one has to consider the various phases that can be formed, in thermodynamic equilibrium, out of the constituents (Kröger, 1964; Qian et al., 1988; Chetty and Martin, 1992). For instance, μ_{Zn} is bounded from above by the energy of a Zn atom in Zn metal: $\mu_{Zn}^{max} = \mu_{Zn(bulk)}$. Indeed, if one would try to raise μ_{Zn} above this level, Zn metal would be preferentially formed. Similarly, μ_{Se} has an upper bound imposed by bulk Se.

We already pointed out [Eq (3)] that the sum of the Zn and Se chemical potentials has to add up to the energy of a two-atom unit of ZnSe. This condition can be expressed as

$$\mu_{Zn} + \mu_{Se} = \mu_{ZnSe} = \mu_{Zn(bulk)} + \mu_{Se(bulk)} + \Delta H_f(ZnSe) \tag{4}$$

where $\Delta H_f(ZnSe)$ is the heat of formation of ZnSe (ΔH_f is negative for a stable compound). Eq (4) shows that imposing an upper bound on the Se chemical potential leads to a lower bound on the Zn chemical potential, given by $\mu_{Zn}^{min} = \mu_{Zn(bulk)} + \Delta H_f(ZnSe)$. We therefore note that the Zn (as well as the Se) chemical potential can vary over a range given by the heat of formation of ZnSe, $\Delta H_f(ZnSe)$. A lucid discussion of similar arguments, in the context of surface reconstructions, was provided by Qian et al. (1988). Following the notation of García and Northrup (1995) a parameter λ can be introduced which varies between zero (Zn rich) and one (Se rich):

$$\mu_{Se} = \mu_{Se(bulk)} - (1 - \lambda)\Delta H_f(ZnSe). \tag{5}$$

To find an upper bound on the chemical potential of the impurity we must explore the various compounds that the impurity can form in its interactions with the systems. For Li, e.g., a possible upper bound on μ_{Li} is of course imposed by Li (bulk) metal. However, the most stringent constraint arises from the compound Li_2Se, which leads to the following upper bound on the chemical potentials:

$$\begin{aligned} 2\mu_{Li} + \mu_{Se} &= \mu_{Li_2Se} \\ &= 2\mu_{Li(bulk)} + \mu_{Se(bulk)} + \Delta H_f(Li_2Se). \end{aligned} \tag{6}$$

Practical applications of the bounds on the chemical potentials will be discussed in Parts V and VI.

In the examples discussed so far, the substances that impose bounds on the chemical potentials are all in the solid phase. Our focusing on *energies*, as opposed to *free energies*, which contain entropy contributions, can then be justified by observing that in equations such as Eq (4) the entropy contributions will largely cancel on both sides. If one of the substances occurs in the form of a gas, neglecting entropy would not seem justified: indeed, entropy contributions for atoms or molecules in a gas can be very large. However, as discussed in Sections III.2.a and III.4, invoking equilibrium between a gas and the growing solid is probably not justified anyway. The *energy* of the atoms or molecules in the gas can still form a useful bound on the chemical potential, if we envision the species to be adsorbed on the surface of the solid.

3. General Expression for Formation Energy

Before writing down a general expression, we have to consider in a bit more detail how the quantity which we called $\mathscr{E}(Li_{Zn}^-)$ in the previous subsection is actually obtained from a calculation. In practice, a calculation for a Li impurity is carried out by considering a chunk of ZnSe crystal (a so-called "supercell"), containing n^{Zn} Zn atoms and n^{Se} Se atoms, and replacing one of the Zn atoms with a Li atom. The quantity $\mathscr{E}(\text{Li}_{Zn}^-)$ is then obtained by taking the calculated total energy for a supercell containing the impurity, $E_{tot}(\text{Li}_{Zn}^-)$, and subtracting the total energy for the same size supercell containing pure bulk ZnSe, $E_{tot}(\text{bulk})$, i.e.,

$$\mathscr{E}(\text{Li}_{Zn}^-) = E_{tot}(\text{Li}_{Zn}^-) - E_{tot}(\text{bulk}) \\ = E_{tot}(\text{Li}_{Zn}^-) - n^{Zn}\mu_{Zn} - n^{Se}\mu_{Se} \quad (7)$$

where we have used the fact that the energy of the bulk supercell can be expressed in terms of the chemical potentials of the host atoms. Now we can rewrite Eq (2) as

$$E_{form}(\text{Li}_{Zn}^-) = E_{tot}(\text{Li}_{Zn}^-) - (n^{Zn} - 1)\mu_{Zn} - n^{Se}\mu_{Se} - \mu_{Li} - E_F \quad (8)$$

We now write down a general expression for the total energy $E_{tot}(D_i)$ for a defect or impurity D_i. The supercell we are considering contains n_i^{Zn} Zn atoms and n_i^{Se} Se atoms. We continue to use Li as a sample impurity; in the general case, the supercell for the impurity calculation could contain n_i^{Li} Li atoms. The formulas are valid, of course, for a general impurity, or for a native defect (in which case $n_i^{Li} = 0$). The defect formation energy, $E_{form}(D_i)$ is defined as

$$E_{form}(D_i) = E_{tot}(D_i) - n_i^{Zn}\mu_{Zn} - n_i^{Se}\mu_{Se} - n_i^{Li}\mu_{Li} - n_i^e E_F \quad (9)$$

where n_i^e is the number of excess electrons in the defect.

4. Thermodynamic Equilibrium

Eq (1) is valid under conditions of thermodynamic equilibrium. In nonequilibrium situations, the concentration of a defect or impurity can exceed the equilibrium value; the nonequilibrium concentration would be frozen into the lattice. In order to decide whether equilibrium conditions apply or not, it is important to examine how equilibrium is established. This will depend on the mobility of various defects at the temperatures of interest (Laks et al., 1991). Indeed, at a given temperature, equilibrium is established through diffusion of various defects and impurities in the system. If the mobility of the various species is sufficiently high, equilibration will occur. It should be emphasized that this does not require that an equilibrium exists between the growing semiconductor and the gases in the growth chamber (in an MBE-type environment). The equilibrium we are invoking here is between the bulk of the growing semiconductor, any impurities or defects present in the bulk, and the various species that are present on the surface in adsorbed or chemisorbed form.

For ZnSe, specifically, it is known that the mobilities of various native defects are very high (Watkins, 1990), even at modest temperatures. It is therefore likely that the assumption of thermodynamic equilibrium is justified.

5. Charge Neutrality — Self-consistent Solutions

Expressions based on Eq (9) can be written down for all configurations of the impurity, in their various charge states, as well as for all native defects in the semiconductor. Once the formation energy is known, the concentration of a specific defect or impurity can be obtained from Eq (1). At this point, all concentrations are still functions of the chemical potentials (μ_{Zn} or μ_{Se}, and μ_{Li}), as well as of the Fermi level (E_F). The chemical potentials, as explained above, are independent parameters; it therefore makes sense to express results as functions of these chemical potentials. The Fermi level, however, is not an independent variable, since it is determined by the condition of charge neutrality:

$$\text{Net charge} = 0 = p - n - \sum_{i} n_i^e [D_i], \quad (10)$$

where p and n are the hole and electron densities, respectively. These free carrier densities are determined from the standard semiconductor equations.

The charge conservation equation provides for an interaction between the concentrations of all charged defects through their influence on the Fermi level. For example, a positively charged defect produces extra free electrons that raise the Fermi level; the higher Fermi level, in turn, increases the concentrations of all negatively charged defects and lowers the concentrations of all positively charged defects. As pointed out by Zhang and Northrup (1991), this "negative feedback" reduces the sensitivity of the final results to possible inaccuracies in the first-principles energies, to be discussed in Part IV. Using this prescription, all of the defect formation energies, and hence the concentrations [D_i], are unique functions of the chemical potentials of the host and impurity atoms, and the temperature T.

The fact that the expressions for defect and impurity concentrations are coupled has stymied some investigators, sometimes causing them to simplify the problem by only considering a subset of defects. The self-consistent solution of the set of equations is actually quite straightforward, when properly formulated. As discussed by Laks *et al.* (1991, 1992) the problem is essentially finding the root of polynomial, which can be done quickly and easily using standard algorithms.

6. First-principles Calculations

a. First-principles Studies of Defects

The formalism described above already provides insight in the mechanisms that determine doping levels. To obtain more concrete results, however, explicit values for the various energies that occur in the equations are necessary. Such energies cannot directly be obtained from experiment; it is possible, however, to calculate them using state-of-the-art first-principles methods.

First-principles calculations have had a major impact on the understanding of defects and impurities in semiconductors. The first applications of electronic structure methods to defect problems focused on obtaining defect wave functions and levels in the band gap (Baraff and Schlüter, 1978; Bernholc *et al.*, 1978). With the advent of the capability to calculate total energies, it became possible to investigate the atomic structure of the defect; i.e., the stable position in the host lattice, the relaxation of the surrounding atoms, as well as the energy along a migration path (Baraff and Schlüter, 1983; Bar-Yam and Joannopoulos, 1984; Car *et al.*, 1985). Calculations of hyperfine parameters also enabled more direct comparisons with experimental observations (Overhof *et al.*, 1989; Van de Walle and Blöchl, 1993; the latter reference includes results for defects in ZnSe). More recently, formalisms have been developed to use the total energy of the defect to calculate its concentration, under the assumption of thermodynamic equilibrium (Zhang and Northrup, 1991; Laks *et al.*, 1991). The same formalism can also be applied to the calculation of impurity solubilities (Northrup and Zhang, 1993; Van de Walle *et al.*, 1993).

It is a strength of the formalism that it treats native point defects (vacancies, self-interstitials, and antisites) and dopant impurities on an equal footing, allowing us to investigate whether native defects can form a significant source of compensation. We also note that the methods and computational approaches are, in principle, general in nature, and can be applied to any defect or impurity in any semiconductor. Since some of the more interesting problems related to defect formation and doping occur in wide-band-gap semiconductors, these materials have become a focus of intense theoretical research activity, in spite of the fact that they tend to be computationally more difficult to treat.

b. Computational Approach

State-of-the-art first-principles calculations for defects or impurities in semiconductors are based on density-functional theory in the local-density approximation (Hohenberg and Kohn, 1964; Kohn and Sham, 1965) and *ab initio* pseudopotentials (Hamann *et al.*, 1979). As discussed above, the calculations are performed in a supercell geometry: the defect or impurity is placed at the center of a certain volume of host material, and this "supercell" is periodically repeated. Normally, introducing a single impurity or defect in the semiconductor would totally break the transational symmetry; the supercell approach preserves periodicity, thereby allowing the use of computational tools such as Fast Fourier Transforms. Convergence as a function of supercell size must always be checked. In the work of Laks *et al.* (1992), 32-atom supercells were found to provide adequate accuracy. Relaxations of the host atoms around the defect or impurity are also included.

This computational approach has been successfully used to study properties of defects or impurities in semiconductors such as Si (see, e.g., Car *et al.*, 1985) or GaAs (see, e.g., Northrup and Zhang, 1993). Applying the approach to ZnSe creates computational difficulties due to the d states of the Zn atoms. The calculations typically employ a plane-wave basis set, used to expand wave functions and potentials (essentially expanding in Fourier series). Since only electronic states corresponding to valence electrons need to be described in a pseudopotential approach, a plane-wave basis set with a relatively modest cutoff often suffices. The $3d$ states of Zn, however, turn out to play an important role in the structural properties of the material (Wei and Zunger, 1988). One also notes that the electronic levels corresponding to these d states fall within the valence bands of ZnSe. These d states can therefore not simply be relegated to the core of the atom, but should be included at the same level as valence electrons. The problem is that the d states are quite localized in nature (compared to the more extended valence states), and are hard to describe with a plane-wave basis set. In the work of Laks *et al.* (1991, 1992) and Van de Walle *et al.* (1993), a mixed-basis set has therefore been used, which contains both plane waves and localized functions aimed at accurately describing the d states. More details about this and other aspects of the calculations on ZnSe are given in the paper by Laks *et al.* (1992).

An accurate treatment of the d states is important for obtaining reliable results for ZnSe. Results from calculations that do not explicitly deal with the d states, and treat them as part of the core, should be inspected with great caution. An intermediate level of treatment has been used by some groups, namely the so-called "nonlinear core correction," based on the work of Louie *et al.* (1982). Here the d states are still treated as part of the core, but the fact that their density significantly overlaps with the valence-electron density is taken into account in the calculation of exchange and correlation. This approach offers a substantial improvement over the simple "d in the core" assumption, at reasonable computational cost. However, it produces results which in some cases still deviate from the full treatment. The extent to which this approach can be relied on is therefore not fully known.

IV. Native Defects

1. First-principles Investigations of Native Defects

A comprehensive investigation of native defects in ZnSe was reported by Laks *et al.* (1991, 1992). The main conclusion from that work was that native defects in this wide-band-gap semiconductor do *not* have significantly lower formation energies than defects in other semiconductors. Their concentrations, therefore, are usually quite limited, and they do not necessarily pose a threat to doping of the semiconductor.

The work of Laks *et al.* (1991, 1992) also highlighted the fact that the prevalence of native defects depends sensitively on the stoichiometry of the material (Zn-rich vs. Se-rich), and on the doping (n-type vs. p-type). A large deviation from stoichiometry (if not accommodated by extended defects or precipitates) would necessarily lead to large concentrations of native defects. In thermodynamic equilibrium, however, when the host-atom chemical potentials are subject to the bounds discussed in Section III.2.c, the deviations from stoichiometry are very limited and the concentrations of native defects modest. Some cases where native-defect concentrations may become high enough to be observable, or to become a threat to doping, are discussed in Sections V.4.c and VII.1.

A table with complete information about all the native defects that were investigated can be found in Laks *et al.* (1992). Here we limit ourselves to presenting salient information on the types of native defects which dominate in ZnSe under various conditions; see Table I. We use the following notation: V_{Zn} is a Zn vacancy; $Zn_i(T_{Se})$ is a Zn interstitial at a tetrahedral interstitial site surrounded by Se atoms; and Zn_{Se} is a Zn-on-Se antisite, i.e., a Zn atom sitting on a site which is nominally a Se site in the lattice. The same notation applies, *mutatis mutandis*, to Se-type defects.

TABLE I

DOMINANT NATIVE DEFECTS IN ZnSe, AS A FUNCTION
OF DOPING TYPE AND STOICHIOMETRY[a]

	p-type	n-type
Zn-rich	Zn_i/V_{Se}	Zn_{Se}
Se-rich	Se_{Zn}	V_{Zn}

[a]Based on results from Laks et al., 1992, and García and Northrup, 1995.

Jansen and Sankey (1989) also addressed the issue of native defects in ZnSe and ZnTe with a first-principles approach, but treating the Zn d electrons as frozen-core states, and using a more approximate basis set. Their results exhibit the same trends as those reported by Laks et al. (1992). However, Jansen and Sankey interpreted their results as supporting the model of strong compensation by native defects. This conclusion seems inconsistent with their numerical results. For instance, Jansen and Sankey found that native-defect concentrations in p-type ZnSe are several orders of magnitude lower than in n-type ZnSe, whereas experimentally n-type ZnSe is easier to obtain than p-type. Furthermore, their results were obtained for a very high temperature ($T = 1658$ K), which far exceeds any temperature that ZnSe or ZnTe would ever be exposed to. Their calculated energies are such that at more typical growth temperatures their native-defect concentrations would be even lower than those reported by Laks et al. (1992).

2. THE Zn INTERSTITIAL

An interesting check on the reliability and accuracy of the calculations is provided by the case of the Zn interstitial in ZnSe, for which experimental information is available. This was the first isolated native interstitial defect directly observed in a semiconductor (Rong and Watkins, 1987), and it was studied extensively using optically detected magnetic resonance. The hyperfine parameters produced by this type of experiment are very sensitive to the symmetry and structure of the defect, including relaxations of the host atoms. The hyperfine parameters have also been calculated for the structure obtained by Laks et al. (1992); the results were reported in Van de Walle and Laks (1990) and Van de Walle and Blöchl (1993) and reveal good agreement with experiment. This confirms both the experimental identification of the defect and the accuracy of the calculated relaxations.

3. ANTISITE DEFECTS

Laks et al. (1992) reported that the neutral Se_{Zn} antisite defect exhibits a large lattice relaxation, in which the Se atom moves by about 1 Å along the $\langle 111 \rangle$ direction toward a tetrahedral interstitial site. This relaxation is similar to that found for the As_{Ga} antisite defect in GaAs (Chadi and Chang, 1988a; Dabrowski and Scheffler, 1988), which has been associated with the $EL2$ defect.

4. THE Se VACANCY

García and Northrup (1995) have recently shown that the Se vacancy in ZnSe, in the +2 charge state (V_{Se}^{2+}), undergoes an unusually large lattice relaxation. Indeed, they found that the Zn atoms surrounding the Se vacancy move outward by 0.51 Å, lowering the energy by 1.6 eV. Taking this relaxation into account, García and Northrup found the Se vacancy to be the lowest-energy point defect in ZnSe. In the earlier work by Laks et al. (1991, 1992), such a large relaxation of V_{Se}^{2+} was not taken into account. Very recent calculations performed in the same framework as those of the 1991 work of Laks et al. indeed confirm the presence of the large lattice relaxation for V_{Se}^{2+} (Van de Walle and Neugebauer, 1995). The corresponding energy lowering makes the Se vacancy energetically slightly more favorable than the Zn interstitial, which had previously been found to be the dominant point defect. Both the Se vacancy and the Zn interstitial are therefore listed in Table I as prevailing defects in Zn-rich p-type ZnSe.

The Se vacancy may occur in large enough concentrations to allow experimental observation. (We note that Ando et al. [1993] have reported a deep hole trap which exhibits a strong temperature dependence of the hole capture/emission rates, indicative of a center with large lattice relaxation.) However, the isolated Se vacancy would still not constitute a major form of compensation of the material. García and Northrup (1995) have pointed out that compensation by an isolated point defect such as the Se vacancy may lead to a reduction of the hole concentration, but cannot cause a saturation of the free carrier concentration. The potential role of the Se vacancy (as well as other defects) in compensation will further be discussed in Section V.4.c and VII.1.

5. SELF-ACTIVATED CENTERS

Table I shows that the dominant defect occurring in Se-rich n-type ZnSe is the Zn vacancy (V_{Zn}). This result is consistent with the experimental observation that as-grown bulk ZnSe samples are highly compensated, and must be annealed in a Zn-rich atmosphere to be made well conducting. The cause of the compensation is large numbers of "self-activated" centers, which are donor–V_{Zn} pairs (Watts, 1977). This shows that Zn vacancies are a prominent defect in as-grown n-type ZnSe. Furthermore, analysis of the Zn-Se phase diagram suggests that ZnSe grown under equilibrium conditions from a melt is Se-rich. Thus, the results for Se-rich n-type ZnSe provide a natural explanation for the occurrence of self-activated centers.

Vacancy-type defects were observed in Ga-doped ZnSe, using a positron-annihilation technique (Miyajima et al., 1991). The concentration of negatively charged vacancies, believed to be Zn vacancies, increased with increasing Ga concentration in the film.

6. Broken-bond Defects

Chadi and Chang (1989b) suggested that a P or As acceptor in ZnSe could undergo a large lattice relaxation, effectively breaking the bond between the impurity and a neighboring Zn atom (see Fig. 2); this mechanism will be addressed in Section V.3. Chadi (1995) subsequently suggested that such a bond-breaking mechanism could take place even in the absence of an impurity, involving only bonds in the bulk of the semiconductor. More recently, Park and Chadi (1995) have reported that a configuration involving a single broken bond would not be stable; they proceeded to propose a new configuration, involving two broken bonds.

Park and Chadi calculated that this "double broken bond" defect, which is illustrated in Figure 3, has a low formation energy in the presence of free holes, and acts as a double donor. Since significant lattice distortions occur in the neighborhood of this defect, its formation energy is lower if it occurs in the vicinity of a nitrogen impurity; indeed, the nitrogen atom is significantly smaller than the host atoms, and its incorporation creates a certain amount of lattice strain which can be relieved by the nearby double broken bond defect. We will return to this issue in Section V.4.c.

7. Critical Examination of Native Defect Compensation as a Generic Compensation Mechanism

In light of the first-principles results discussed in Section IV.1, it is illuminating to reexamine the notion that native-defect compensation increases with the width of the band gap. The standard argument for this trend was stated in Section II.1; let us make the argument a bit more

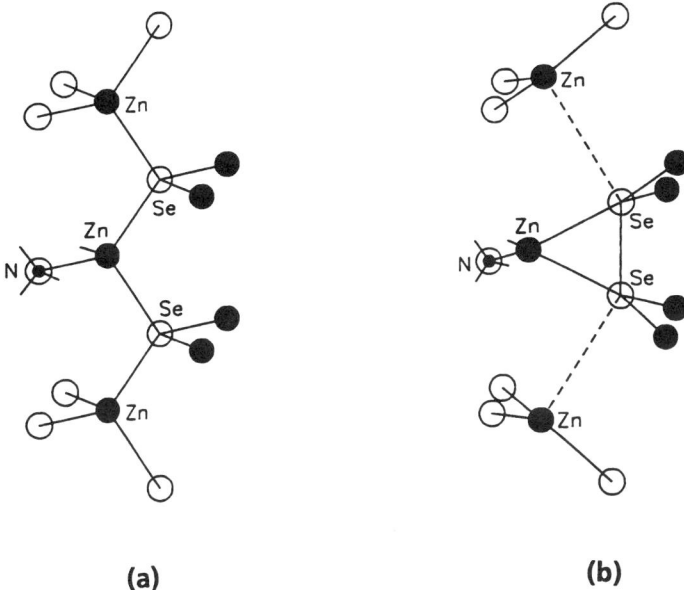

FIG. 3. Configuration of the "double broken bond" defect in ZnSe proposed by Park and Chadi (1995). The ideal ZnSe lattice is shown in (a); the formation of a double broken bond state resulting from the breaking of two Zn–Se bonds and the formation of a Se–Se bond is shown in (b). The double broken bond state is shown to occur near a nitrogen impurity, which would lower the formation energy of the defect (from Park and Chadi, 1995).

quantitative here. Assume that a prototypal native defect, with donor character, would occur in p-type material. When neutral, this defect introduces one electron into a state in the gap, and the formation energy in the neutral charge state is equal to E^0. The energy gained by transferring an electron from the level in the gap ($E^{+/0}$) (Fig. 4) to the Fermi level should increase with the width of the band gap, if the donor level is near the conduction band, and the Fermi level is near the valence band. The net energy to form the compensating defect is therefore $E^0 - (E^{+/0} - E_F)$, and this quantity indeed seems to decrease as the band gap increases.

However, this conclusion only holds if one assumes (as was often implicitly done) that E^0 does not depend on the width of the band gap — an assumption for which there is no real justification. Alternatively, it can be pointed out that $E^{+/0}$ and E_0 are not independent of one another. Indeed, the level in the gap, $E^{+/0}$, is defined as a transition level, meaning it corresponds to the Fermi-level position for which the formation energies of the positive and neutral charge states would be equal: $E^{+/0} = E^0 - E^+$. The net

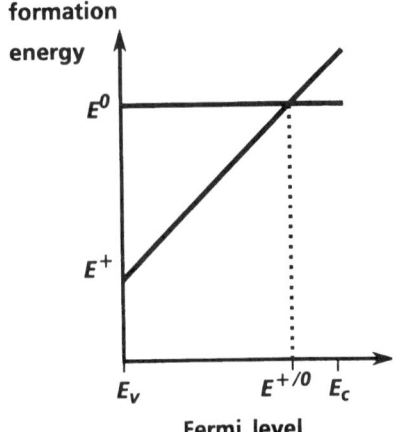

FIG. 4. Defect formation energy as a function of Fermi level, for a donor-type defect with a level in the gap at energy $E^{+/0}$. The formation energies for the neutral and positive charge states are shown; the gap level $E^{+/0}$ is the value of the Fermi level for which the two charge states have the same energy.

energy required to create the compensating defect is thus $E^0 - (E^{+/0} - E_F) = E^+ + E_F$, which is basically the formation energy of the defect in the positive charge state (E^+ being the formation energy when E_F is at the top of the valence band). We therefore see that the net energy required to create the compensating defect does not depend on the energy of the neutral defect at all! A generic native-defect compensation mechanism would require that the formation energy of the positive charge state, E^+, decreases as the width of the band gap increases. The existence of such a trend has not been established. Therefore, there is no basis for invoking a generic native-defect compensation mechanism in wide-band-gap semiconductors.

V. p-Type Doping

1. LITHIUM IN ZnSe

    a. *Configurations of Li in the Lattice*

    The case of lithium provides an excellent opportunity to illustrate the physics of a number of processes related to doping and compensation, offering a prime example for application of the formalism outlined in Section

III. Various possible configurations and charge states of the lithium impurity in the lattice have been analyzed (Van de Walle et al., 1993). Li behaves as an acceptor when located on the Zn site. The substitutional acceptor, Li_{Zn}^-, induces virtually no relaxation of the surrounding host atoms. For the lithium interstitial (Li_i^+), which is a shallow donor, the T_d site surrounded by Se atoms (T_d^{Se}) is 0.2 eV lower in energy than the T_d^{Zn} site. An investigation of other interstitial positions indicated that the barrier for migration of the interstitial is less than 0.5 eV (i.e., a Li interstitial can move readily, even at room temperature). Finally, Li on a substitutional Se site was also investigated, but found to have a prohibitively large formation energy.

b. Contour Plots of Li Concentration

As pointed out in Section III.2, the results for formation energies of the various configurations are functions of the chemical potentials; in this case, the Li chemical potential, μ_{Li}, as well as the chemical potential of one of the host constituents, say μ_{Zn}. This dependence can be highlighted by presenting the results in the form of contour plots, with the chemical potentials μ_{Zn} and μ_{Li} as the variables. Instead of presenting the formation energies themselves, it is more informative to show results for physical observables such as impurity concentrations and Fermi level positions. As explained in Section III.5, the Fermi energy is not an independent variable, since it is determined by charge neutrality. Figure 5a shows a contour plot for the total concentration of Li in ZnSe, at $T = 600$ K, which is a typical temperature in MBE growth of ZnSe:Li (Haase et al., 1990).

Looking at the contour lines in Figure 5a, we note that the total Li concentration ([Li]) increases with increasing μ_{Li}; indeed, it becomes more favorable for the impurity to dissolve in the semiconductor as the energy of the Li reservoir rises. Similarly, [Li] increases with *decreasing* μ_{Zn}, which is the energy of the reservoir to which Zn needs to be removed in order to accommodate Li on Zn sites.

c. Competition between Interstitials and Substitutionals

In order to understand the shape of the contour lines in more detail, we need to consider that Li not only occurs as a substitutional acceptor, but also as an interstitial donor. The formation energy for Li in a substitutional location was given in Eq (2). For the interstitial site, where Li is a shallow donor, we have

$$E_{form}(Li_i^+) = \mathscr{E}(Li_i^+) - \mu_{Li} + E_F. \tag{11}$$

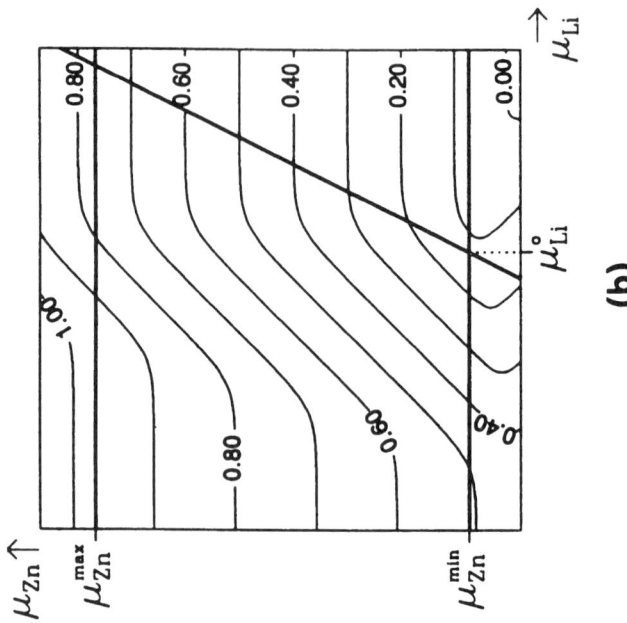
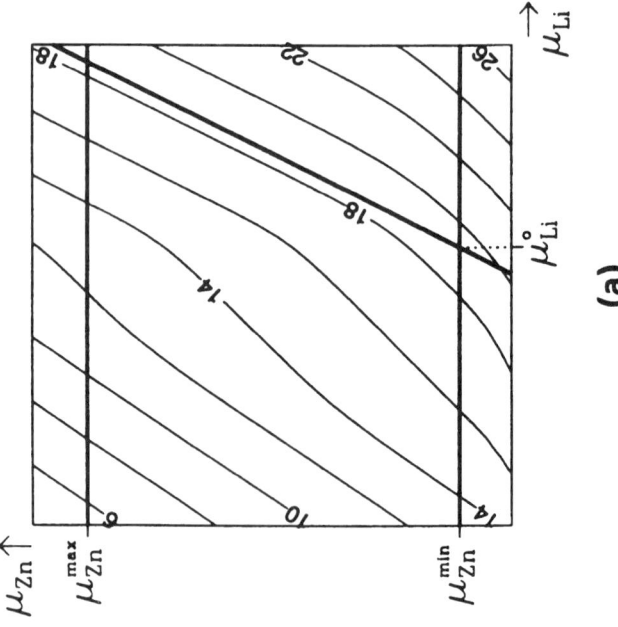

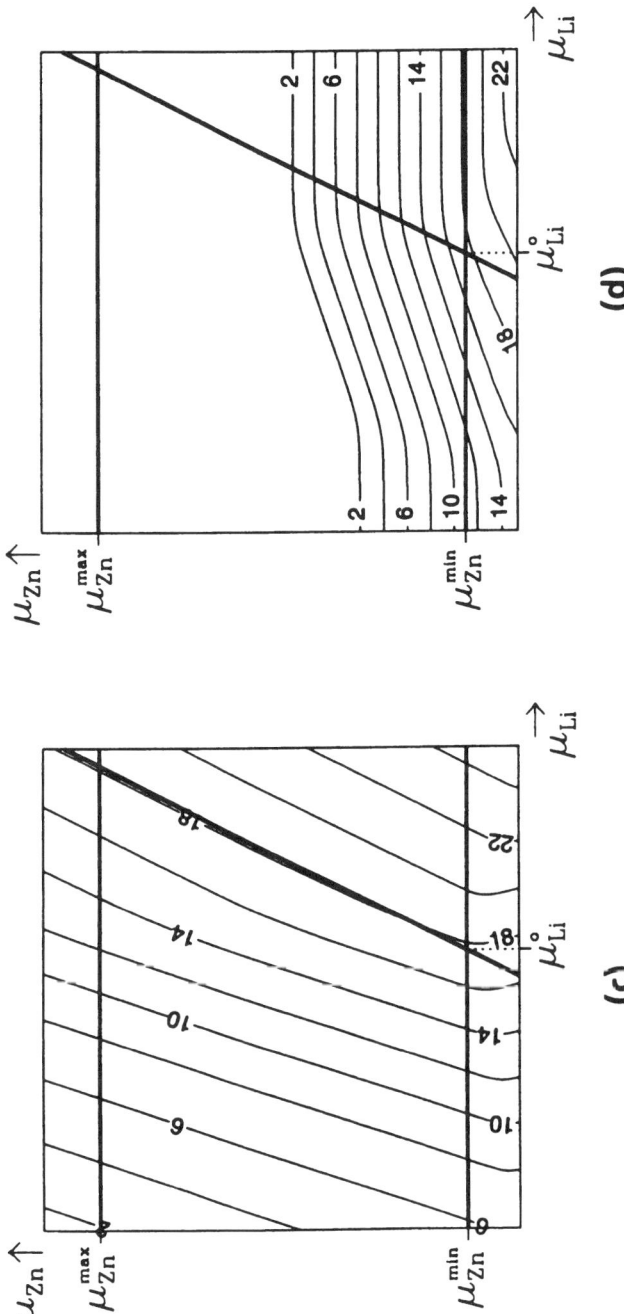

FIG. 5. Contour plots of (a) $\log_{10}$ [Li], where [Li] is the total Li concentration in cm^{-3}; (b) Fermi level (in eV, referred to the top of the valence band); (c) $\log_{10}$ [Li$_i$], where [Li$_i$] is the interstitial Li concentration in cm^{-3}; and (d) $\log_{10}$ [Se$_{Zn}^{++}$], the Se antisite concentration in cm^{-3}; at 600 K in ZnSe:Li, as a function of Zn and Li chemical potentials. Solid lines indicate bounds on μ_{Zn} and μ_{Li}.

143

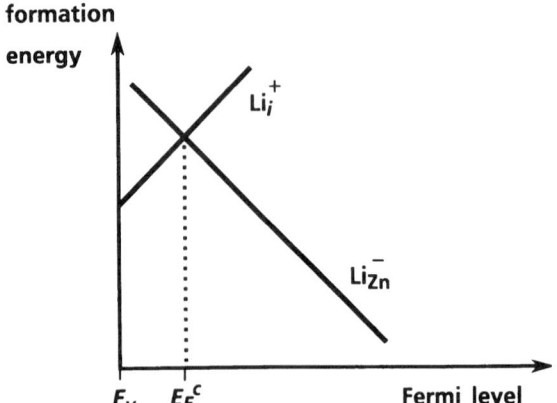

FIG. 6. Formation energies of acceptor (Li_{Zn}^-) and donor (Li_i^+) species as a function of Fermi level. As the p-type doping level increases, the Fermi level moves closer to the valence band, and saturates at the critical Fermi-level position, E_F^c, where the formation energy of the donor equals that of the acceptor.

$\mathscr{E}(Li_i^+)$ is the calculated energy of an interstitial Li at its most stable site, which is at the tetrahedral interstitial site surrounded by Se atoms (T_d^{Se}). The dependence of the formation energy on Fermi level position is schematically illustrated in Figure 6, which shows that as the Fermi level moves down (i.e., as the material becomes increasingly p-type), the formation energy of the acceptor species rises, whereas the formation energy of the donor species goes down. This predicts the existence of a limiting Fermi level position (maximum hole concentration), which can be obtained by equating the two formation energies. Attempts to push the Fermi level lower would result in preferential formation of donors, which would push the Fermi level back up. Incorporation of additional Li leaves the Fermi level unchanged, as each substitutional acceptor is immediately compensated by an interstitial donor.

The position of the Fermi level (at 600 K) is shown in Figure 5b. For very low Li concentrations, the Fermi level is pinned by native defects. Li interstitials are responsible for the flattening of the contour lines on the right-hand side of the plot. For a fixed value of μ_{Zn}, the Fermi level saturates as μ_{Li} is raised, even though the total Li concentration still increases (Fig. 5a). If no interstitials could form, the contour lines for the Fermi level would continue with the same slope as in the left-hand side of the plot. The interstitials cause compensation and limit the achievable hole concentration. The presence of Li interstitials has been experimentally observed (Haase *et al.*, 1990, 1991; Marshall, 1993). A contour plot of the Li interstitial concentration is shown in Figure 5c.

The position at which the Fermi level saturates due to interstitial compensation still depends on the Zn chemical potential, as can be noted in Figure 5b. These results differ markedly from those of Sasaki et al. (1991), where it was concluded that compensation by Li interstitials would always dominate; the authors of that work did not recognize that the level of compensation depends on the Zn chemical potential, and hence on the growth conditions. The dependence on chemical potentials explains the experimental observation that the degree of compensation by Li interstitials varies widely in different samples (Haase et al., 1990, 1991). The results also provide a guideline for optimizing the growth conditions: low values of μ_{Zn} lead to lower compensation, as well as higher Li_{Zn} concentrations.

d. Bounds on Chemical Potentials

In order to determine solubility limits, we need to use the information about bounds on the chemical potentials discussed in Section III.2.c. The bounds on the Zn chemical potential are shown as the horizontal lines in Figure 5. For Li, the chemical potential is limited by formation of the compound Li_2Se. Formation of Li_2Se on the growing ZnSe surface in MBE has actually been experimentally observed in the case of heavy Li doping (Zhu et al., 1992). The compound Li_2Se leads to the line with slope $+2$ in Figure 5, which was defined in Eq (6). The point where this line intersects the lower bound on μ_{Zn} is given by $\mu_{Li}^o = \mu_{Li(bulk)} + \frac{1}{2}\Delta H_f(Li_2Se)$. The heats of formation for the various compounds which are calculated from first principles turn out to be close to the experimental values.

e. Ability to Dope ZnSe with Li

The contours presented in Figure 5, together with the bounds on the chemical potentials, provide important insights in the ability to dope ZnSe with Li. We note that, over much of the range of the Li and Zn potentials, the maximum Li concentration is slightly higher than 10^{18} cm^{-3}. The fact that the slope of the contours in this region coincides with the slope of the Li_2Se boundary in Figure 5a is accidental, caused by the fact that in this region the removal of one Zn atom leads to the incorporation of two Li atoms (one substitutional and one interstitial). The highest Li concentration (and lowest Fermi level, i.e., highest hole concentration) occurs in the lower right-hand corner of the accessible region, for $\mu_{Zn} = \mu_{Zn}^{min}$ and $\mu_{Li} = \mu_{Li}^o$. At this point of highest Li incorporation, fewer than 3% of the Li atoms occur in the form of interstitials.

Some additional conclusions can be drawn. First, even though all native point defects were explicitly included in the calculations, their concentrations are very small over the whole of the accessible range in Figure 5. The effect of native defects is noticeable for low μ_{Zn} values, causing bending of the contour lines; however, their concentration would only become important if $\mu_{Zn} < \mu_{Zn}^{min}$, which is physically not allowed. The dominant native defect is the Se_{Zn} antisite, which is a donor. Figure 5d shows a contour plot of the Se_{Zn}^{++} concentration. At the point of highest Li incorporation, the concentration of Se_{Zn} is two orders of magnitude smaller than the Li concentration. Clearly the native defect concentration is too low to play any significant role in compensation. However, the concentration may be high enough to be detectable experimentally.

f. Complex Formation

So far we have only discussed isolated point defects and isolated impurities. In principle we should also consider complexes. Although the formalism presented in Section III is general enough to include any possible complexes, an exhaustive treatment is computationally prohibitive. Inspection of expressions for formation energies actually shows that a complex will only occur in appreciable concentrations (i.e., concentrations on the order of or larger than those of the individual defects out of which it is formed) if the binding energy exceeds the individual formation energy of the components of the complex. This consideration makes it less likely that complexes would play an important role.

In the context of Li doping, it is interesting to consider a complex consisting of a Li interstitial and a Li substitutional. Formation of such complexes seems plausible, since the interstitial is quite mobile, and the acceptor and donor are coulombically attracted (Neumark and Catlow, 1984). The calculated binding energy of this complex is ~ 0.3 eV (Van de Walle *et al.*, 1993). This value is small enough not to lead to appreciable concentrations of this type of complex. Moreover, any complexes would be dissociated at a growth temperature of 600 K. If we assume, however, that the concentration of Li substitutional and Li interstitial atoms is determined at the growth temperature, and remains fixed as the sample is cooled down, then the concentration of $Li_{Zn}-Li_i$ pairs may increase as the temperature is lowered. The presence of such complexes has indeed been observed experimentally, by a technique based on dissociation of the complex, followed by drift of the Li_i in an electric field (Marshall, 1993). The resulting change in the doping profile can be detected in a C-V or small-signal AC-transmittance measurement.

2. Sodium in ZnSe

Sodium is another column-I impurity which has been considered as an acceptor dopant in ZnSe (Cheng *et al.*, 1989). A full investigation of ZnSe:Na, including contour plots similar to the ones shown for Li above, was reported in Van de Walle *et al.* (1993). Sodium behaves qualitatively similar to lithium, but exhibits important quantitative differences. The relevant bound on the Na chemical potential is imposed by the compound Na_2Se. The most important result is that the calculated solubility of substitutional Na is more than three orders of magnitude lower than that of Li.

Experimental doping attempts with Na have been unsuccessful (Cheng *et al.*, 1989); the theoretical results indicate that the solubility limit is the culprit, rather than, e.g., compensation due to foreign impurities in the source.

Electronically, Na behaves similarly to Li; however, its slightly larger size causes it to fit less comfortably on the Zn site in ZnSe. This size mismatch is likely the main reason for the higher formation energy (and hence lower solubility) of Na in ZnSe (Laks *et al.*, 1993). This argument indicates that other column-I impurities, such as K, would fare even worse.

3. Phosphorous and Arsenic in ZnSe

It has long been known that P and As, which should act as acceptors when located on the Se site in ZnSe, do not lead to low-resistive *p*-type material. The reasons for this failure are still not completely clear. Experimentally, it has been found that P (Yao and Okada, 1986) and As (Okajima *et al.*, 1986) form both shallow and deep acceptor levels in ZnSe. A number of theoretical studies have been carried out, which we summarize here.

Chadi and Chang (1989b) proposed the following explanation: while P and As indeed reside on the substitutional Se site when they are in the negative charge state, they would undergo a large relaxation off the substitutional site when they are put in a positive charge state (Fig. 2). The energy of this state would be low enough for the relaxation to occur spontaneously, and As and P would therefore spontaneously "self-compensate." The mechanism is very similar to the one proposed by Chadi and Chang (1988b, 1989a) to explain the *DX* center in AlGaAs alloys. In a subsequent paper (Chadi, 1991) the importance of the positive charge state was downplayed, in favor of a stronger emphasis on the lattice relaxation that can occur in the neutral (paramagnetic) charge state. The latter provides a link with early electron-spin-resonance (ESR) experiments by

Watts et al. (1971) and Reinberg et al. (1971); the ESR is only sensitive to paramagnetic charge states, and observed a state with C_{3v} symmetry. Park and Chadi (1995) have also investigated the pressure dependence of the binding energies of the various configurations of P and As, and linked the results to the experimental observations of Strachan et al. (1994) and Li et al. (1994).

Kwak et al. (1993, 1994, 1995) have used state-of-the-art first-principles techniques to study the properties of substitutional and interstitial P, as well as P-related antisites (P_{Zn}). First of all, they find that substitutional P_{Se} in the negative charge state has T_d symmetry, and undergoes only a very small lattice relaxation in the neutral charge state; this relaxation is smaller than the one reported by Chadi (1991), and, according to Kwak et al., too small to agree with the experimental data of Watts et al. (1971) (although hyperfine parameters were not calculated). Kwak et al. find a larger lattice relaxation for the positive charge state, but the value is still much smaller than the one reported by Chadi and Chang (1989b) for this state. One reason for the discrepancy between the results of Chadi and those of Kwak et al. may be the limited size of the supercell used in Chadi's studies, which may induce spurious relaxations.

Kwak et al. (1994, 1995) attribute the failure of P to produce good p-type doping to the introduction of P_{Zn} antisite defects, which act as triple donors and compensate the material. They find that the formation energy of interstitial phosphorous is always higher than that of substitutional P_{Se}, and of the P_{Zn} defect, for any value of the Se chemical potential.

Kwak et al. propose that the P interstitial is responsible for the deep acceptor level observed by Watts et al. (1971) and Reinberg et al. (1971). They argue that the properties of the P interstitial, which is located near a hexagonal interstitial site, are consistent with the observed luminescence and ESR spectrum; the agreement with ESR is mostly based on qualitative arguments. It would be interesting to actually calculate the hyperfine parameters for the proposed configuration (using the procedure described by Van de Walle and Blöchl, 1993), to enable a more direct comparison with experiment. Lacking such an identification, the precise nature of the phosphorous-induced deep level remains uncertain.

4. NITROGEN IN ZnSe AND ZnTe

a. Substitutional Nitrogen

An investigation of N in ZnSe, along the same lines as for Li discussed in Section V.I above, was reported in the paper by Van de Walle et al. (1993).

Nitrogen on a substitutional Se site (N_{Se}) is a shallow acceptor. For N_{Se}^-, the surrounding Zn atoms undergo a significant inward relaxation, reducing the calculated Zn–N distance to 2.1 Å, very close to the Zn–N distance in the compound Zn_3N_2 (Wyckoff, 1964). Regarding charge states other than the negative, Chadi and Chang (1989b) proposed that even if symmetry-lowering relaxations would occur, they would not interfere with the shallow acceptor character of the dopant. Kwak et al. (1993) subsequently reported that N_{Se} retains T_d symmetry in all charge states.

The bounds on the N chemical potential are due to formation of N_2 molecules and the Zn_3N_2 compound, leading to a calculated solubility for N (Van de Walle et al., 1993) which is about one order of magnitude higher than for Li, consistent with experimental results. The failure of nitrogen doping starting from N_2 is due to the large kinetic barrier for breaking up the molecule; a plasma source or other technique for obtaining N in an atomic state, or at least N_2 in an excited state, is required (Park et al., 1990). Once one succeeds in incorporating atomic (as opposed to molecular) nitrogen into the lattice, N should act as a good acceptor, allowing hole concentrations high enough for useful device applications.

It is interesting to note that Chadi (1994b) found that nitrogen forms a localized state in MgSe, as opposed to a delocalized state in ZnSe. This could explain the experimental results of Han et al. (1994), who have observed a decrease in doping efficiency in N-doped ZnMgSSe alloys, as well as a persistent conductivity effect.

b. Other Configurations of N in the Lattice

The possibility of incorporation of nitrogen on interstitial sites has also been examined. In the paper by Van de Walle et al. (1993), nitrogen interstitials located on T_d sites in the lattice were examined and found to be prohibitively high in energy; nitrogen on substitutional Zn sites was similarly found to be unfavorable. These conclusions were confirmed in recent calculations by Kwak (1995). Chadi and Troullier (1993) proposed split-interstitial configurations for nitrogen sharing a substitutional Se site with the Se atom; they did not, however, report energies of these configurations relative to the N_{Se} acceptor. Chadi (1994b) also suggested the possibility of formation of N_2, which would assume a split-interstitial configuration (2 N atoms sharing a Se site). Cheong et al. (1995), finally, examined a number of different configurations and found that neutral N_2 molecules would dominate in Se-rich material (in addition to N_{Se}, of course), whereas in Zn-rich material a split-interstitial complex consisting of two N atoms occupying a Se site would dominate.

c. Complexes between Native Defects and Substitutional Acceptors

Zn interstitials may form complexes with substitutional nitrogen acceptors (Van de Walle *et al.*, 1993). Such complexes are formed if a positive binding energy exists between the native defect and the acceptor; if this binding energy is large enough, the overall formation energy of the complex may be lower than that of the individual point defect. For this to happen, the binding energy of the complex needs to exceed the formation energy of the individual native defect.

Recent calculations for the (Zn_i-N_{Se}) complex (Van de Walle and Neugebauer, 1995) yield a binding energy of $\sim 0.6\,eV$, large enough to make the (Zn_i-N_{Se}) complex the prime candidate for a compensating center in ZnSe. García and Northrup (1995) have addressed this issue for the case of As as the substitutional acceptor. They find that a (Zn_i-As_{Se}) complex has a large binding energy, approximately 1.2 eV. The calculations by Van de Walle and Neugebauer (1995) produced a binding energy of 0.6 eV for the (Zn_i-As_{Se}) complex, i.e., the same as for (Zn_i-N_{Se}).

García and Northrup (1995) also carried out calculations for a $(V_{Se}-As_{Se})$ complex, finding a binding energy of 0.3 eV — a rather low value, ostensibly due to the fact that the constituents of this complex are only second-nearest neighbors in the lattice. $V_{Se}-N_{Se}$ complexes have been discussed in the literature; Hauksson *et al.* (1992) and Murdin *et al.* (1993) proposed this complex to explain the compensation observed in their samples.

According to the calculations, (Zn_i-N_{Se}) complexes are more likely to occur than $V_{Se}-N_{Se}$ complexes. If they indeed occur in appreciable quantities, it should be possible to detect them experimentally. One possible technique of observation would involve dissociation of the complex, followed by drift of the Zn_i in an electric field — similar to an experiment described in Section V.1.f for complexes between Li interstitial donors and Li_{Zn} acceptors (Marshall, 1993). Other techniques, such as vibrational-mode spectroscopy, could also be very fruitful. The impact of complexes on compensation and doping limits will be further discussed in Section VII.1.

Finally, we mention the "double broken bond" state proposed by Park and Chadi (1995), which was discussed in Section IV.6 and depicted in Figure 3. Park and Chadi pointed out that the formation energy of this defect would be lowered if it were formed in the vicinity of a N acceptor — effectively creating a complex between the acceptor and the defect; the overall complex acts as a single donor. Park and Chadi found that the formation energy of the complex was lower in MgSe than in ZnSe; this trend is consistent with the observed difficulty in *p*-type doping of ZnMgSSe alloys (Han *et al.*, 1994). Park and Chadi also suggested that the double broken bond defects would explain the persistent conductivity effect observed by Han *et al.*

5. Oxygen in ZnSe

A number of groups have reported that oxygen gives rise to a shallow-acceptor state in ZnSe (Akimoto *et al.*, 1989; Shahzad *et al.*, 1991). This is surprising, given that oxygen is an isovalent impurity if it sits on the substitutional Se site. Akimoto *et al.* (1989) proposed that the acceptor character of oxygen was induced by its high electronegativity, causing a large charge transfer from the host lattice to the O atom. This charged oxygen could then bind a hole by a long-range Coulomb force.

The only theoretical work on this problem, to our knowledge, has been performed by Chadi (1994b), who suggested that the electrical activity arises from an interstitial oxygen atom which forms a strong bond with a Zn atom, effectively breaking a Zn–Se bond and attaching to the Zn from the antibonding direction. The resulting onefold coordinated O was found to behave as a shallow acceptor.

We venture here to propose another, very speculative explanation for the electrical activity associated with oxygen: electrical activation of the isovalent oxygen impurity by hydrogen. Hydrogen is known to activate isovalent impurities in elemental semiconductors, such as a Si impurity in Ge (Haller *et al.*, 1980; Denteneer *et al.*, 1989). It is conceivable that hydrogen could bind to an otherwise electrically inactive O substitutional impurity in ZnSe, and in the process create an acceptor-like center. As discussed in Section II.5, hydrogen is likely to occur as a contaminant during growth, possibly even introduced by the oxygen source.

VI. *n*-Type Doping

1. Al and Ga in ZnSe and ZnTe

Chadi (1994a) has addressed *n*-type doping in ZnSe, as well as ZnTe, MgSe, and MgTe, in light of the possibility of formation of DX-like centers. The results are consistent with the observations that ZnTe can be doped *p*-type more easily than *n*-type, whereas ZnSe, CdSe, ZnS, and CdS are more easily doped *n*-type than *p*-type. Chadi's investigation was aimed at addressing the possibility of using MgZnSe and ZnSeTe alloys for band-structure engineering, while maintaining the ability to dope both *p*-type and *n*-type.

Chadi used first-principles methods to examine Ga, Al, and Cl impurities in ZnSe and ZnTe. While the calculations are state-of-the-art, rather small supercells (18 atoms) were used, which could result in larger error bars on some of the calculated values. Chadi found that Al and Ga form stable deep centers in ZnTe, analogous to DX centers in AlGaAs alloys. DX centers

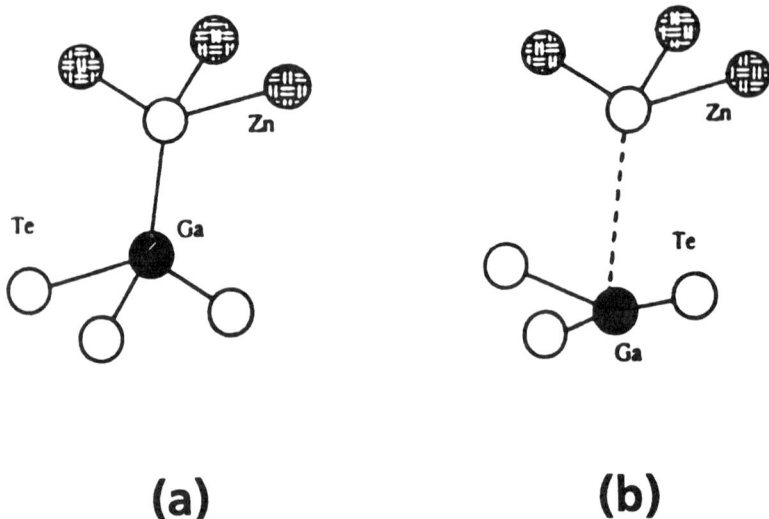

FIG. 7. (a) The normal substitutional state of a Ga donor impurity in ZnTe. (b) The localized deep donor DX state, in which the Ga atom has undergone a large displacement (from Chadi, 1994a).

were found to be high-energy states in ZnSe, however. The formation of the DX^- state in ZnTe is accompanied by a large displacement (1.85 Å) of the Ga or Al donor along the antibonding [111] direction, as shown schematically in Figure 7. Chadi also reported a configuration-coordinate diagram showing activation energies for various optical and thermal excitations. The same trend that was observed for occurrence of DX centers between ZnSe and ZnTe was also observed for MgSe versus MgTe. Strong experimental evidence for DX centers in ZnCdTe has been found by Khachaturyan et al. (1989).

Chadi also pointed out that the calculated difference in the energy of DX states in ZnSe and ZnTe correlates with the conduction-band offset between the two materials. Such a correlation, namely that the positions of the valence-band maximum and conduction-band minimum, on an absolute energy scale, are a determining factor in doping, had been suggested previously (Ren et al., 1988; Dow et al., 1991). Indeed, it has been observed that the valence band in ZnSe lies much lower in energy than the valence band in ZnTe—and ZnSe is harder to dope p-type than ZnTe. While intriguing, the correlation with band-edge positions by itself does not offer any further insight as to the cause of the doping limitations. For p-type doping, Dow et al. (1991) proposed that a defect-induced p-like deep level lies within the valence band in ZnTe, but emerges into the band gap in ZnSe.

Chadi's suggestion about the relevance of conduction-band positions in the formation of DX centers in ZnSe and ZnTe is similar in nature.

2. Cl in ZnSe and ZnTe

For Cl, Chadi (1994a) found that the donor wavefunction in ZnSe is delocalized, whereas it is strongly localized on the s states of the Zn neighbors in ZnTe, a difference which was attributed to the larger lattice constant of ZnTe. Chadi pointed out, however, that calculations using larger supercells are required to confirm this result.

VII. Comparison between Theory and Experiment

We have already made connection with experimental results throughout the preceding sections. Here we focus on a few additional topics and cases where specific theoretical predictions can and should be checked experimentally.

1. Compensation due to Native Defects and Native-defect Complexes

At this point in time, nitrogen seems to be the most successful acceptor species in ZnSe. In spite of the successes, nitrogen doping still has its problems. One issue is the large ionization energy. This is a problem common to most acceptors in wide-band-gap materials, and it seems hard to circumvent. Also, it has been suggested (Mensz, 1994) that during device operation a larger fraction of the N_{Se} acceptors will actually be ionized than expected from purely thermal ionization, leading to more favorable doping behavior than expected from simple inspection of the ionization energy.

The other problem with nitrogen is more severe, namely the fact that the achievable doping levels are still limited. Various groups have published experimental results for carrier concentrations as a function of the total nitrogen concentration in the layer (Nishikawa *et al.*, 1994; Ito *et al.*, 1992; Qiu *et al.*, 1991). (We presume that the quantity referred to as $N_A - N_D$ in these papers is actually the hole concentration, as measured in C-V measurements. In order to obtain the actual acceptor concentration, a further analysis would need to be carried out, taking into account that, due to the large ionization energy of the N acceptor, only a fraction of the

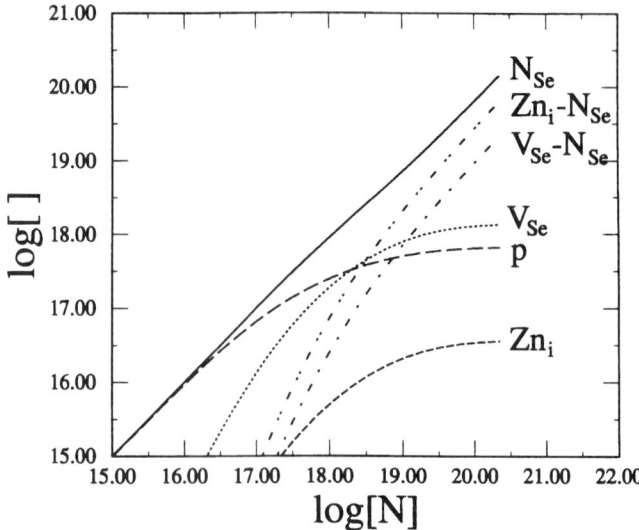

FIG. 8. Concentrations (per cm^3) of various species in N-doped ZnSe, as a function of total nitrogen concentration, at 600 K. The Se chemical potential is chosen to yield maximum solubility of N_{Se} [corresponding to $\lambda = 0.06$, Eq. (5)]. The curves are labeled according to the type of impurity or defect they correspond to; p indicates the hole concentration (from Van de Walle and Neugebauer, 1995).

dopants are ionized.) When the nitrogen concentration approaches 10^{18} cm^{-3}, the carrier concentration appears to saturate, and then decreases as [N] is further increased.

The question is: what mechanism causes the hole concentration to saturate, then drop, as the nitrogen concentration is increased? Is it related to a compensation mechanism? Or is it somehow due to limited solubility? In order to address this issue, we need to investigate the effects of compensation in a bit more detail, looking at compensation due to isolated native defects as well as due to complexes between native defects and acceptors.

Figure 8 illustrates the dependence of the concentration of the acceptor species and the various compensating centers on the total nitrogen concentration at 600 K. The nitrogen concentration is changed by varying the nitrogen chemical potential, μ_N, up to the upper bound discussed in Section V.4.a; the curves are plotted up to the maximum achievable nitrogen concentration (i.e., the solubility limit). In Figure 8 we also assume that the selenium chemical potential is fixed at the value that optimizes the incorporation of nitrogen on substitutional Se sites (i.e., maximizes the solubility) (see Van de Walle *et al.*, 1993); this value corresponds to $\lambda = 0.06$ [Eq (5)], i.e., quite close to the Se-poor (Zn-rich) limit.

4 DOPING OF WIDE-BAND-GAP II–VI COMPOUNDS—THEORY

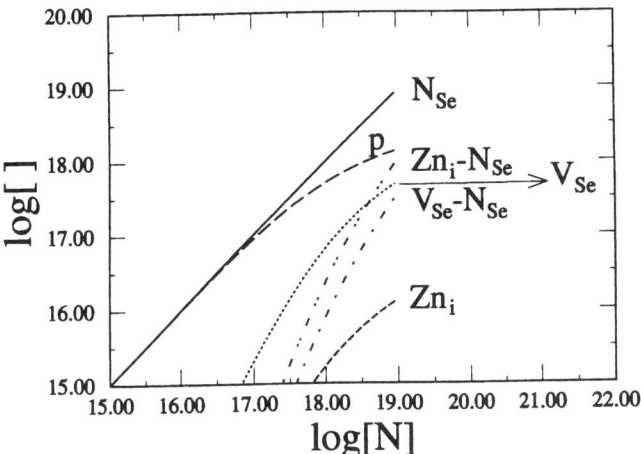

FIG. 9. Same as Figure 8, except that the Se chemical potential is chosen corresponding to $\lambda = 0.15$ (from Van de Walle and Neugebauer, 1995).

The formation energies for the native defects, impurities, and complexes used in Figure 8 are those derived in Laks et al. (1992), Van de Walle et al. (1993), and Van de Walle and Neugebauer (1995); an estimated vibrational entropy of $5\,k_B$ has been included for each native defect and native-defect complex (note that, without this entropy contribution, the concentrations of the compensating defects would be almost two orders of magnitude lower than in Figure 8). Under the conditions described here, the Zn_i–N_{Se} complexes are the dominant compensating defects, causing a noticeable reduction in the hole concentration, which is also plotted in Figure 8.

Figure 8 reflects the concentrations of various species at 600 K. If a sample is grown at this temperature, and subsequently cooled to room temperature, the concentrations of the defects may change, because various species (for instance, the Zn interstitial) are quite mobile. The room-temperature concentrations will depend on the mobility of the defects, the binding and dissociation energies of the complexes, and on the cooling rate.

Compensation by (Zn_i–N_{Se}) complexes causes the saturation of the hole concentration observed in Figure 8. Such a saturation of the carrier concentration as the nitrogen concentration is increased has been experimentally observed (Nishikawa et al., 1994; Ito et al., 1992). However, it bears pointing out that the results displayed in Figure 8 are quite sensitive to the Se chemical potential (i.e., to the details of the growth conditions). To illustrate this point, Figure 9 shows the results obtained for a value of the Se chemical potential ($\lambda = 0.15$) which is only slightly different from the choice made in Figure 8 ($\lambda = 0.06$). Note that the maximum concentration of nitrogen is now lower, since we have moved away from the point of maximum solubility. More importantly, however, note that the concentration of compensating defects has decreased — by as much as an order of magnitude, for similar N concentrations. The hole concentration is correspondingly increased, and no longer saturates at a plateau value. The fact that the hole concentration is lower than the N_{Se} concentration is not solely due to compensation; because of the high ionization energy (110 meV; see Shahzad et al., 1990), only a fraction of N_{Se} acceptors are actually ionized, pushing the hole-concentration curve below the N_{Se} curve.

If we continue to increase λ (i.e., make the system more Se-rich) we find that the degree of compensation is further reduced; at the expense, however, of N incorporation, due to the lower solubility. Our calculations show that if λ is raised above 0.3, the N_{Se} concentration falls below 10^{17} cm^{-3}. The fact that the effects of compensation are reduced as λ is increased (i.e., as the system becomes more Se-rich) was also noted by García and Northrup (1995); however, the values of λ considered by García and Northrup are much greater than in the work presented here. For values of λ as high as discussed in the work by García and Northrup (1995) we find that the calculated N incorporation would become very low. García and Northrup did not explicitly consider the formation energy of the N_{Se} species, and they allowed incorporation of the acceptor species above the solubility limit. Whether incorporation of the acceptor species occurs under equilibrium or nonequilibrium conditions has not been decisively established. Fan et al. (1994) interpreted their SIMS and transport data on N-doped ZnSe and ZnTe as consistent with the notion of equilibrium incorporation, governed by solubility limits. More investigations are necessary, however.

2. DISCUSSION OF DOPING SATURATION EFFECTS

A saturation in the hole concentration would be consistent with a compensation mechanism, as described above and illustrated, for instance, in Figure 8. However, a precipitous decrease in the carrier concentration for high [N], as observed by Nishikawa et al. (1994) and by DePuydt et al. (1995) cannot be attributed to compensation; some other mechanism needs to be invoked. One possibility is that above a critical impurity concentration the crystal quality degrades, leading to a decline in doping efficiency. A mechanism by which increased impurity concentrations can affect crystal quality is discussed in Section VII.3.

Since the experimental observations reveal a decrease in the hole concentration at high acceptor concentrations, it is not clear whether the observed "plateau" in the carrier concentration (around $[N] \approx 10^{18}$ cm^{-3} for nitrogen) should be interpreted as a saturation effect that is due to compensation, or rather as the onset of an approaching downturn in the hole concentration. We have pointed out that the occurrence of compensation may also be a rather sensitive function of the growth conditions, as illustrated by Figures 8 and 9, which differ only by a slightly different choice of Se chemical potential. Plots of carrier concentration vs. dopant incorporation therefore offer, in our opinion, no conclusive evidence regarding compensation. Other experiments, in which compensating defects are actually observed and identified, are required for that purpose. And we emphasize that the theoretical results suggest that compensation is not unavoidable — a careful choice of growth conditions may be able to suppress the formation of compensating species.

3. Nucleation of Misfit Dislocations

ZnSe is commonly grown on GaAs substrates, which provide a good but not perfect lattice match: the lattice constants differ by about 0.25%. This mismatch leads to the formation of misfit dislocations once the critical-layer thickness (about 1500 Å) is exceeded. It has been observed that the formation of dislocations is influenced by the doping of the ZnSe layer. Kuo *et al.* (1993) have reported an increase in the density of threading dislocations in ZnSe grown on GaAs substrates when the N concentration exceeds a certain critical value. Laks *et al.* (1993) suggested that this increase is triggered by the fact that the dopant incorporation approaches the solubility limit; excess quantities of the dopant impurity may be incorporated as microprecipitates, which can act as nucleation sites for dislocation loops. This proposed explanation is consistent with the observation by the 3M group (DePuydt, 1995) that the "critical acceptor concentration," above which the hole concentration decreases (and the crystal quality degrades), depends on the type of acceptor. The measured values are 10^{16}, 10^{18}, 10^{19} cm^{-3} for Na, Li, and N doping, respectively; the trend in these values agrees very well with our calculated solubility limits for these acceptors.

4. Comparison between Different II–VI Materials

It has long been known that ZnTe is much easier to dope *p*-type than ZnSe. This can be attributed to the difference in solubility of acceptor impurities in these two materials. Indeed, the solubilities of Li, Na, and N were calculated in both semiconductors (Laks *et al.*, 1993; Van de Walle and

Laks, 1995), and it was found that the solubility limits are systematically higher (by at least an order of magnitude) in ZnTe than in ZnSe.

Han et al. (1993) and Fan et al. (1994) have recently accomplished high-level p-type doping in MBE growth of ZnTe. They correlated secondary-ion mass spectroscopy (SIMS) measurements of N concentrations in ZnSe and ZnTe with transport data, and concluded that in both cases close to 100% of N atoms were incorporated on substitutional Se lattice sites, and served as active acceptors. This result indicates that, at least for the growth conditions employed by Fan et al., the p-type doping level is not so much compensation limited, as it is limited by solubility of the N acceptor. In addition, they found that, for similar growth conditions, the N concentration in ZnTe exceeded that in ZnSe by almost an order of magnitude — in agreement with theoretical values for dopant solubilities (Van de Walle and Laks, 1995).

Finally, we mention that Laks and Pantelides (1995) have investigated the solubility of Li and Na in BeSe, MgSe, CaSe, and BeTe. They find solubilities which are lower than in ZnSe, except for BeTe.

5. SOLUBILITY-LIMITING PHASES

Experimental observations could address the role of solubility in limiting the doping. As discussed in Section V.4.a, the incorporation of N in ZnSe (Van de Walle et al., 1993) as well as in ZnTe (Van de Walle and Laks, 1995) is limited due to the formation of Zn_3N_2. It would be interesting to check experimentally whether any Zn_3N_2 can be detected on the surface of an MBE-grown sample, under conditions that favor high N doping. For the case of Li doping, evidence for the formation of islands of Li_2Se (see Sections III.2.c and V.1.d) was found in experiments using reflection high energy electron diffraction (RHEED) (Zhu et al., 1992).

6. EFFECT OF N INCORPORATION ON THE LATTICE CONSTANT

One expects that the incorporation of N_{Se} in the ZnSe lattice would be accompanied by a shrinking of the lattice constant, since N is a much smaller atom than Se. This effect was indeed experimentally observed, as discussed by Petruzzello et al. (1993). The effect of the shorter bond length on the lattice constant of the material can be evaluated, as outlined by Van de Walle and Laks (1995). It turns out the theoretical prediction is close to the observed value. Since the size mismatch between the N impurity and the host is even larger in the case of ZnTe:N, significant effects on the lattice constant are expected in that case as well.

VIII. Conclusions and Future Directions

In this chapter I have attempted to review the theoretical understanding of doping of wide-band-gap semiconductors; my goal was not just to list results for specific cases, but to provide a framework for understanding the mechanisms that govern doping.

It is fair to say that our understanding, both from experiment and from theory, of doping of ZnSe is quite advanced. For p-type doping, Li is the best candidate among the column-I impurities; Na has too low a solubility, and the trend is likely to get worse as one goes to heavier column-I elements. Among column-V elements, nitrogen currently seems the best choice. In addition, we know what types of native defects may become important under certain conditions, and we know about potential complex formation. Whether compensation by native-defect complexes is a determining factor in nitrogen doping of ZnSe, and to what extent dopant solubility plays a role, are issues which need further investigation. For As and P, we concluded that there are still significant gaps in our understanding. There are definite indications that these impurities can give rise to shallow levels, in addition to deep levels; the nature of the latter is still unclear. If there would be any way of suppressing the deep-level formation, these elements might turn out to be promising candidates for p-type doping. One issue might be the lack of a suitable source for phosphorous doping—just like the early failures of N doping needed to be overcome by the development of a plasma source. We think further investigation, particularly for P, is warranted.

The level of knowledge about doping of other wide-band-gap II–VI materials is far lower. This is disconcerting, because the design of light-emitting devices requires the ability to do band-structure engineering, necessitating the use of other compounds, and alloys, in addition to ZnSe. It has already been found that p-type doping of ZnMgSSe is harder to accomplish than in ZnSe. Fundamental research into the nature of doping mechanisms in a variety of materials, as well as in alloys, will be required.

Further investigations into the behavior of hydrogen in the II–VI's would also be fruitful. There seems to be a consensus that hydrogen plays an important role in passivating acceptor impurities, causing the failure of MOCVD to produce p-type material. Nonetheless, and in spite of the expense that has been involved in pursuing MOCVD growth of ZnSe, basic research about hydrogen in the II–VI's has been scarce.

At the computational level, we noted that vibrational entropies are not explicitly being calculated at the present time. We pointed out that such entropy contributions are small enough not to affect any qualitative conclusions, but they could affect the quantitative results. Accurate calculations of vibrational entropies are gradually becoming feasible and will no doubt be carried out for various systems in the near future. In addition, theoretical and computational developments that allow going beyond density-functional theory would be fruitful for addressing issues such as location of transition levels in the band gap.

The progress that has been made in research and device applications of II–VI materials since the late 1980s has been extremely impressive. Further advances are still needed to support the technology; such progress will crucially depend on our fundamental understanding of the mechanisms that govern the structural and electronic properties of the material. I hope that this chapter, by providing an overview of the state of the art, will contribute to this progress.

ACKNOWLEDGMENTS

Thanks are due to D. B. Laks and S. T. Pantelides for productive collaborations. I also appreciated useful discussions with Alberto García, Jörg Neugebauer, and John Northrup.

REFERENCES

Akimoto, K., Miyajima, T., and Mori, Y. (1989). *Phys. Rev. B* **39**, 3138.
Ando, K., Kawaguchi, Y., Ohno, T., Ohki, A., and Zembutsu, S. (1993). *Appl. Phys. Lett.* **63**, 191.
Baraff, G. A., and Schlüter, M. (1978). *Phys. Rev. Lett.* **41**, 892.
Baraff, G. A., and Schlüter, M. (1983). *Phys. Rev. B* **28**, 2296.
Bar-Yam, Y., and Joannopoulos, J. D. (1984). *Phys. Rev. Lett.* **52**, 1129.
Bernholc, J., Lipari, N. O., and Pantelides, S. T. (1978). *Phys. Rev. Lett.* **41**, 895.
Car, R., Kelly, P. J., Oshiyama, A., and Pantelides, S. T. (1985). *Phys. Rev. Lett.* **54**, 360.
Chadi, D. J., and Chang, K. J. (1988). *Phys. Rev. Lett.* **60**, 2187.
Chadi, D. J., and Chang, K. J. (1988b). *Phys. Rev. Lett.* **61**, 873.
Chadi, D. J., and Chang, K. J. (1989a). *Phys. Rev. B* **39**, 10 063.
Chadi, D. J., and Chang, K. J. (1989b). *Phys. Rev. Lett.* **55**, 575.
Chadi, D. J. (1991). *Phys. Rev. Lett.* **59**, 3589.
Chadi, D. J., and Troullier, N. (1993). *Physica B* **185**, 128.
Chadi, D. J. (1994a). *Phys. Rev. Lett.* **72**, 534.
Chadi, D. J. (1994b). *J. Cryst. Growth* **138**, 295.
Chadi, D. J. (1995). *Proceedings of the 22nd International Conference on the Physics of Semiconductors*, ed. D. J. Lockwood. World Scientific, Singapore. 2311.
Cheng, H., DePuydt, J. M., Potts, J. E., and Haase, M. A. (1989). *J. Cryst. Growth* **95**, 512.
Cheong, B.-H., Park, C. H., and Chang, K. J. (1995). *Phys. Rev. B* **51**, 10610.
Chetty, N., and Martin, R. M. (1992). *Phys. Rev. B* **45**, 6089.
Dabrowski, J., and Scheffler, M. (1988). *Phys. Rev. Lett.* **60**, 2183.
Denteneer, P. J. H., Van de Walle, C. G., and Pantelides, S. T. (1989). *Phys. Rev. Lett.* **62**, 1884.
DePuydt, J. M. (1995). Private communication.

Dow, J. D., Hong, R. D., Klemm, S., Ren, S. Y., Tsai, M. H., Sankey, O. F., and Kasowski, R. V. (1991). *Phys. Rev. B* **43**, 4396.
Fan, Y., Han, J., He, L., Gunshor, R. L., Brandt, M. S., Walker, J., Johnson, N. M., and Nurmikko, A. V. (1994). *Appl. Phys. Lett.* **65**, 1001.
García, A., and Northrup, J. E. (1995). *Phys. Rev. Lett.* **74**, 1131.
Haase, M. A., Cheng, H., DePuydt, J. M., and Potts, J. E. (1990). *J. Appl. Phys.* **67**, 448.
Haase, M. A., DePuydt, J. M., Cheng, H., and Potts, J. E. (1991). *Appl. Phys. Lett.* **58**, 1173.
Haller, E. E., Joos, B., and Falicov, L. M. (1980). *Phys. Rev. B* **21**, 4729.
Hamann, D. R., Schlüter, M., and Chiang, C. (1979). *Phys. Rev. Lett.* **43**, 1494.
Han, J., Stavrinides, T. S., Kobayashi, M., Gunshor, R. L., Hagerott, M. M., and Nurmikko, A. V. (1993). *Appl. Phys. Lett.* **62**, 840.
Han, J., Ringle, M. D., Fan, Y., Gunshor, R. L., and Nurmikko, A. V. (1994). *Appl. Phys. Lett.* **65**, 3230.
Hauksson, I. S., Simpson, J., Wang, S. Y., Prior, K. A., and Cavenett, B. C. (1992). *Appl. Phys. Lett.* **61**, 2208.
Hohenberg, P., and Kohn, W. (1964). *Phys. Rev.* **136**, B864.
Ito, S., Ikeda, M., and Akimoto, K. (1992). *Jpn. J. Appl. Phys.* **31**, L1316.
Jansen, R. W., and Sankey, O. F. (1989). *Phys. Rev. B* **39**, 3192.
Khachaturyan, K., Kaminska, M., Weber, E. R., Becla, P., and Street, R. A. (1989). *Phys. Rev. B* **40**, 6304.
Kohn, W., and Sham, L. J. (1965). *Phys. Rev.* **140**, A1133.
Kröger, F. A. (1964). *The Chemistry of Imperfect Crystals.* North-Holland, Amsterdam. 136, 628.
Kuo, L. H., Salamanca-Riba, L., DePuydt, J. M., Cheng, H., and Qiu, J. (1993). *Appl. Phys. Lett.* **63**, 3197.
Kwak, K. W., King-Smith, R. D., and Vanderbilt, D. (1993). *Phys. Rev. B* **48**, 17 827.
Kwak, K. W., Vanderbilt, D., and King-Smith, R. D. (1994). *Phys. Rev. B* **50**, 2711.
Kwak, K. W., Vanderbilt, D., and King-Smith, R. D. (1995). *Phys. Rev. B* **52**, 11 912.
Laks, D. B., Van de Walle, C. G., Neumark, G. F., and Pantelides, S. T. (1991). *Phys. Rev. Lett.* **66**, 648.
Laks, D. B., Van de Walle, C. G., Neumark, G. F., and Pantelides, S. T. (1992). *Phys. Rev. B* **45**, 10 965.
Laks, D. B., Van de Walle, C. G., Neumark, G. F., and Pantelides, S. T. (1993). *Appl. Phys. Lett.* **63**, 1375.
Laks, D. B., and Pantelides, S. T. (1995). *Phys. Rev. B* **51**, 2570.
Li, M. M., Strachan, D. J., Ritter, T. M., Tamargo, M., and Weinstein, B. A. (1994). *Phys. Rev. B* **50**, 4385.
Louie, S. G., Froyen, S., and Cohen, M. L. (1982). *Phys. Rev. B* **26**, 1738.
Mandel, G. (1964). *Phys. Rev.* **134**, A1073.
Marshall, T. (1993). *Physica B* **185**, 433.
Mensz, P. M. (1994). *J. Cryst. Growth* **138**, 697.
Miyajima, T., Okuyama, H., Akimoto, K., Mori, Y., Wei, L., and Tanigawa, S. (1991). *Appl. Phys. Lett.* **59**, 1482.
Mooney, P. M. (1992). In *Deep Centers in Semiconductors*, 2nd ed., ed. S. T. Pantelides. Gordon and Breach, Philadelphia. 643.
Murdin, B. N., Cavenett, B. C., Pidgeon, C. R., Simpson, J., Hauksson, I. S., and Prior, K. A. (1993). *Appl. Phys. Lett.* **63**, 2411.
Neumark, G. F. (1980). *J. Appl. Phys.* **51**, 3383.
Neumark, G. F., and Catlow, C. R. A. (1984). *J. Phys. C* **17**, 6087.
Nishikawa, Y., Ishikawa, M., Saito, S., and Hatakoshi, G. (1994). *Jpn. J. Appl. Phys.* **33**, L361.

Northrup, J. E., and Zhang, S. B. (1993). *Phys. Rev. B* **47**, 6791.
Okajima, M., Kawachi, M., Sato, T., Hirahara, K., Kamata, A., and Beppu, T. (1986). In *Proceedings of the 18th International Conference on Solid State Devices and Materials*, 647.
Overhof, H., Scheffler, M., and Weinert, C. M. (1989). *Mat. Sci. Forum* **38–41**, 293.
Pankove, J. I., and Johnson, N. M., editors (1991). *Hydrogen in Semiconductors; Semiconductors and Semimetals*, Vol. 34, treatise editors R. K. Willardson and A. C. Beer. Academic Press, Boston.
Park, R. M., Troffer, M. B., Rouleau, C. M., DePuydt, J. M., and Haase, M. A. (1990). *Appl. Phys. Lett.* **57**, 2127.
Park, C. H., and Chadi, D. J. (1995) *Phys. Rev. Lett.* **75**, 1134.
Petruzzello, J., Gaines, J., van der Sluis, P., Olego, D., and Ponzoni, C. (1993). *Appl. Phys. Lett.* **62**, 1496.
Qian, G.-X., Martin, R. M., and Chadi, D. J. (1988). *Phys. Rev. B* **38**, 7649.
Qiu, J., DePuydt, J. M., Cheng, H., and Haase, M. A. (1991). *Appl. Phys. Lett.* **59**, 2992.
Ray, A. K., and Kröger, F. A. (1978). *J. Electrochem. Soc.* **125**, 1348.
Reinberg, A. R., Holton, W. C., de Wit, M. and Watts, R. K. (1971). *Phys. Rev. B* **3**, 410.
Ren, S. Y., Dow, J. D., and Shen, J. (1988). *Phys. Rev. B* **38**, 10 677.
Rong, F., and Watkins, G. D. (1987). *Phys. Rev. Lett.* **58**, 1486.
Sasaki, T., Oguchi, T., and Katayama-Yoshida, H. (1991). *Phys. Rev. B* **43**, 9362.
Shahzad, K., Khan, B. A., Olego, D. J., and Cammack, D. A. (1990). *Phys. Rev. B* **42**, 11 240.
Shahzad, K., Jones, K. S., Lowen, P. D., and Park, R. M. (1991). *Phys. Rev. B* **43**, 9247.
Strachan, D. J., Li, M. M., Ritter, T. M., Tamargo, M., and Weinstein, B. A. (1994). *J. Cryst. Growth* **138**, 318.
Van de Walle, C. G., and Laks, D. B. (1990). In *Proceedings of the 20th International Conference on the Physics of Semiconductors*, ed. E. Anastassakis and J. D. Joannopoulos. World Scientific Publishing, Singapore. 722.
Van de Walle, C. G., and Blöchl, P. E. (1993). *Phys. Rev. B* **47**, 4244.
Van de Walle, C. G., Laks, D. B., Neumark, G. F., and Pantelides, S. T. (1993). *Phys. Rev. B* **47**, 9425.
Van de Walle, C. G., and Laks, D. B. (1995). *Solid State Communications* **93**, 447.
Van de Walle, C. G., and Neugebauer, J. (1995). In *Defect and Impurity Engineered Semiconductors and Devices*, ed. S. Ashok, I. Akasaki, J. Chevallier, and N. M. Johnson. Materials Research Society Symposia Proceedings, Vol. 378, Materials Research Society, Pittsburgh, 467.
Watkins, G. D. (1990). *Defect Control in Semiconductors*, ed. K. Sumino. Elsevier Science Publishers B. V., Amsterdam, 933.
Watts, R. K., Holton, W. C., and de Wit, M. (1971). *Phys. Rev. B* **3**, 404.
Watts, R. K. (1977). *Point Defects in Crystals*. Wiley, New York. 252.
Wei, S.-H., and Zunger, A. (1988). *Phys. Rev. B* **37**, 8958.
Wolk, J. A., Ager, J. W., III, Duxstad, K. J., Haller, E. E., Taskar, N. R., Dorman, D. R., and Olego, D. J. (1993). *Appl. Phys. Lett.* **63**, 2756.
Wyckoff, R. W. G. (1964). *Crystal Structures*, 2nd ed., Vol. 2. Interscience Publishers, New York.
Yao, T., and Okada, Y. (1986). *Jpn. J. Appl. Phys.* **25**, 821.
Zhang, S. B., and Northrup, J. E. (1991). *Phys. Rev. Lett.* **67**, 2339.
Zhu, Z., Mori, H., Kawashima, M., and Yao, T. (1992). *J. Cryst. Growth* **117**, 400.

CHAPTER 5

Optical Properties of Excitons in ZnSe-based Quantum Well Heterostructures

Roberto Cingolani

DIPARTIMENTO DI SCIENZA DEI MATERIALI
UNIVERSITÀ DI LECCE
LECCE, ITALY

I. INTRODUCTION 163
II. MODELING OF EXCITONIC STATES IN II–VI QUANTUM WELLS 164
III. LINEAR OPTICAL PROPERTIES OF QUASI-TWO-DIMENSIONAL EXCITONS 169
   1. Optical Absorption 169
   2. Excitons at the Dimensionality Cross-over 185
   3. Phototransport Processes 189
   4. Temporal Evolution of the Excitonic Transitions 195
IV. NONLINEAR EXCITONIC PROPERTIES 202
   1. Basic Theoretical Concepts 202
   2. Excitons and the One-component Electron Plasma 205
   3. Excitons and the Electron-hole Plasma 211
V. ROLE OF EXCITONS IN THE LASING OF ZnSe-BASED QUANTUM WELLS . . . 215
VI. CONCLUSIONS 223
   REFERENCES 223

I. Introduction

Excitons in ZnSe-based quantum well heterostructures exhibit strong stability as compared to bulk semiconductors or III–V quantum wells, due to the enhancement of the binding energy and the reduction of the exciton–phonon coupling caused by quantum confinement. Due to these effects excitons are expected to play an important role in many-body processes such as lasing and nonlinear absorption of II–VI quantum wells even at room temperature. For example, excitonic gain has been demonstrated in $Zn_{1-x}Cd_xSe/ZnSe$ quantum wells of thickness and composition such that the exciton binding energy was larger than the LO phonon energy ($E_b > \hbar\omega_{LO}$) (Ding et al., 1992). A detailed study of excitons in ZnSe multiple quantum wells (MQWs) is thus important to understand the optical properties of these wide gap heterostructures, also in view of their application to blue-green optoelectronic devices. In this chapter we overview the most recent experimental and theoretical work about the optical properties of excitons in ZnSe-based quantum wells. In Section II we briefly summarize the basic theoretical concepts used to model quasi-two-dimensional excitons in quantum wells. In Section III, we discuss the linear optical properties of excitons, including the electrostatic and thermal stability, the strength of the quantum size effect in quantum wells and thin films, the phototransport properties and the temporal evolution of the excitonic transitions. In Section IV we treat the nonlinear optical properties of excitons. After a short theoretical introduction, we discuss the interaction of excitons with a single-carrier and a two-carrier plasma. Finally, we discuss the role of excitons in the lasing processes of II–VI quantum wells in Section V. Our conclusions are drawn in Section VI.

II. Modeling of Excitonic States in II–VI Quantum Wells

We concentrate on II–VI quantum wells consisting of binary compounds ZnSe, CdSe, ZnS and their solid solutions, ZnCdSe, ZnSSe. The combination of these materials spans the whole green-blue spectral range (Fig. 1), though with the complication of lattice mismatch which strongly affects the electronic states and the valence band offset.

The shift of the conduction (c), heavy-hole (HH), and light-hole (LH) band edges due to the hydrostatic and uniaxial deformation of the strained semiconductor is given by (Pollak and Cardona, 1968):

$$\delta E_c = 2 \cdot \alpha_c \cdot \left(\frac{C_{11} - C_{12}}{C_{11}}\right) \cdot \sigma$$

$$\delta E_{HH} = \left(-2 \cdot \alpha_v \cdot \left(\frac{C_{11} - C_{12}}{C_{11}}\right) + b \cdot \left(\frac{C_{11} + 2C_{12}}{C_{11}}\right)\right) \cdot \sigma$$

$$\delta E_{LH} = \left(-2 \cdot \alpha_v \cdot \left(\frac{C_{11} - C_{12}}{C_{11}}\right) - \frac{1}{2} b \cdot \left(\frac{C_{11} + 2C_{12}}{C_{11}}\right)\right) \cdot \sigma$$

$$-\frac{\Delta}{2} + \frac{1}{2} \cdot \sqrt{\Delta^2 + \frac{9}{4}\left(2b \cdot \left(\frac{C_{11} + 2C_{12}}{C_{11}}\right) \cdot \sigma\right)^2 - 2\left(b \cdot \left(\frac{C_{11} + 2C_{12}}{C_{11}}\right) \cdot \sigma\right) \cdot \Delta}$$

(1)

where α_c and α_v are the conduction and valence band hydrostatic deformation potentials, respectively, C_{ij} are the elastic constants, σ is the strain, b is

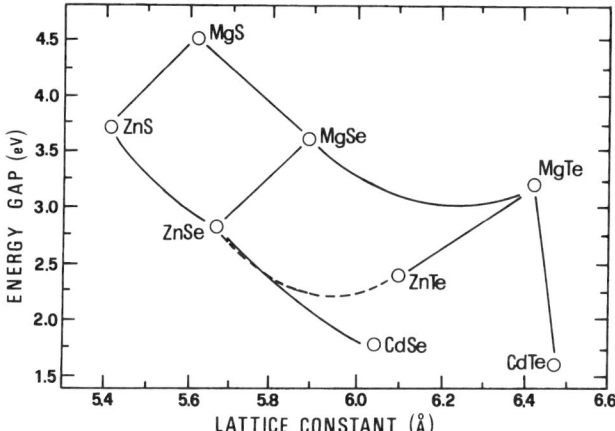

FIG. 1. Energy gap at 300 K versus lattice constant for a few relevant II–VI semiconductors.

the uniaxial deformation potential, and Δ the spin orbit splitting. Actually, the way in which the total deformation potential α is shared between the conduction and valence bands ($\alpha = \alpha_c + \alpha_v$) is unknown. This is usually determined empirically, by comparing experimental data and calculations assuming different band-offsets and α_c/α_v ratio.

The calculation of the confinement energies and excitonic states requires a detailed knowledge of the band structure parameters. Those used in this chapter are summarized in Table I. For the solid solutions, the effective masses and dielectric constants are scaled linearly with the stoichiometric fraction, whereas the band gap is usually scaled taking into account some bowing (Tamargo et al., 1991). The electronic states are evaluated following the usual envelope function model, in the effective mass approximation (Bastard, 1991). The main problem is the conduction to valence band offset ratio assumed in the calculations. There is presently some spread of data in the literature concerning the ZnSe/CdSe material system, whereas the negligible conduction band offset of the ZnS/ZnSe system is well documented. The available data and their differences will be commented in Section III.1.

The exciton binding energy is a very important parameter to be evaluated in II–VI compounds. The enhancement of the excitonic stability due to quantum confinement and reduced phonon coupling is one of the most interesting properties of II–VI quantum wells for optoelectronics. What is important for technology is the correct description of the well width and composition dependence of the exciton binding energy. This in turn depends

TABLE I
MAIN STRUCTURAL AND ELECTRONIC PARAMETERS OF BINARY II-VI COMPOUNDS

	ZnSe	CdSe	ZnS
$E_g(eV)$	2.821^a	1.765^a	3.840^b
$m_e(m_0)$	0.16^a	0.13^a	0.34^c
$m_{hh}(m_0)$	0.6^a	0.45^a	$1.76^d/0.61^e$
$m_{lh}(m_0)$	0.145^b	—	—
ε	8.8^b	9.3^b	8.32^b
$a(\text{Å})$	5.6676^a	6.077^a	5.4060^b
$C_{11}(N/m^2)$	$8.26 \cdot 10^{10a}$	$6.67 \cdot 10^{10a}$	$10.40 \cdot 10^{10f}$
$C_{12}(N/m^2)$	$4.98 \cdot 10^{10a}$	$4.63 \cdot 10^{10a}$	$6.50 \cdot 10^{10f}$
$\alpha(eV)^g$	—	—	-4.53^h
$b(eV)$	-1.2	0.8	-1.25^i
$\Delta(eV)$	0.43^b	—	0.070^b

The conduction to valence band offsets adopted in the calculations are 80:20 for the ZnCdSe/ZnSe material system, and about 5:95 for the ZnSe/ZnSSe heterostructures, including the effect of strain.

For the ternary alloys, the effective masses and dielectric constant are evaluated by linear interpolation from the bulk values of the corresponding binary compounds. The energy gap is also linearly scaled unless otherwise stated.

^{a}Lozykowsky and Shastri, 1991.
^{b}Landolt and Bornstein, 1982.
^{c}Yokogawa et al., 1994.
^{d}Lawaetz, 1971.
^{e}Lippens and Lanoo, 1989.
^{f}Singh and Singh, 1987.
$^g\alpha = -\dfrac{C_{11}+2C_{12}}{3} \cdot \dfrac{\partial E_g}{\partial P}$, where the $\dfrac{\partial E_g}{\partial P}$ was taken after Thomas et al., 1992.
^{h}Ves et al., 1990.
^{i}Shahzad et al., 1988.

on the offset which is affected by the strain. The exciton binding energy is usually evaluated by a variational calculation. In the effective mass approximation, the in-plane motion of the electron and hole is transformed into a center-of-mass motion of the exciton and in the relative motion of the constituent carriers. Neglecting the kinetic energy term of the in-plane motion, the exciton Hamiltonian can be written as:

$$H = \sum_{i=e,h}\left[-\frac{\hbar^2}{2m_i}\frac{\partial^2}{\partial z_i^2} + V_i(z_i)\right] - \frac{\hbar^2}{2\mu}\left(\frac{\partial^2}{\partial x^2} + \frac{\partial^2}{\partial y^2}\right) - \frac{e^2}{\varepsilon\sqrt{\rho^2 + (z_e - z_h)^2}} \quad (2)$$

where m_e, m_h are the effective masses of the electron and hole, μ is the reduced exciton mass, z_e and z_h are the coordinates perpendicular to the plane of the layers for the electron and hole, ρ is the in-plane relative position given by $\rho = \sqrt{x^2 + y^2}$, x and y are the relative coordinates of electrons and holes, ε is the dielectric constant, $V_e(z_e)$ and $V_h(z_h)$ are the actual confinement potentials for electrons and holes, respectively. The lowest exciton energy is calculated variationally with a simple trial function

$$\Psi(\rho, z_e, z_h) = \psi_e(z_e)\psi_h(z_h)\phi_{e-h}(\rho) \tag{3}$$

where $\psi_e(z_e)$ and $\psi_h(z_h)$ are the exact electron and hole wavefunctions in the finite quantum well, respectively. ϕ is the wavefunction of the in-plane radial motion, given by a 1S-orbital

$$\phi_{e-h}(\rho) = \left[\frac{2}{\pi\lambda^2}\right]^{1/2} \exp\left[-\frac{\rho}{\lambda}\right] \tag{4}$$

where λ is the trial parameter representing the radius of the exciton orbit. This separable trial function is strictly correct only for narrow well structures (Bastard *et al.*, 1982; Miller *et al.*, 1985), however it can be used to obtain an estimate of the exciton binding energy also for wider wells, its limiting value being the bulk exciton Rydberg.

The two-dimensional exciton radius is determined by maximizing variationally the exciton binding energy

$$E_b(\lambda) = -\frac{\hbar^2}{2\mu\lambda^2} + \frac{e^2}{4\pi\varepsilon}\langle\Psi|1/\rho|\Psi\rangle \tag{5}$$

The calculated exciton binding energy strongly depends on the degree of confinement of the electronic wavefunctions. Type I quantum wells like ZnCdSe/ZnSe exhibit binding energies ranging between the bulk value (about 20 meV) up to 40 meV or more, depending on the well depth and width (Fig. 2). Conversely, structures with vanishing potential offset, like the ZnSe/ZnS MQWs (very small conduction band offset) have bulk-like excitonic properties (Galbraith, 1992). In Section III.1 we will compare the calculated exciton binding energies with the available experimental data.

Another important figure of merit of the ZnSe-based MQWs is the considerable enhancement of the optical absorption strength. For light propagating perpendicular to the basal plane of the quantum well, this is given by (Sugawara, 1992)

$$\alpha \cdot L_x \simeq \frac{|P_{cv}|^2}{E_b\lambda^2}|\langle\psi_e(z_e)|\psi_h(z_h)\rangle|^2 \cdot L(\hbar\omega - E_b) \tag{6}$$

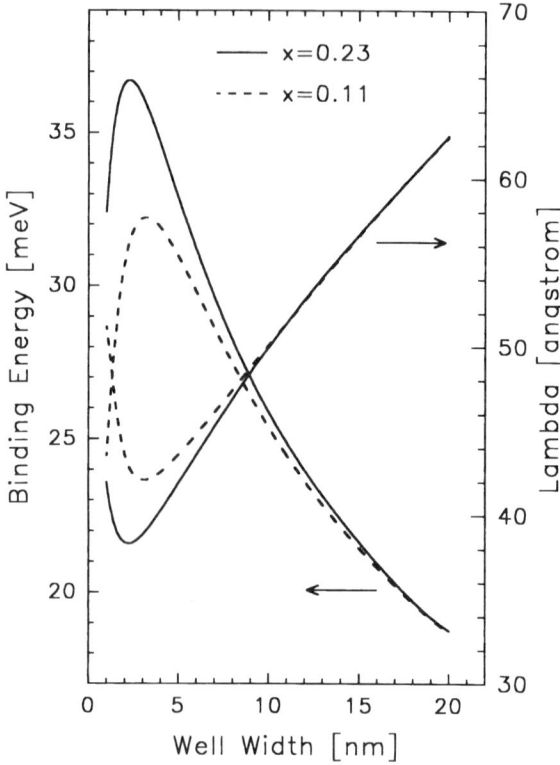

FIG. 2. Variational calculation of the exciton binding energy (left-hand scale) and transverse extent of the exciton wavefunction (λ on the right-hand scale) of $Zn_{(1-x)}Cd_xSe/ZnSe$ quantum wells of different composition.

where P is the momentum matrix element and L is a Lorentzian function modeling the exciton density of states. Calculations performed with available material parameters (Table I) clearly show that the integrated absorption strength is 2–3 times larger than the corresponding value of III–V (GaAs) quantum wells. This finding, together with the extremely large exciton binding energy, suggests that excitons play a very important role in the linear and nonlinear optical response of ZnSe-based quantum wells. It is therefore very important to have a precise description of the excitonic properties, in order to understand some of the relevant optoelectronic phenomena of interest for modern technology: namely, lasing, optical modulation, and nonlinear switching.

III. Linear Optical Properties of Quasi-two-dimensional Excitons

1. Optical Absorption

Representative low-temperature absorption spectra of selected $Zn_{1-x}Cd_x$-Se/ZnSe MQWs are shown in Figure 3. The samples investigated here were grown by molecular beam epitaxy (Bratina *et al.*, 1993) on GaAs substrates. The width and composition of the ZnCdSe quantum wells of these samples span the range from bulk-like (3D) behavior to quasi-two-dimensional (2D) limit. Changes in composition x cause variations in the depth of the quantum well and in the value of the elastic strain within the $Zn_{1-x}Cd_xSe$ layers, which are all expected to be pseudomorphically strained to the ZnSe buffer. The combination of these compositional and configurational par-

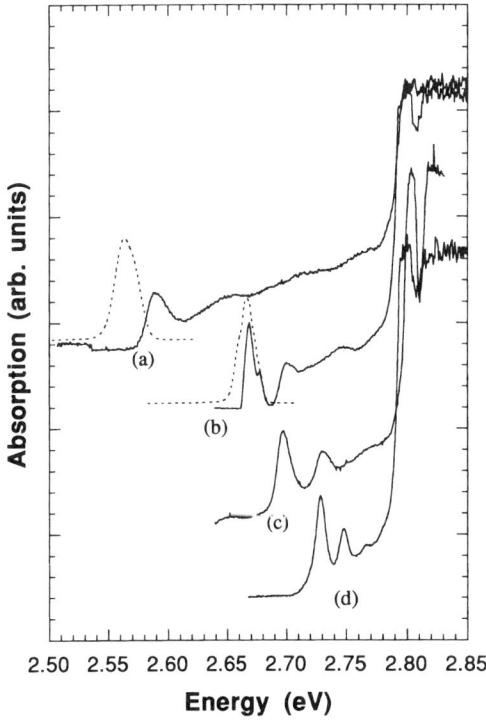

FIG. 3. Absorption (continuous lines) and photoluminescence (dashed lines) spectra of different $Zn_{(1-x)}Cd_xSe/ZnSe$ multiple quantum wells at 10 K. The well width and composition are: (a) $x = 0.23$, $L_w = 3$ nm, (b) $x = 0.11$, $L_w = 7$ nm, (c) $x = 0.16$, $L_w = 3$ nm, (d) $x = 0.11$, $L_w = 3$ nm.

ameters permits a fine tuning of the excitonic properties of the ZnCdSe MQWs, which can be investigated by systematic optical and structural studies.

For transmission measurements, the heterostructures are glued to a sapphire plate and the GaAs substrate is mechanically polished down to a thickness of about 80 µm. The samples are then masked with a photoresist leaving the substrate exposed through 100-µm-diameter windows, and the GaAs selectively removed by wet chemical etching using a $H_3PO_4:H_2O_2:H_2O$ solution at 30°C followed by a $H_2SO_4:H_2O_2:H_2O$ solution at 40°C. The strong excitonic feature around 2.8 eV is due to the ZnSe barrier and buffer layer absorption. At lower energies all samples exhibit distinct excitonic peaks superimposed to the QW continuum. The quantum size effect is clearly demonstrated by the blue-shift of the absorption spectrum with decreasing well width. Furthermore, samples with higher Cd content show absorption spectra extended toward the green range, reflecting the increased depth of the quantum well. Narrow wells ($L_W = 3$ nm) exhibit a single confined subband, as expected. The absorption spectra become more structured in the 20-nm-wide MQWs, where at least three electron subbands may be confined in the well. The photoluminescence spectra (dashed lines in Fig. 3) exhibit clear excitonic bands with typical Stokes shift ranging between 2 meV for wide wells with little Cd content, and 20 meV for the narrow wells with high Cd content. This indicates that the exciton localization at compositional or thickness fluctuations is more important in deep (Cd-rich) quantum wells. X-ray diffraction studies (Cingolani *et al.*, 1994a) show distinct satellite peaks due to the superlattice periodicity, with a relevant broadening of the peaks due to disorder. A typical rocking curve measured in the vicinity of the symmetric (002) direction is shown in Figure 4. The quantitative analysis of these x-ray patterns based on the dynamical diffraction theory (Tapfer and Ploog, 1986) provides the well width and composition in good agreement with the nominal parameters. In particular the stoichiometric Cd content is found to vary by ±1% around the growth value. A more detailed analysis of the satellite peaks line-shape, which is beyond the scope of this chapter, reveals some inhomogeneous strain fluctuation presumably caused by a smooth long-range modulation of the quantum well thickness having a somewhat "wavy" profile. Both these effects produce relevant potential modulation in the quntum well, causing carrier and/or exciton localization. This localization affects dramatically the optical properties of II–VI quantum wells. We should mention that a quantitative understanding of the microscopic disorder in II–VI quantum wells is still lacking and much work has to be done in order to clarify this important issue. So far there are a number of phenomenological approaches to the disorder-induced broadening of

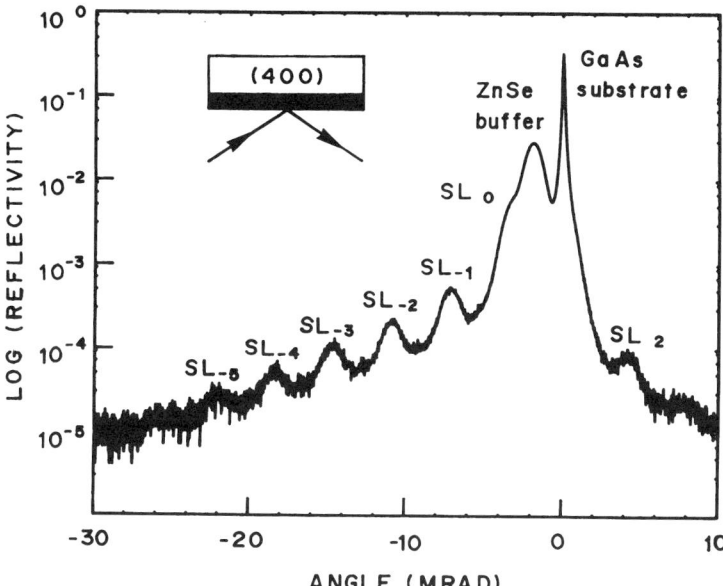

FIG. 4. Double crystal x-ray diffraction pattern of a $Zn_{0.9}Cd_{0.1}Se/ZnSe$ multiple quantum well structure recorded near the (400) reflection of the GaAs, using the $CuK\alpha_1$ radiation. The well width is 2.8 nm, the barrier width is 21.7 nm.

the optical spectra. These are based on a statistical distribution of disorder (either compositional or thickness fluctuation), which is assumed to cause a Gaussian broadening of the single particle states (Young et al., 1994; Baranowskii et al., 1993), usually referred to as inhomogeneous broadening. Furthermore, the formation of ternary alloy at the interface of thin binary/binary quantum wells has been experimentally investigated by Zhu et al. (1993) for the CdSe/ZnSe system and by Yao et al. (1991) for the ZnS/ZnSe system.

The temperature dependence of the absorption spectra provides information on the thermal stability of excitons in these quantum wells. In Figure 5a and b we show absorption spectra in the 10–300 K temperature range for MQWs with $x = 0.16$ and $L_W = 3$ nm, and $x = 0.11$ and $L_W = 20$ nm, respectively. As recently reported by various authors (Ding et al., 1990a; Ding et al., 1992; Ding et al., 1993; Pelekanos et al., 1992a; Cingolani et al., 1995; Liaci et al., 1995, Pellegrini et al., 1995) deep quantum wells (high Cd content) have strongly confined excitons, with large binding energies and clear exciton features observed up to room temperature (Fig. 5a). On the contrary, shallow quantum wells (lower Cd content) do not exhibit room

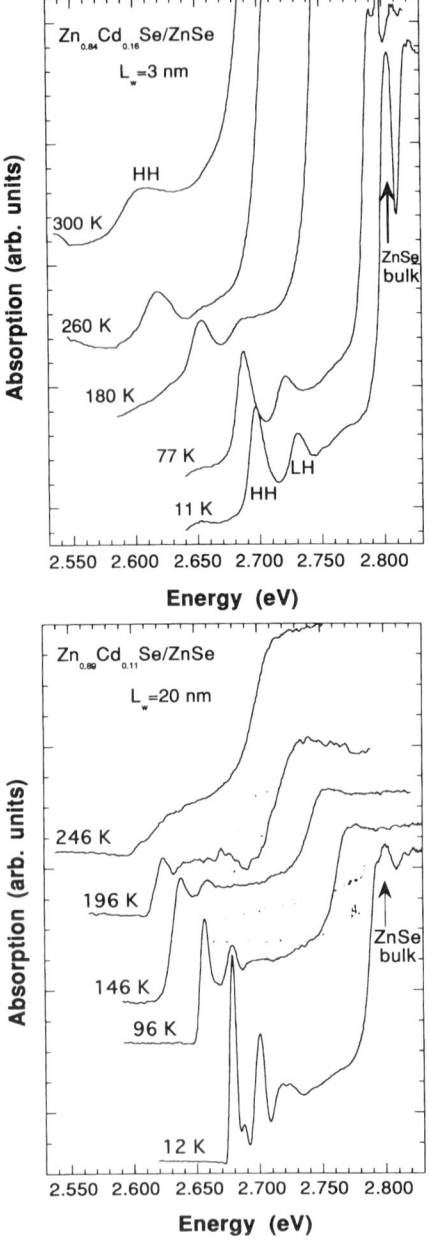

FIG. 5. (a) Temperature-dependent absorption spectra of a $Zn_{0.84}Cd_{0.11}Se/ZnSe$ multiple quantum well sample with $L_w = 3$ nm. (b) Temperature-dependent absorption spectra of a $Zn_{0.89}Cd_{0.11}Se/ZnSe$ multiple quantum well sample with $L_w = 20$ nm.

temperature excitonic absorption (Fig. 5b). The thermal stability of the exciton depends on the ratio of the exciton binding energy (E_b) to the LO phonon energy ($\hbar\omega_{LO} = 31.8$ meV in ZnSe), and on the actual strength of the exciton–phonon coupling. The latter parameter can be approximately estimated from the temperature-dependent absorption linewidth of the ground level HH exciton. In order to get information on this important parameter and, more in general, on the excitonic eigenstates and binding energy, one has to model the absorption line-shape with a statistical model. The absorption spectra are fitted to Gaussian line-shapes (Cingolani et al., 1995; Pellegrini et al., 1995), with linewidths mainly reflecting the inhomogeneous broadening of the states—i.e., statistical fluctuations in the ternary alloy composition, strain and layer thickness (Γ_{inh})—and the homogeneous term due to lifetime broadening caused by the exciton–phonon interaction. The full width at half maximum (FWHM) of the exciton can be written as:

$$\Gamma(T) = \Gamma_{inh} + \frac{\Gamma_{LO}}{[(\exp(\hbar\omega_{LO}/k_B T) - 1)]} \quad (7)$$

The homogeneous term, proportional to the LO phonon population, becomes important at high temperature and is weighted by the electron–phonon coupling constant Γ_{LO}. The step-like continuum of the quantum well density of states is simulated by a step function, convoluted with the same inhomogeneous Gaussian broadening used to reproduce the exciton absorption resonance, and with the inclusion of the Coulomb enhancement factor at the band edge (Sommerfeld factor) (see for instance Chemla et al., 1984), which is found to play a minor role in these II–VI materials. The binding energy is obtained from the spectral separation between the exciton line and the continuum edge. This simple criterion is justified by the lack of any feature associated with the 2s state of the exciton in the absorption spectra, due to the unavoidable broadening induced by local inhomogeneities of the ternary alloy well.

An example of the result of this line-shape analysis is shown in Figure 6 for a $Zn_{0.87}Cd_{0.13}Se/ZnSe$ MQW with $L_W = 3$ nm. The overall calculated absorption spectrum (solid line) obtained from a best fit to the data (solid circles) reproduces fairly well the experimental spectrum. The individual contributions of the heavy-hole (HH) and light-hole (LH) excitonic features and the HH continuum are also shown (dashed lines).

The analysis of the temperature-dependent absorption spectra of ZnCdSe/ZnSe MQWs (Pelekanos et al., 1992a; Cingolani et al., 1995) by means of Eq (7) (neglecting the contribution of acoustic-phonon scattering [Lee et al.,

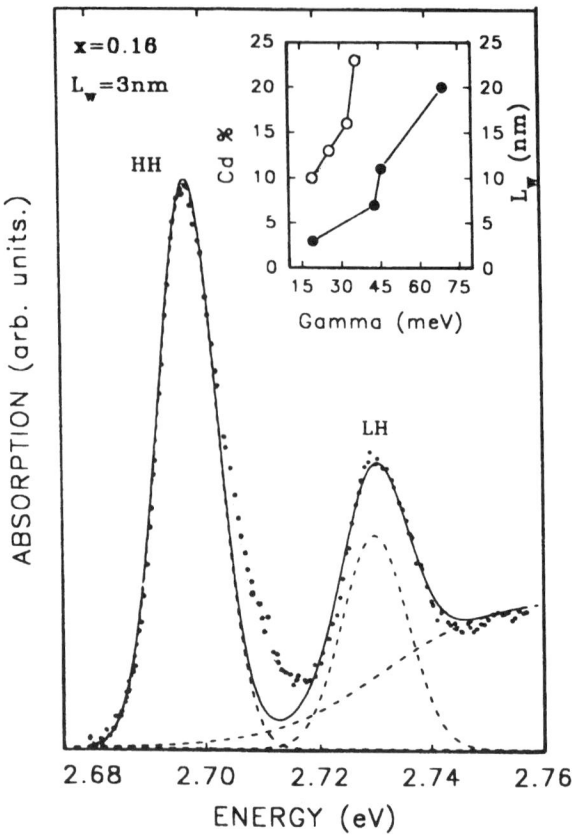

FIG. 6. Line-shape fitting of the absorption spectrum at 10 K of a $Zn_{0.84}Cd_{0.16}Se/ZnSe$ multiple quantum well with $L_w = 3$ nm. The overall best fit (continuous line) to the experimental data (solid circles) is shown together with the individual heavy-hole (HH) and light-hole (LH) exciton resonances and HH continuum (dashed lines). Inset: exciton–phonon coupling constant (Γ_{LO}) versus the well width L_w^e at $x = 0.11$ (solid circles, right-hand scale) and versus Cd concentration (empty circles, left-hand scale), as obtained from the temperature-dependent exciton linewidth.

1986]) provides the Γ_{LO} values displayed in the inset of Figure 6. The exciton–phonon coupling in narrow QWs (3 nm) is found to be smaller than in bulk ZnSe, whereas for wide wells (20 nm) it is larger than in the bulk, in agreement with the theoretical work of Young et al. (1994). The important implication of this result is that the reduced exciton–phonon coupling in narrow wells favors the exciton stability, leading to a dominant excitonic role in the optical processes of II–VI quantum wells under strong injection or high temperatures. We also observe in the inset of Figure 6 that, for a given well width (3 nm), the exciton–phonon coupling appears to increase with increasing Cd content. Apparently, the increased exciton confinement occurring in deep $Zn_{1-x}Cd_xSe/ZnSe$ QWs is partly balanced by the increased exciton–phonon coupling occurring in the Cd-rich alloys. This result is in contrast with the expected reduction of the exciton–phonon coupling occurring in CdSe with respect to ZnSe (Rudin et al., 1990). A totally different situation is found for the ZnSe/ZnS system, in which the weak confinement does not affect strongly the phonon coupling. Strong exciton–phonon interaction is indeed observed in the optical spectra, as described at the end of this subsection.

In wide wells with large Cd content ($x > 0.2$) the broadening becomes comparable to the subband splitting, and the interband continuum can hardly be resolved in the experimental spectra. For example, in the investigated composition range ($0.1 < x < 0.3$) a $\pm 1\%$ fluctuation in the Cd content of the alloy results in a fluctuation of ± 11 meV in the energy gap (Tamargo et al., 1991). The corresponding unintentional modulation of the quantum well depth would cause per se a fluctuation of 1–3 meV in the quantization energies of the carriers in the well.

The experimental exciton binding energies are summarized in Figure 7 for the different MQWs (solid circles and triangles for $x = 0.11$ and $x = 0.23$ QWs, respectively), together with the results of the variational calculations described in Section II (solid lines). For shallow MQWs, the calculated exciton binding energy varies from the bulk value (about 19 meV for $L_W = 20$ nm) up to 32 meV for $L_W = 3$ nm. For even lower well widths the calculated exciton binding energy decreases, due to the increased penetration of the exciton wavefunction in the barriers. Deeper quantum wells ($x = 0.23$) exhibit a similar behavior, but the maximum exciton binding energy is increased to about 37 meV for $L_W = 2$ nm. Similar values are reported by other authors (Pelekanos et al., 1992b; Liaci et al., 1995; Pellegrini et al., 1995). Some discrepancy is found in the wide well region ($L_W = 20$ nm), and is likely to reflect the limited accuracy of the trial wavefunction Eq. (4) when the 3D exciton limit is approached (Miller et al., 1985; Bastard, 1991; Liaci et al., 1995).

These results clearly show that exciton confinement in $Zn_{1-x}Cd_xSe/ZnSe$ MQWs can be tuned independently by varying the well composition or thickness. For a given well width of 3 nm, the increase of the well depth (i.e., the increase of the Cd content from typically $x = 0.1$ to $x = 0.25$) results in a 15% enhancement of the exciton binding energy, reflecting the enhanced localization of the carrier wavefunctions in the well. This is shown more clearly in Figure 8, where we plot calculated (solid and dashed lines) and experimental (solid circles) ground state exciton binding energies as a

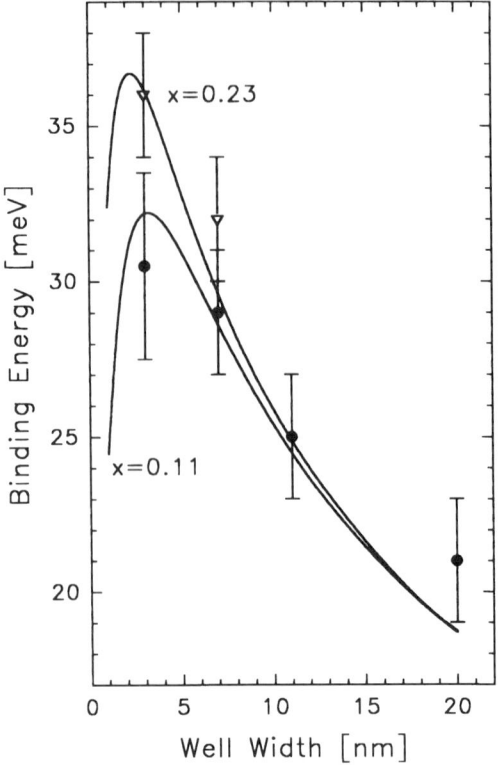

FIG. 7. Experimental exciton binding energy of $Zn_{(1-x)}Cd_xSe/ZnSe$ quantum wells with $x = 0.11$ (circles) and $x = 0.23$ (triangles) as a function of the well width. The curves are the results of variational calculations taking a linear interpolation of the dielectric constants $[\varepsilon = 8.8(1 - x) + 9.3x]$.

function of Cd content. The exciton binding energy is clearly seen to vary from just below to above the LO phonon energy in the well composition range examined.

The data of Figures 7 and 8 suggest that maximum exciton stability $(E_b > \hbar\omega_{LO})$ cannot be achieved in shallow quantum wells ($x \simeq 0.1$). Conversely, a strongly stabilized exciton exists in the deeper quantum wells ($x \geqslant 0.2$), at least for sufficiently narrow well widths, as shown, for example, by the presence of a clear excitonic resonance in the room temperature absorption spectrum of Figure 5a. Under the condition $E_b > \hbar\omega_{LO}$, excitons are thus expected to play an important role in real devices operating at high temperatures and strong injection rates, namely lasers and modulators.

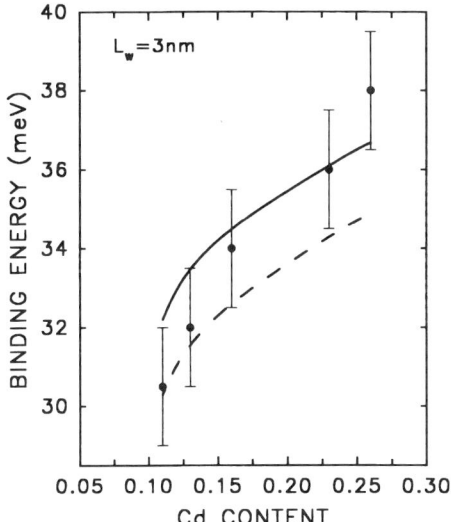

FIG. 8. Compositional dependence of the exciton binding energy of $Zn_{(1-x)}Cd_xSe/ZnSe$ with well width $L_w = 3$ nm. The curves represent the results of variational calculations taking a linear interpolation of the dielectric constants (continuous) or the CdSe bulk value $\varepsilon = 9.3$ (dashed).

We have now all of the information required to evaluate the interband transition energies of ZnCdSe/ZnSe MQWs. In Figure 9 we compare transition energies calculated in the envelope function approximation including strain (see Section II) with the experimental values obtained by adding the binding energy data in Figure 7 to the experimental energy position of HH and LH resonances (solid and open circles in Figure 9, respectively). The calculated HH interband transitions with quantum number $n = 1,2,3$ (solid lines) and LH (dashed lines) match within 10 meV the experimental values (solid and open circles), for both $Zn_{0.89}Cd_{0.11}Se/ZnSe$ (topmost section) and $Zn_{0.77}Cd_{0.23}Se/ZnSe$ MQWs. Such a difference is comparable to the experimental uncertainty expected for fluctuations of the order of $\pm 1\%$ in the well composition, and/or monolayer fluctuations in the width of narrow wells. Similar agreement is also found in Figure 10, where we plot the Cd dependence of the ground level HH and LH excitonic transition energies for a series of MQWs with $L_W = 3$ nm and well composition x spanning the 0.10–0.26 range. We should mention that most of the band parameters of ZnSe, zincblende CdSe, and related ternary alloy are not well known (Landolt and Bornstein, 1982). The remarkable overall agreement between calculated and measured transition energies supports the

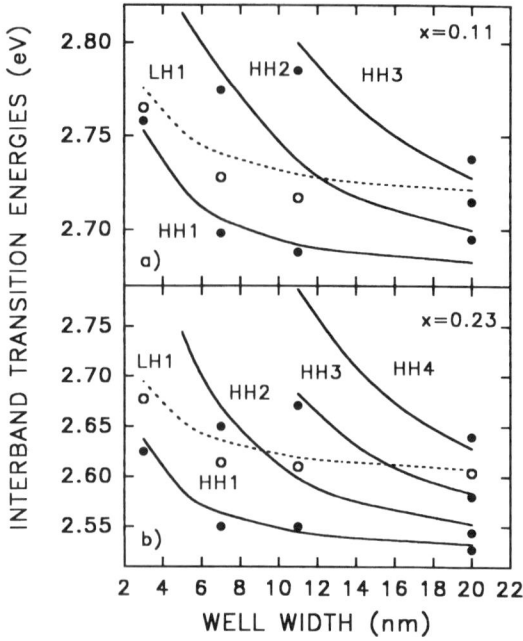

FIG. 9. Interband heavy-hole (HH, solid circles) and light-hole (LH, empty circles) transition energies of $Zn_{(1-x)}Cd_xSe/ZnSe$ quantum wells with $x = 0.11$ and $x = 0.23$ at 10 K. The curves represent the results of envelope function calculations in the effective mass approximation, including strain. The continuous curves denote HH transitions with quantum number $n = 1,2,3...$, whereas the dashed curves represent the $n = 1$ LH transition. The experimental values are obtained by adding the exciton binding energies of Figure 7 to the experimental energy position of the HH and LH excitonic resonances measured in absorption.

choice of band parameters (Table I) and exciton binding energies used here. Many theoretical and experimental studies of excitons in ZnCdSe/ZnSe MQWs have used quite different values for the conduction and valence band effective masses, deformation potentials, and band offsets (Ding et al., 1990b; Ding et al., 1992; Pelekanos et al., 1992b; Lozykowsky and Shastri, 1992; Alonso et al., 1992; Ding et al., 1993; Chung et al., 1993; Ren et al., 1994; Young et al., 1994; Cingolani et al., 1995; Pellegrini et al., 1995; Liaci et al., 1995) leading to slightly different results. The major problem in the eigenstate calculations is the interplay between the band offset expected for the bulk, unstrained materials and the effect of tetragonal strain field on the band extrema in determining the actual total band discontinuities. Recent data reported in the literature about ZnCdSe/ZnSe indicate that the conduction band offset is much larger than the valence band offset, and

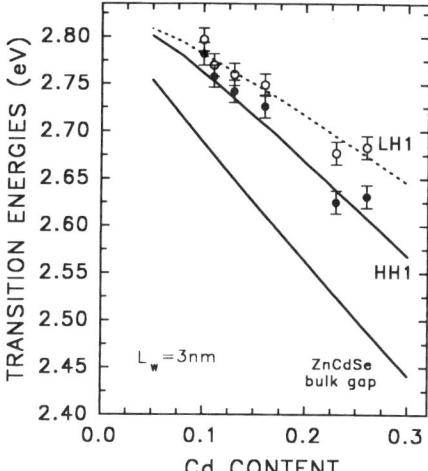

FIG. 10. Compositional dependence of the interband transition energies of $Zn_{(1-x)}Cd_xSe/$ZnSe quantum wells with $L_w = 3$ nm at 10 K. The solid and dashed lines correspond to the calculated HH and LH transition energy, respectively. A constant light-hole exciton binding energy of 20 meV has been assumed in the calculations. The energy gap of the unstrained bulk ternary alloy $Zn_{(1-x)}Cd_xSe$ is also plotted for comparison (bottommost solid line).

ranges between 0.65 and 0.85 of the band gap difference. The offset value is usually obtained by comparing the energy of the higher index transitions measured by absorption (Ding et al., 1990a; Cingolani et al., 1995; Pellegrini et al., 1995) or reflectivity (Alonso et al., 1992; Liaci et al., 1995) with envelope function calculations. The spread of these determinations thus depends on the choice of the effective masses, and structural parameters for the strain evaluation, and, in most cases, on the poor assessment of the well width and composition. On the other hand, the theoretical prediction of the offset (Chung et al., 1993; Ren et al., 1994) suffers the typical indetermination of band structure calculations which is of the order of several tens of meV.

Nevertheless, the different choice of band offset ratio and/or α_c and α_v (see Section II) only slightly affects the agreement between calculated and experimental transition energies for the higher index excitonic eigenstates of technologically interesting $Zn_{1-x}Cd_xSe$ quantum wells (with $x < 0.3$), due to large effective masses and limited potential depth. For the specific case of Figure 9, we performed several calculations for different values of the conduction band offset (in the 0.7–0.9ΔE_g range) and with $\alpha_c = 0$, (ii) $\alpha_c = 2/3\alpha$, and (iii) $\alpha_c = \alpha$. The combination of parameters that was found to better reproduce the experimental results was a conduction band offset equal to about $0.8\Delta E_g$ and $\alpha_c = 2/3\alpha$.

Another important issue is the evaluation of the binding energy of the light-hole exciton. In fact, in these heterostructures the combination of strain and the relatively low valence band offset leads to a vanishing confinement potential for the LH states, largely independent of well width and composition. In particular, the depth of the LH potential well varies between 3 meV at $x = 0.1$ and 15 meV at $x = 0.3$, causing the light-hole states to be weakly confined or resonant with the continuum. This explains why the LH excitonic features readily disappear from the absorption spectra with increasing temperature. In the calculations of Figure 9 we have used the model of Galbraith (1992) for quantum wells of vanishing band offset, to evaluate a binding energy of about 20 meV for the LH exciton in ZnCdSe/ZnSe MQWs, irrespective of the well width and composition. More accurate results for the light-hole exciton binding energy can be found in the work of Liaci et al. (1995).

A totally different scenario is found for the companion ZnSe/ZnS material system. In this case the relevant property is that the conduction band offset is about 10% of the total band gap discontinuity. This value is well documented either theoretically (Quiroga et al., 1990·; Bertho and Jouanin, 1993) or experimentally (Mohammed et al., 1987; Shahzad et al., 1988; Shen et al., 1992; Gil et al., 1994; Cingolani et al., 1994b) for ZnSe/ZnSSe heterostructures with different sulfur content. The conduction band offset is reduced to a few meV by the strain, resulting in an almost flat conduction band in heterostructures consisting of ternary ZnS_xSe_{1-x} barriers (with $x < 0.3$). The major consequences of this characteristic are (i) the formation of almost delocalized conduction-band states, (ii) a bulk-like exciton binding energy, and (iii) eventually the occurrence of type II band alignment in the presence of large strain or moderate external hydrostatic pressures (Yamada et al., 1991; Lomascolo et al., 1994; Gorczyca and Christensen, 1993).

The excitonic properties of these materials are strongly affected by the weak electronic confinement, and are primarily determined by the quantization of the valence band states. Other materials in which a similar situation occurs are the CdTe/ZnCdTe or ZnSe/ZnMgSe structures, having little or no valence band offset and strong electron confinement. Several authors have carried out calculations to model these situations. Wu and Nurmikko (1988) have used a modified band edge potential to confine the unconfined particle by the Coulomb force. Galbraith (1992) used a variational technique assuming an anisotropic exciton in which the electron is bound to the hole represented by a delta function. This technique, though giving a weak enhancement of the exciton binding energy (in agreement with the experiments), results in an unphysical independence of the binding energy on the well width. The use of the simple model outlined in Section II provides a more correct description of the exciton in these materials, in which the weak electron confinement is treated by using explicitly the heavy-hole and electron wavefunctions. The precise determination of the conduction band offset is crucial for the determination of the eigenstates and exciton binding energy. This has been achieved by photoluminescence (PL) experiments under high hydrostatic pressure. The application of hydrostatic pressure is expected to cause significant relative changes of conduction band offset in these superlattices. This is due to the difference in the band gap pressure coefficients of ZnSe (72 meV/GPa)(Hwang et al., 1994) and ZnS (64 meV/GPa). At some critical pressure the Γ point of ZnSe is expected to shift at higher energy than that of the ZnSSe alloy, so that electrons are confined in the barrier and holes remain in the well, resulting in a pressure-induced type I-type II transition. Evidence for this transition has been reported for ZnSe/ZnS (Yamada et al., 1991) and for ZnSe/ZnSSe superlattices (Lomascolo et al., 1994). Above the critical pressure the transition was monitored through (i) the strong decrease of the PL intensity, (ii) the appearance of different emission bands in the PL spectra, and (iii) by the change of the phonon sideband energies, reflecting the change from ZnSe LO phonons to ZnSSe LO phonons. These results are summarized in Figure 11 for a 2.8 nm/2.8 nm $ZnSe/ZnS_{0.18}Se_{0.82}$ superlattice measured at low temperature. The type I-type II transition occurs around 4.5 GPa, when the integrated emission intensity drops abruptly. At the critical pressure value the Γ points of the well and the barrier are aligned, corresponding to a conduction and offset of 6–7 meV. At the same pressure a change in the splitting of the PL phonon sidebands is observed, consistent with the change from the ZnSe LO phonon to the $ZnS_{0.18}Se_{0.82}$ LO phonon assisted recombination ($\hbar\omega_{LO}^{ZnSe} = 31.8$ meV and $\hbar\omega_{LO}^{ZnSSe} = 29.3$ meV, respectively).

The precise evaluation of the conduction band offset allows us to evaluate the small electron confinement energy and the exciton binding energy of the $ZnSe/ZnS_{0.18}Se_{0.82}$ superlattices. These values are compared to the experimental absorption and photoluminescence excitation data, in a way similar to the ZnCdSe/ZnSe case. In Figure 12 we show the transmission spectrum of a ZnSe/ZnSSe SL, consisting of 80 periods of 3.7 nm ZnSe/3.7 nm $ZnS_{0.8}Se_{0.82}$. The line-shape analysis reveals a distinct edge of the continuum, from which a binding energy of about 22 meV is estimated. The systematic analysis of symmetric superlattices with well widths comprised in the range 2 nm $< L_x <$ 15 nm, reveals the expected weak dependence of the exciton binding energy on the well width (Fig. 13). Unlike the case of the ZnCdSe MQWs, the exciton binding energy increases only by about 25% in the narrowest quantum wells, clearly showing the important role of the electron delocalization in heterostructures with negligible conduction band offset. Similar absorption experiments have been performed in ZnS/ZnSe

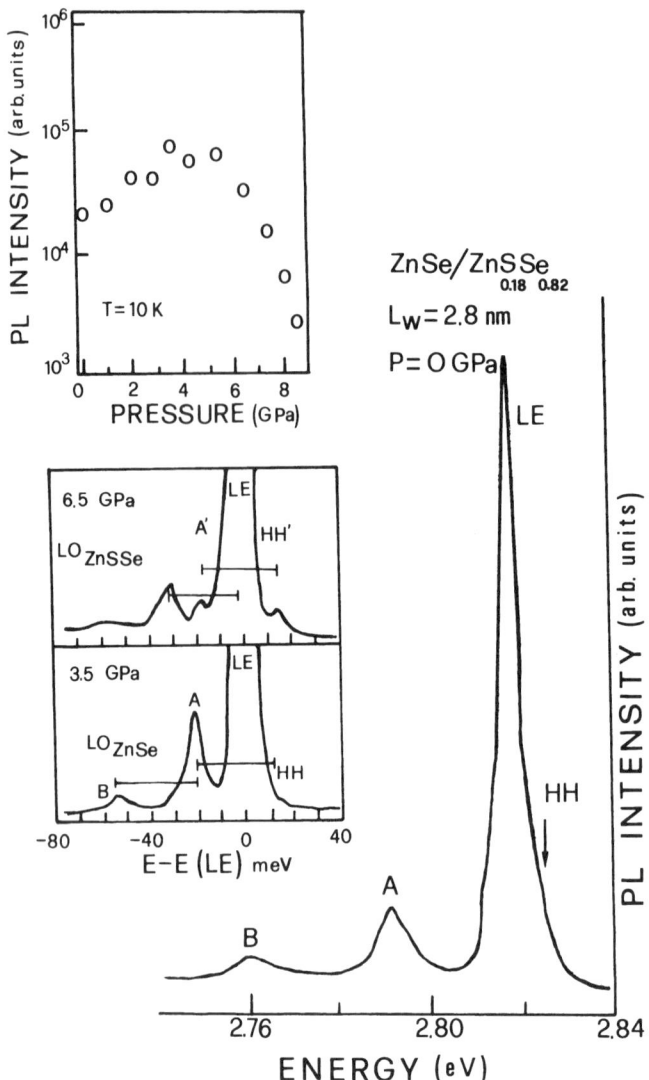

FIG. 11. Low-temperature continuous-wave luminescence spectrum recorded from a symmetric $ZnS_{0.18}Se_{0.82}/ZnSe$ superlattice of well and barrier width equal to 2.8 nm. LE is the emission of excitons localized at flat islands at the interfaces, A and B are the LO-phonon replicas of the heavy-hole exciton (HH). *Top inset:* LE emission intensity versus the hydrostatic pressure. The type I–type II transition is monitored through the drop in intensity occurring at 4.5 GPa. *Bottom inset:* Fine structure of the luminescence spectra recorded just below and above the pressure-induced type I–type II cross-over. Below threshold the phonon-assisted transitions (A,B) involve ZnSe LO phonons (31.8 meV). Above threshold, the phonon replicas involve ZnSSe LO phonons.

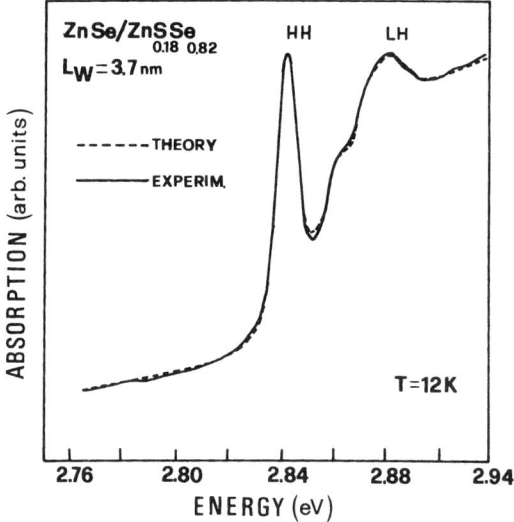

FIG. 12. Experimental and calculated absorption spectrum of a symmetric $ZnS_{0.18}Se_{0.82}/$ZnSe superlattice (well and barrier width equal to 3.7 nm) at 12 K.

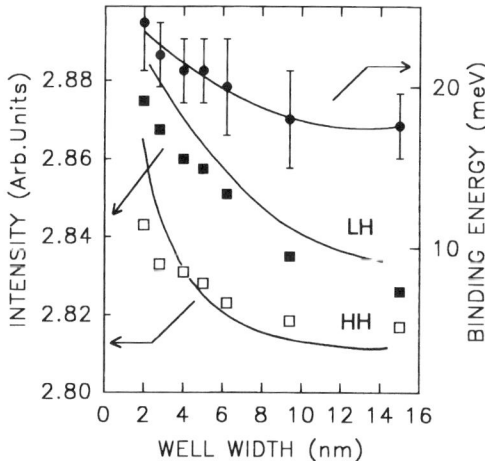

FIG. 13. Well width dependence of the exciton binding energy (circles, right-hand scale) and of the excitonic transition energies (squares, left-hand scale) at 10 K in symmetric $ZnS_{0.18}Se_{0.82}/$ZnSe superlattices. The curve interpolating the binding energy symbols is a guide for the eye. The curves reproducing the excitonic energies have been calculated according to the envelope function model, assuming a free-standing superlattice (total thickness about 1 μ).

(Shen et al., 1992; Hohnoki et al., 1994) and CdSe/ZnSe heterostructures (Shan et al., 1993; Yang et al., 1993).

The inclusion of the exciton binding energy, of the experimentally estimated conduction band offset, and of the strain corrections in a Krönig-Penney model, gives a satisfactory description of the well width dependence of the HH and LH energies in Figure 13 (Cingolani, 1994b). The total exciton energy increases by less than 60 meV in the investigated well width range, because the confinement is dominated by the hole quantization. The discrepancy between experiment and theory is affected by the lack of information on the effective masses and deformation potentials. Furthermore, the assumption of totally free-standing superlattices used in the strain calculations is ideal (the superlattices used in this study had a total thickness of about 1 μm), and might be not totally correct in real samples.

Before concluding this section we recall that no reduction of the electron-phonon coupling is expected in ZnSe/ZnSSe superlattices with respect to bulk, due to the weak confinement. This is reflected not only in the weak

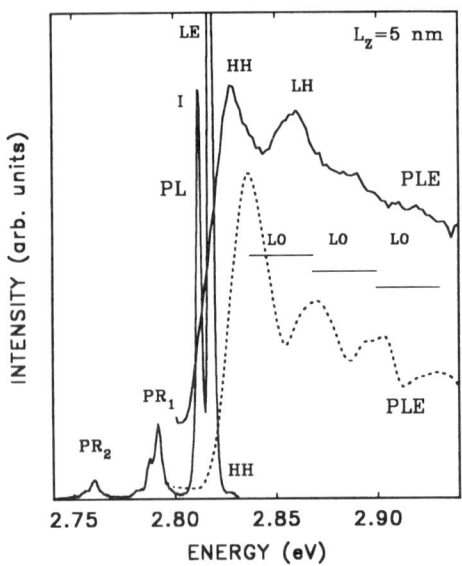

FIG. 14. Photoluminescence (PL) and photoluminescence excitation spectra (PLE) of a 5 nm/5 nm $ZnS_{0.18}Se_{0.82}$/ZnSe superlattice at 10 K. The PL spectrum shows sharp localized-state emission (LE,I) and phonon replicas (PR). The PLE spectrum recorded under low excitation power (dashed curve) exhibits distinct oscillation with periodicity of one LO phonon, characteristic of hot-exciton generation. The PLE measured under higher excitation power recovers the density of states line-shape with distinct heavy-hole (HH) and light-hole (LH) excitonic resonances.

enhancement of the exciton binding energy, but also in the appearance of strong hot-exciton features in the photoluminescence excitation (PLE) spectra. This situation is depicted in Figure 14, where we compare the PLE spectra of a 5 nm/5 nm ZnSe/ZnS$_{0.18}$Se$_{0.82}$ superlattice recorded under different excitation intensities. At very low power (of the order of few μW cm^{-2}), the formation of hot excitons can be observed in the PLE spectra, through the characteristic oscillatory behavior of period equal to the ZnSe LO-phonon energy (Cingolani et al., 1994b). This is due to the enhancement of the excitonic absorption in resonance with the phonon-assisted transition in the continuum. With increasing the excitation intensity (of the order of 1 mW cm^{-2}), intercarrier scattering becomes a dominant relaxation mechanism, and the usual density of states line-shape is recovered. In this case the PLE spectrum reproduces the usual absorption shape with clear excitonic resonances (continuous curve in Fig. 14).

2. EXCITONS AT THE DIMENSIONALITY CROSS-OVER

Excitons in a quantum well are treated assuming that the lateral barriers influence primarily the wavefunctions of the individual carriers rather than the exciton wavefunction itself, resulting in the so-called strong confinement regime. Under this condition one can reasonably assume that the eigenvalue of the Coulomb interaction (i.e., the exciton binding energy) is smaller or at most comparable to the quantization energy gained by the carriers. In III–V materials this condition is easily fulfilled, due to the large extent of the exciton envelope function (typically $a_0 > L_w$, where a_0 is the exciton Bohr radius and L_w is the well width), and to the light carrier masses (Cingolani and Ploog, 1991). On the contrary, in ZnSe-based quantum wells the exciton Bohr radius is rather small ($a_0 \leqslant 4$ nm) whereas the effective masses of carriers are quite large. The condition $a_0 \leqslant L_w$ is easily verified for most common well width values (Fig. 2). Such condition is referred to as the intermediate confinement regime, in which the lateral boundaries influence the exciton envelope function as a whole, rather than the individual carrier wavefunctions, resulting in the so-called quantization of the center of mass (CM) motion of the exciton. This situation is at the crossover between the three-dimensional and the two-dimensional behavior for the excitonic wavefunction.

From the theoretical point of view, we assume that the exciton has a center of mass motion and moves in real space due to its kinetic energy. Coupled with the radiation field it forms an exciton polariton (Hopfield and Thomas, 1963), whose spatial dispersion affects the optical spectra of absorption, photoluminescence, and reflectivity. The fundamental equation of the exciton–photon interaction, which gives the polariton dispersion curves $E(k)$, is obtained by equating the dielectric function (assuming Lorentzian excitonic resonances) and the photon dispersion law:

$$\varepsilon_\infty + \frac{4\pi\alpha_o E_o^2}{E_o^2 - E^2 + \beta k^2 + i\gamma E} = \frac{\hbar^2 c^2}{E^2} k^2 \tag{8}$$

where ε_∞ is the dielectric constant, $4\pi\alpha_o$ is the oscillator strength, E_o is the free exciton energy, $\beta = \hbar^2 E_o/M_{exc}$ with M_{exc} the effective exciton mass, γ is the exciton broadening, k is its wave vector. The momentum of the CM along the growth direction (k_z) is quantized because the polariton wave is strongly reflected at the well–barrier interfaces. Standing waves then exist for $L_w = n\lambda/2$, where n is a nonzero integer and λ is the polariton wavelength. Consequently, the continuous polariton curve $E(k)$ transforms into a series of discrete points at $k_n = n\left(\dfrac{\pi}{L_w}\right)$ with $n = 1,2,3\ldots$. This is true for well width values down to about two times the exciton Bohr radius a_0.

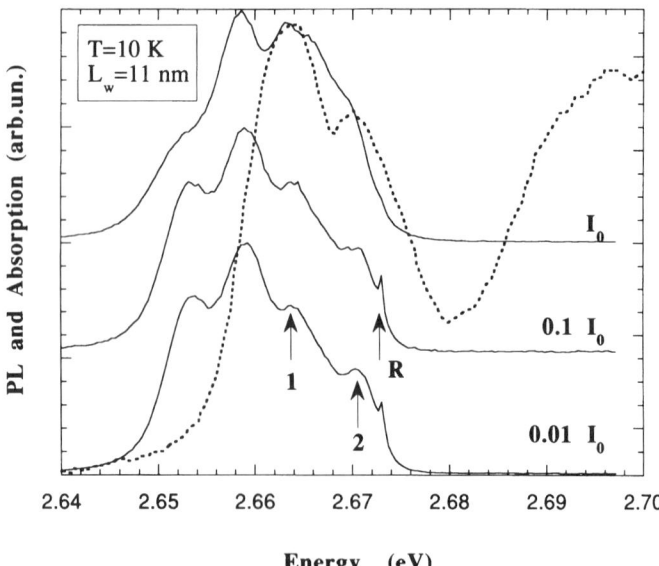

FIG. 15. Photoluminescence spectra at different excitation intensities (continuous lines) and absorption spectrum (dashed line) of a $Zn_{0.89}Cd_{0.11}Se/ZnSe$ MQWs with $L_w = 11$ nm, recorded at 10 K. $I_0 = 16$ W cm^{-2}. The arrows indicate the center of mass states. The peak labeled R is the Raman line of the exciting laser.

Under this condition, the energy dispersion of the exciton–polariton becomes

$$E_n^{exc} = E_o + \frac{\hbar^2}{2M_{exc}} \left(\frac{\pi}{L_w}\right)^2 n^2 \tag{9}$$

In Figure 15 we show the PL and absorption spectra of a $Zn_{0.89}Cd_{0.11}Se/ZnSe$ MQWs of well and barrier width 11 nm and 20 nm, respectively. In the PL spectrum we can distinguish up to four peaks, the lowest energy doublet being extrinsic as deduced from the saturated intensity independence. The comparison between luminescence and absorption spectra confirms the excitonic nature of the other features, which extend over about 10 meV on the high-energy tail of the excitonic resonance. This energy range is smaller than the LO phonon energy in the ternary alloy (about 30 meV). This prevents fast relaxation of the excited states into the lowest level, resulting in the simultaneous recombination of the excited CM states.

In Figure 16 we compare the absorption spectra of samples with different well widths $7\,nm < L_w < 20\,nm$ (Greco et al., 1996). This thickness range

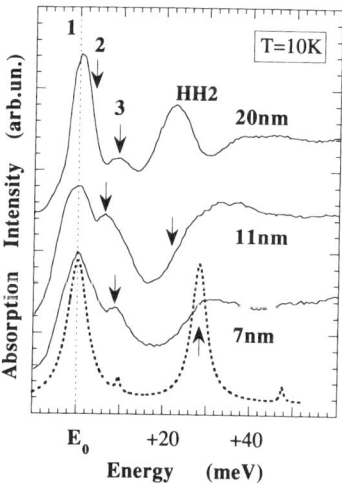

FIG. 16. Comparison of the absorption spectra of $Zn_{0.89}Cd_{0.11}Se/ZnSe$ MQWs with different well widths (continuous lines) at 10 K. The arrows indicate center of mass states. The dashed line is the absorption spectrum calculated according to the "thin-slab" variational model discussed in the text (for the 7 nm MQWs sample). The broadening and the continuum of the quantum wells have been neglected. HH2 labels the $n = 2$ exciton resonance of the 20-nm MQWs structure.

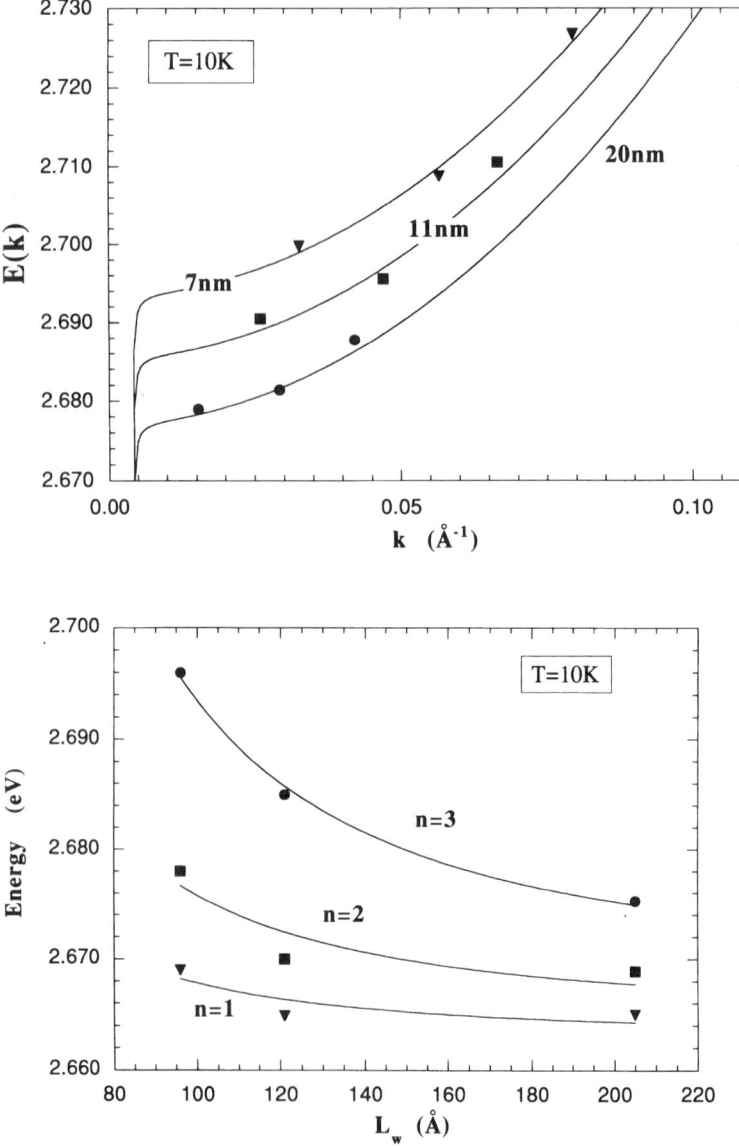

FIG. 17. (a) Energy-momentum dispersion curve of the exciton–polariton calculated by means of the variational model [Eq (10)] and quantized center of mass states (symbols) obtained from the 10 K absorption spectra of different ZnCdSe/ZnSe MQWs. (b) Well width dependence of the center of mass eigenstates (symbols) at 10 K compared with the energy calculated according to Eq. (9) by using the variational L_{eq}. The $n = 2$ and $n = 3$ curves have been suitably scaled to compare with the data.

corresponds to the condition $2.5\,a_o < L_w < 5\,a_o$, i.e., to the cross over between two- and three-dimensional crystal. As expected from the L_w dependence of the quantization condition (Eq. 9) we observe an increasing splitting of the CM states with decreasing the well thickness. The bulk-like dispersion [Eq (9)] does not reproduce the experimental data unless an unphysical mass enhancement is assumed for the narrower wells. In this case the true exciton–polariton dispersion has to be taken into account. This can be done by using the variational wavefunctions of D'Andrea and Del Sole (1992), valid in the thin-film regime. In this model the exciton is confined in an infinitely deep quantum well of equivalent width $L_{eq} = L_w + 2/P$, where $1/P$ represents the penetration of the exciton wavefunction in the barrier. $1/P$ and the exciton Bohr radius are varied in order to minimize the energy of the CM states. The CM eigenstates and the dispersion curve can be obtained after the solution of the transcendental equations (D'Andrea and Del Sole, 1990):

$$k_n \tan(k_n L_{eq}/2) + P \tanh(P L_{eq}/2) = 0$$
$$\frac{\tan(k_n L_{eq}/2)}{k_n} - \frac{\tanh(p L_{eq}/2)}{P} = 0 \qquad (10)$$

for even and odd states. Using the variational wavefunctions of D'Andrea and Del Sole (1990) and Eq. (10), we compute the normal incidence optical response of the structures to fit the experimental absorption spectra. The results of the calculations are shown by the dashed line in Figure 16. As we are interested only in the energy–momentum values of the CM states, the actual inhomogeneous broadening and the absorption continuum of the quantum well are neglected in the calculations.

The energy–momentum dispersion for the investigated well widths is plotted in Figure 17a together with the measured CM states. The K_n values are obtained after the variational calculation of the absorption spectra. The polariton curve flattens with decreasing well widths. The well width dependence of the CM states is plotted in Figure 17b (Greco et al., 1996). The experimental data are well reproduced by the theory, provided the geometrical well width L_w is replaced by the variational L_{eq} in Eq (9). These data demonstrate that the optical response of ZnCdSe/ZnSe MQWs around the cross-over between two- and three-dimensional crystal is dominated by the quantization of the center of mass motion of the exciton–polariton.

3. PHOTOTRANSPORT PROCESSES

In this section we discuss the phototransport processes in II–VI quantum wells. These effects are very important for the realization of optoelectronic devices, like lasers and nonlinear optical modulators, exploiting the enhanced excitonic properties of II–VI heterostructures. The control of doping for the fabrication of *p-i-n* heterostructures, and the deposition of high-quality ohmic contacts are presently a matter of intense investigation, and presumably much work has still to be done before reaching the technological standard of the III–V materials. These technological limitations prevent the precise determination of the built-in electric field of real devices. On this basis, the present achievements in the field of phototransport have to be considered as very preliminary. Different technological approaches have been pursued to realize II–VI quantum well diodes. In particular either Schottky-barrier diodes or *p-i-n* heterostructures have been fabricated, with ZnCdSe/ZnSe and ZnSe/ZnSSe quantum wells. The effect of tunneling and thermoionic emission at the metal/*p*-ZnSe interface on the current voltage characteristics of ZnSe/ZnSSe diodes has been investigated in detail by Suemune (Suemune, 1993). A comparative study of the optical and electrical properties of *p-i-n* diodes and Schottky-barrier diodes based on MOCVD grown ZnSe/ZnSSe samples has demonstrated a considerable improvement in the current-voltage characteristics of these diodes with respect to corresponding MBE-grown ZnCdSe/ZnSe superlattices (Fujii *et al.*, 1994; Suemune *et al.*, 1994). Furthermore, well-resolved excitonic resonances have been observed in the low-temperature (77 K) photocurrent spectra of ZnSe/ZnSSe superlattices (Suemune and Yamanishi, 1993).

Photocurrent spectroscopy has been successfully investigated in ZnCdSe/ZnSe quantum wells grown in *p-i-n* configuration (Wang *et al.*, 1993a; Cingolani *et al.*, 1994c; Cingolani *et al.*, 1994d). All these experiments demonstrate that in samples with proper configuration, the photocurrent spectra closely reproduce the actual absorption spectrum of the heterostructure. In Figure 18 we show the layer sequence of a *p-i-n* diode embedding ZnCdSe MQWs in the intrinsic region grown on a *p*-type GaAs substrate (Cingolani *et al.*, 1994c). The band profile clearly evidences the advantage of the present configuration. Electron–hole pairs photogenerated at the surface (*n*-type layer) are readily separated by the internal electric field. The minority carriers experience two favorable potential discontinuities, allowing efficient collection of holes at the bottom contact. Conversely, the majority carriers are retained in the photogeneration region by the double potential barrier. The narrow gap *p*-type layer, besides avoiding the complex problem of *p*-doping of the II–VI substrate, is thus exploited to enhance the minority carrier collection of the heterostructure, i.e., the strength of the electrical signal induced by the absorption of light.

In Figure 19 we show typical photocurrent (PC) spectra measured at different temperatures. The spectra exhibit sharp excitonic features and the distinct edge of the ZnSe barrier. At low temperature carrier freeze-out

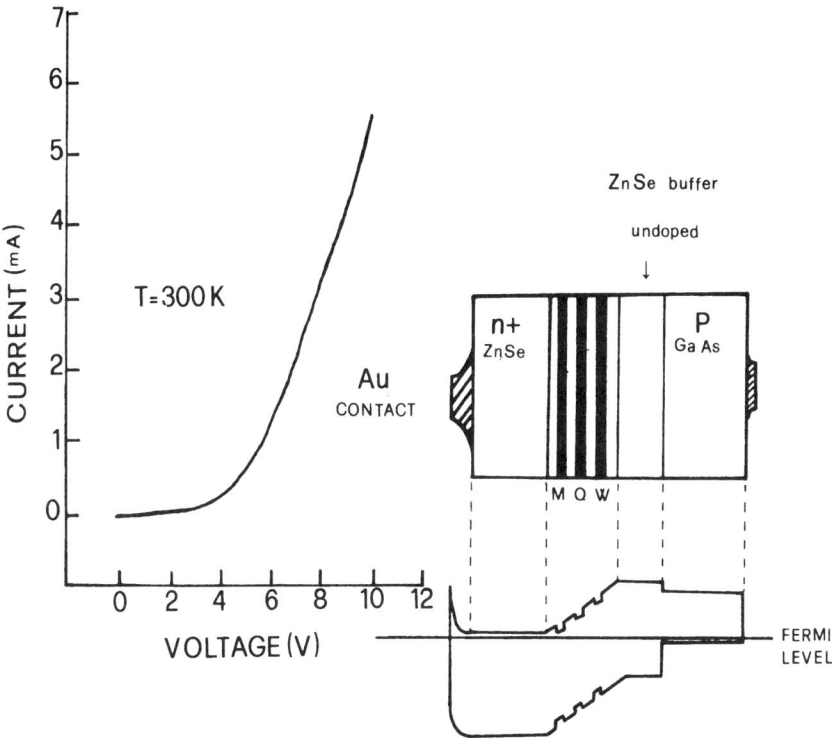

FIG. 18. *Right:* Schematics of the *p-i-n* heterostructure embedding $Zn_{(1-x)}Cd_xSe/ZnSe$ and corresponding band profile in real space, obtained from the self-consistent solution of the Schrödinger and Poisson equation. *Left:* Room temperature current-voltage characteristic of a structure containing 10 quantum wells with $x = 0.10$ and $L_w = 3$ nm.

occurs, and the photocurrent is rather low. With increasing temperature, carriers are thermally activated and the PC line-shape closely reproduces the absorption spectrum (dashed lines in Fig. 19). Following the discussion of Section III.1, the fundamental exciton resonance is seen to disappear at high temperatures in the shallower quantum wells ($\hbar\omega_{LO} > E_b$), whereas it remains stable up to room temperature in the deeper QWs.

For a quantitative treatment of the photocurrent process one has to write the continuity equation (Moss *et al.*, 1973)

$$\frac{dJ_z}{dz} = eI\alpha e^{-\alpha z} - \frac{\Delta p}{\tau} \qquad (11)$$

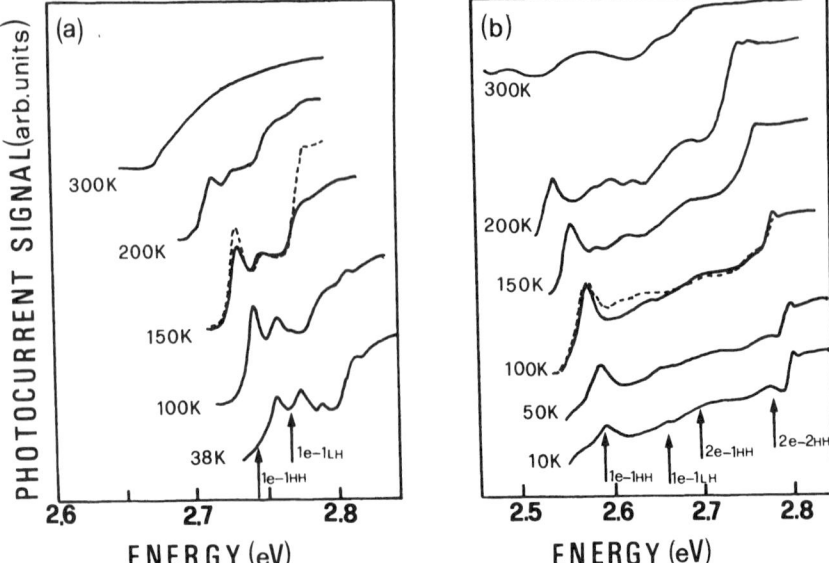

FIG. 19. (a) Temperature-dependent photocurrent spectra recorded from $Zn_{0.9}Cd_{0.1}Se/$ ZnSe multiple quantum wells embedded in p-i-n heterojunction. The well width is $L_w = 3$ nm. The dashed line represents the measured absorption spectrum. (b) Same as in Figure 19a, but for a $Zn_{0.75}Cd_{0.25}Se/ZnSe$ MQWs heterostructure.

and the transport equation

$$J_z = -eD\frac{d}{dz}(\Delta p) + e\mu\Delta pE \qquad (12)$$

Eq (11) consists of two terms: the photogeneration rate depending on the absorption coefficient α (I is the photon flux) and the nonradiative recombination losses τ (where Δp is the photogenerated excess of minority carriers). The latter includes the surface recombination $eS\Delta p$ (where S is the surface recombination velocity). This loss mechanism is particularly relevant in semiconductor heterostructures with many interfaces. Values as high as $S = 10^5$ cm sec^{-1} can be reached at the air/semiconductor interface, which reduce by about a factor 1000 or more at the semiconductor/semiconductor interface (Gowar, 1984). The total current density flowing across the heterostructure is given by Eq (12). The major contribution comes from the diffusion term (where D is the diffusion coefficient) and by the drift term, which is negligible in the absence of external polarization (μ is the carrier mobility and E is the electric field).

Eqs (11) and (12) contain the main ingredients for the understanding of the photocurrent spectra of Figure 19. When the energy of the exciting photons exceeds the barrier energy, the light is absorbed directly in the ZnSe top-layer ($=2.82\,\text{eV}$ at 10 K). The minority carriers are thus generated in a thin layer at the surface of the n-type ZnSe contact, from where they have to cross the intrinsic region and the p-type region before reaching the contact. Under these conditions there are two main carrier losses mechanisms: (i) the nonradiative recombination along the carrier path which is longer than the carrier diffusion length, and (ii) the surface recombination at the sample surface. The photogenerated excess of minority carriers (first term of Eq (11)) is partly balanced by the losses. The PC signal is thus not proportional to the absorption as the nonradiative recombination affects the internal carrier collection efficiency. Indeed the ZnSe region of the PC spectra, though showing a distinct band gap edge with excitonic resonance, does not reproduce quantitatively the strength of the absorption measured by transmission experiments (see the dashed lines in Fig. 19).

When the photon energy is in the transparency region of the ZnSe barrier, the carriers are generated directly in the quantum wells, with a homogeneous density due to the small total thickness of ZnCdSe. Here they diffuse across the ZnSe buffer layer toward the GaAs p-contact. Due to the limited distance from the p-contact (the total thickness of the MQW region is $L = 30\,\text{nm}$) and the lack of free surfaces, the collection efficiency is rather large, resulting in a PC signal proportional to the actual absorption of the quantum well. This means that the loss terms in Eq (11) give a minor contribution to the total photogenerated current. Indeed, in the limit of vanishingly small loss terms in Eq (11), one obtains

$$J_z = e\alpha I \int_0^L e^{-\alpha z} dz = eI(1 - e^{-\alpha z}) \simeq eI\alpha L \qquad (13)$$

Under this condition, the PC signal is directly proportional to the absorption coefficient. This finding is confirmed by the close similarity of the QW absorption and photocurrent spectra in Figure 19. The important role of the surface recombination is further evidenced when the free surface surrounding the metallic contact on top of the sample is chemically removed by deep chemical etching. Indeed, dramatic improvement in the PC response of the structures is obtained after fabricating mesa-structures of about 800 nm depth and area coincident with the metallic contact itself, as shown in Figure 20. The photocurrent signal increases by a factor 10, even without bias, and the excitonic features are strongly sharpened, especially in the bulk ZnSe region. In this case the PC spectrum really reproduces the absorption

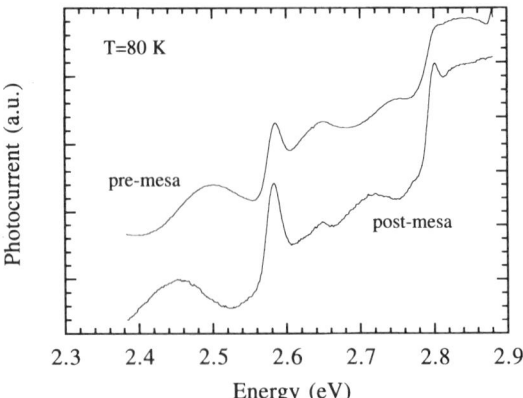

FIG. 20. Photocurrent spectra recorded before and after the fabrication of submillimetric mesas from a $Zn_{0.9}Cd_{0.1}Se/ZnSe$ multiple quantum well structure in *p-i-n* configuration.

spectrum. This is a direct consequence of the reduction of the surface recombination velocity in the illuminated region. This suggests that the surface recombination velocity is the dominant carrier loss mechanism reducing the internal quantum efficiency of these II–VI heterostructures.

Finally we analyze the PC spectra under external polarization. For the specific case of Figure 18, no shift or bleaching of the exciton resonance is observed under reverse bias, because of the modest internal field (of the order of $10^4 \cdot V \, cm^{-1}$) due to the large total thickness of the intrinsic area and to the series resistance effects at the nonohmic contacts. However, dramatic changes in the spectral line-shape of the photocurrent are observed for reverse bias around the breakdown field. These are shown in Figure 21 where we can see that the excitonic peaks observed at zero bias progressively transform into sharp dips with increasing bias. For polarization values corresponding to the breakdown value ($-4.5\,V$) the PC spectra become flat, then the sign reversal of the PC signal occurs for larger negative voltages. Similar effects have been observed in the PC spectra of III–V heterostructures under similar experimental conditions (Marshall et al., 1989; Leavitt and Bradshaw, 1991; Grahan et al., 1992; Yokuda et al., 1994). At low field the transit time of the carriers across the intrinsic region containing thick barriers can be longer than the carrier lifetime, resulting in a poor carrier collection at the bottom contact. On the contrary, at higher fields or in the presence of thin barriers, the transit time and the lifetime of the carriers can be comparable. These effects manifest themselves with a sudden change of the PC line-shape, as the one observed in Figure 21. It is worth noting that the sign reversal of the PC resonances disappears

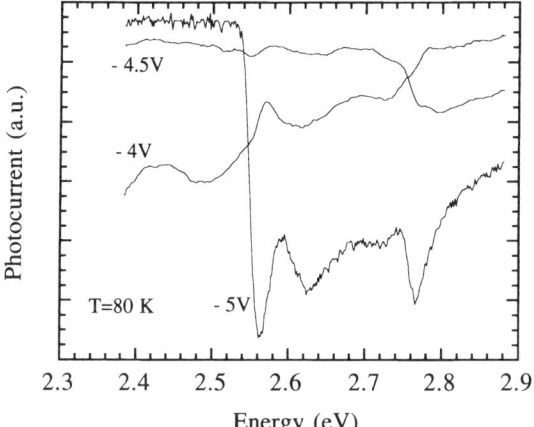

FIG. 21. Sign-reversal of the photocurrent occurring around the breakdown voltage in a $Zn_{0.9}Cd_{0.1}Se/ZnSe$ p-i-n structure.

when the mesa structures are fabricated. This is again a consequence of the reduction of the surface recombination, which allows a good carrier transport, independent of the transit time effects in the intrinsic region.

4. Temporal Evolution of the Excitonic Transitions

In this section we discuss the transient properties of excitons confined in ZnSe-based quantum wells. In optical experiments excitons are formed with some excess energy by the off-resonant pumping. The excess energy is relaxed in a few hundreds fs by the exciton–LO phonon interaction, resulting in a quasi-equilibrium distribution of thermalized excitons. After this short transient, excitons undergo a number of different interactions, namely localization at potential fluctuations, scattering with electrons or other excitons, and eventually recombine radiatively on a time scale of the order of few hundred ps.

The femtosecond dynamics of excitons has been measured in ZnSe/ZnSSe (Stevens *et al.*, 1994; Kuroda *et al.*, 1994) and in ZnCdSe/ZnSe (Ding *et al.*, 1993; Neukirch *et al.*, 1994; Tokizaki *et al.*, 1994) heterostructures. The main features of the transient pump and probe transmission spectra are the small blue shift of the excitonic resonance occurring 2–3 ps after the excitation, due to the hard-core exciton repulsion (exchange effect), and the presence of induced absorption bleaching within the inhomogeneously broadened ex-

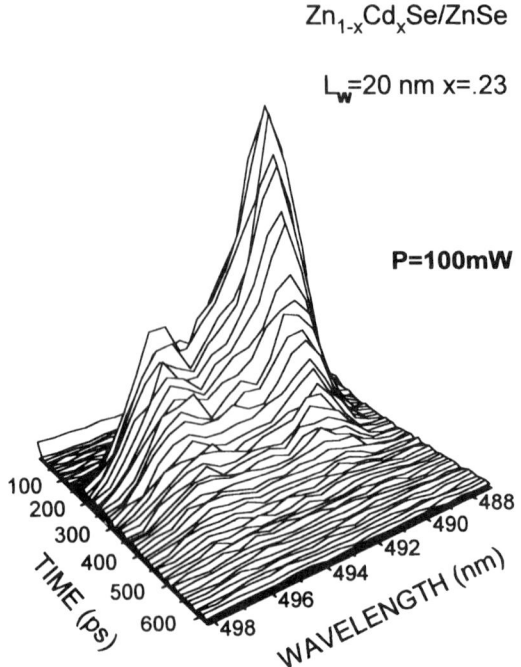

FIG. 22. Time-resolved luminescence spectrum of a $Zn_{0.77}Cd_{0.23}Se/ZnSe$ multiple quantum well with $L_w = 20$ nm at 10 K. The integrated power was 100 mW (excitation wavelength 390 nm, pulse width 2 ps, repetition frequency 82 MHz). The $n = 1$ and $n = 2$ exciton bands can be clearly resolved around 495 and 491 nm, respectively.

citon linewidth. In ZnCdSe/ZnSe MQWs the typical exciton–phonon interaction time is found to be of the order of 90 fs, whereas intraband relaxation of the photoinjected carriers occurs within 2–3 ps. After that, exciton formation occurs (Tokizaki et al., 1994).

On the time scale of the exciton lifetime the behavior of the II–VI quantum wells does not differ appreciably from that of III–V structures. The exciton dynamics is indeed dominated by the temporal evolution of the polariton states (Andreani et al., 1992; Citrin, 1994). In Figure 22 we display the time-resolved luminescence spectrum of a ZnCdSe/ZnSe MQW structure of well width 20 nm and Cd content $x = 0.23$. The $n = 1$ and $n = 2$ excitonic subbands are clearly resolved with typical decay times of the order of 200 and 150 ps respectively, consistent with the results of Tsutsumi et al., 1994. Somewhat shorter decay times are measured in similar samples (Gutowsky et al., 1995) probably due to relevant localization problems.

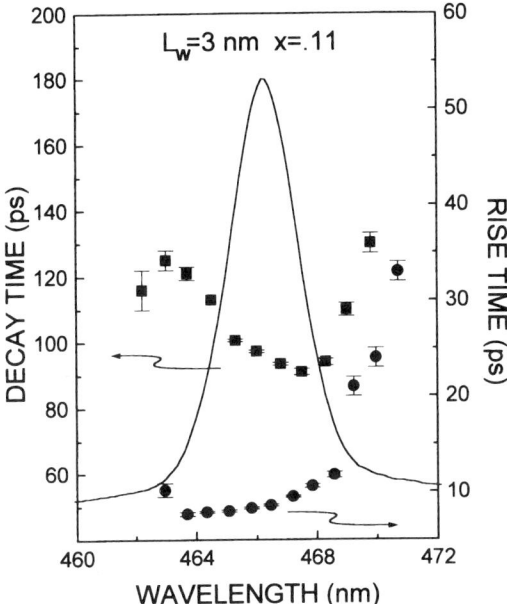

FIG. 23. Spectrally resolved rise time (dots and right-hand scale) and decay-time (squares and left-hand scale) of a $Zn_{0.89}Cd_{0.11}Se/ZnSe$ multiple quantum well with $L_w = 3$ nm. Note the increase of the rise time occurring in the low-energy tail of the exciton band due to trapping and localization.

The time-resolved luminescence experiments provide important information on the localization of excitons within the inhomogeneously broadened density of states caused by disorder in the ternary alloy ZnCdSe matrix. As a general trend the temporal evolution of the luminescence from localized excitons exhibits rise time and decay time longer than the free exciton. This is clearly seen in the low-energy tail of the exciton resonance of ZnCdSe/ZnSe quantum wells under above-barrier excitation (Fig. 23). The long rise time indicates the trapping of excitons at local potential fluctuations. This temporal evolution changes with the excitation intensity, as shown in Figure 24. With increasing the power density both the rise time and the decay time get shorter, as expected for the free exciton luminescence. This change reveals the saturation of the available density of localization states and the recovery of the intrinsic free-exciton recombination. A further increase of excitation generates a rather high carrier density resulting in the almost constant luminescence intensity during the first 150 ps, followed by a longer decay time. Under this condition the excitonic population is fed by the carrier plasma as long as excess carriers are available to form new excitons,

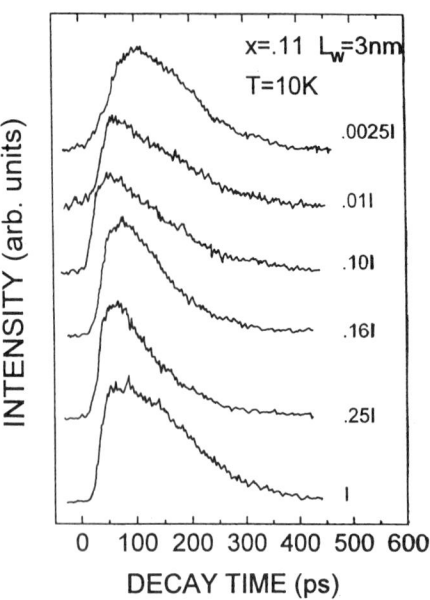

FIG. 24. Temporal evolution of the photoluminescence intensity in a $Zn_{0.89}Cd_{0.11}Se/ZnSe$ of well width $L_w = 3$ nm at 10 K. The detection wavelength was fixed at the peak of the exciton band. The maximum integrated power is $I = 80$ mW.

resulting in the "plateau" region on the time-resolved curve. The important point is that the localization dynamics is found to change with the quantum well stoichiometry in ternary alloy (ZnCdSe/ZnSe) quantum wells. Cd-rich samples suffer important compositional and strain fluctuations which are clearly evidenced by the stronger excitonic-absorption linewidth as compared to shallow QWs. Therefore, the recover of the intrinsic free-exciton recombination, monitored through the shortening of the rise and decay time, is often not observed in deep ZnCdSe MQWs. This is clearly shown in Figure 25 where we compare the intensity dependence of the spectrally resolved decay and rise time of ZnCdSe MQWs of different composition. As already discussed in Figure 24, in the shallow quantum well the short rise time characteristic of the free-exciton recombination is recovered by increasing the power density (Fig. 25a). On the contrary, a clear localization dynamics is observed in the whole excitation intensity range for the deep quantum wells (Fig. 25b). A detailed rate-equation analysis of the time resolved luminescence traces provides the density of localization centers existing in the quantum well (Lomascolo et al., 1996). This is found to vary between $9 \cdot 10$ cm^{-2} and $4 \cdot 10^{10}$ cm^{-2} for Cd content in the range

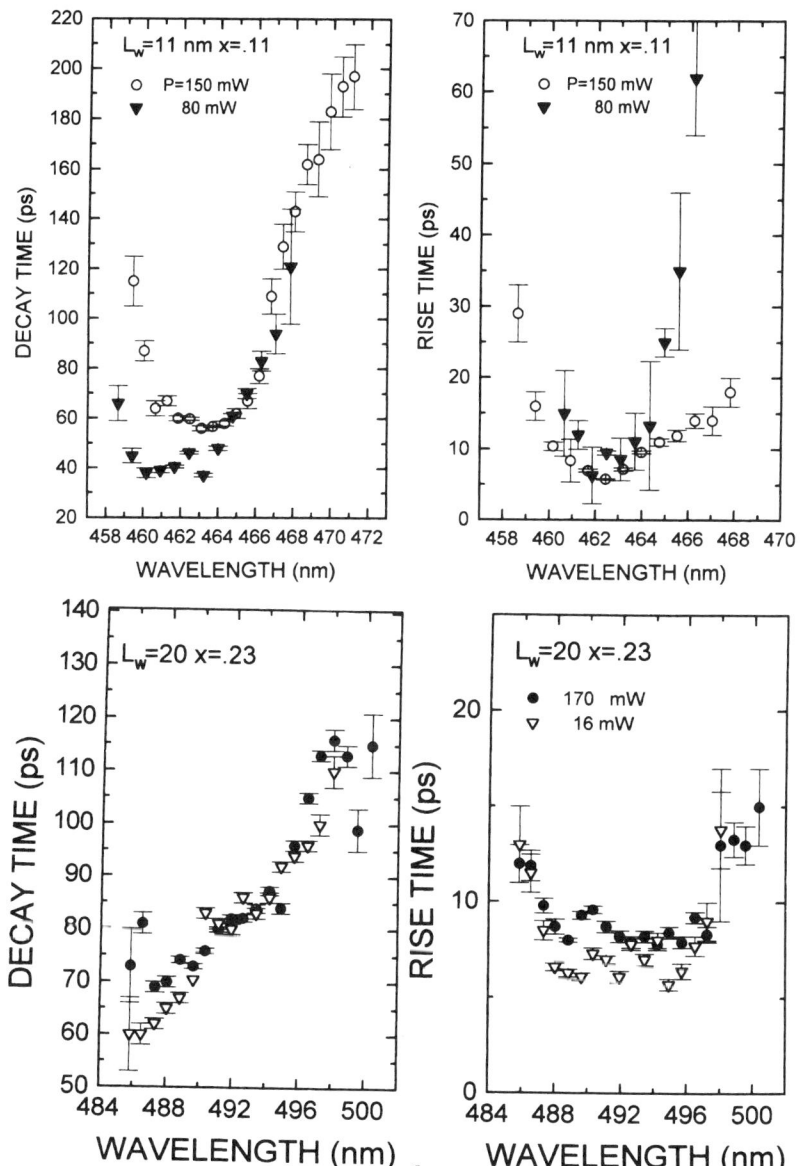

FIG. 25. (a) Intensity dependence of the decay time (left panel) and rise time (right panel) measured from a $Zn_{0.89}Cd_{0.11}Se/ZnSe$ heterostructure of well width $L_w = 11$ nm at 10 K under different excitation power. Note the shortening of the rise time, indicating the saturation of the localization centers occurring at high photogeneration rates. (b) The same as in Figure 25a, but for a $Zn_{0.77}Cd_{0.23}Se/ZnSe$ multiple quantum well of width $L_w = 20$ nm. No saturation of the localization centers is observed under intense photoexcitation.

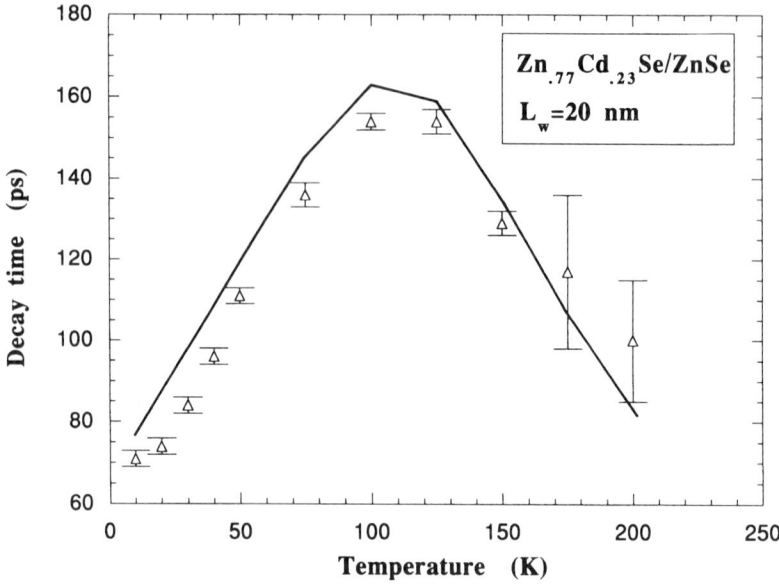

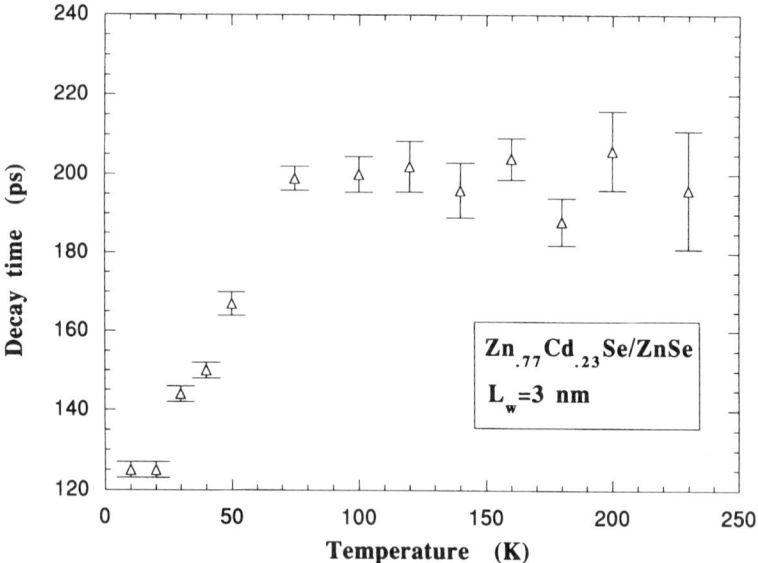

FIG. 26. (a) Temperature dependence of the exciton decay time in $Zn_{0.89}Cd_{0.11}Se/ZnSe$ multiple quantum wels ($L_w = 7$ nm). The curve is calculated according to Eq. 14. (b) Same as in Figure 26a, but for a $Zn_{0.77}Cd_{0.23}Se/ZnSe$ multiple quantum well sample.

0.1 < x < 0.23. This result is very important for the understanding of the lasing mechanism. In fact, stimulated emission is found to occur in the sample of Figure 25b, at injection densities for which the localization of excitons is still dominant. This supports the model of optical gain from localized exciton states proposed by Ding et al. (1992). Conversely, the lasing threshold of the shallow quantum wells (Fig. 25a) occurs at intensity such that the localization is saturated, indicating that the exciton localization is not the main lasing mechanism in shallow ZnCdSe MQWs (Cingolani et al., 1994a). Similar conclusions are found to be valid for the ZnSe/ZnSSe heterostructures (Stevens et al., 1994). In this case the localization of excitons primarily occurs at the barrier/well interfaces due to the formation of extended islands.

The temperature dependence of the decay time of the exciton luminescence provides more information on the polariton confinement in the quantum well. The polariton lifetime in quantum wells is expected to grow linearly with the temperature (Andreani et al., 1992; Citrin, 1994). This phenomenon competes with the thermal escape of carriers from the quantum wells which reduces the overlap of the electron and hole wavefunctions, resulting in a decrease of the exciton lifetime. This dynamics is shown in Figure 26 for deep and shallow ZnCdSe/ZnSe MQWs. In the shallow quantum well the decay time increases almost linearly with the temperature, then it saturates and starts to decrease around 100 K. At this temperature the less confined particle, presumably the heavy-role, is thermally activated above the barrier and the exciton itself becomes unstable. In the deep quantum well the saturation is observed, but the depth of the quantum well is large enough to prevent thermal escape of the carriers up to room temperature. This spectral behavior has been observed in III–V quantum wells, and has been quantitatively explained by Michler et al. (1992) and by Gurioli et al. (1992). The temperature-dependent exciton decay time is described by

$$\tau(T) = \frac{\tau_w}{1 + \frac{\tau_w}{\tau_b} C^{1/2} \exp\left(-\frac{\Delta E}{kT}\right)} \quad (14)$$

where C is the scattering rate, $\tau_{w(b)}$ is the decay time in the well (barrier), and ΔE is the activation energy of the carriers confined in the well (this is analogous to the work function of the heavy-hole in the investigated structures). The calculated temperature-dependent decay times are compared to the experimental data in Figure 26 (continuous lines). The activation energy obtained from the fit compares quite well with the potential step experienced by the heavy-holes, thus confirming the offset and the strain correction adopted for the evaluation of the quantized states in these heterostructures (see Section III.1).

IV. Nonlinear Excitonic Properties

1. Basic Theoretical Concepts

In this section we discuss the nonlinear optical properties of excitons in II–VI quantum wells. Due to the enhanced excitonic stability, nonlinear optical processes involving excitons are very important in II–VI quantum wells and must be considered relevant even for the design of optoelectronic devices operating with high carrier density and at room temperature. The excitonic nonlinearities originate from different mechanisms, namely: (i) the interaction of excitons with an external electric field and (ii) the many-body interactions (Schmitt-Rink *et al.*, 1989). The former case is based on the charge separation occurring in a polarized quantum well, which causes a decreased overlap of the electron–hole wavefunction (quantum confined Stark effect). This in turn causes a reduction of the exciton oscillator strength, and of the binding energy. There are only a few experiments reported in the literature due to the technological problems to be faced in the fabrication of good *p-i-n* structures (see Section III.3). Electroabsorption experiments have been performed on ZnSe/ZnSSe superlattices by Marquadt *et al.* (1994) who showed a negligible Stark shift of the excitonic resonance. This finding was interpreted as a consequence of the weak exciton polarizability due to the negligible conduction band offset. A distinct quantum confined Stark effect has been revealed under external forward bias either in *p-i-n* or Schottky diodes embedding ZnCdSe/ZnSe MQWs in the depletion layer. In the former case the built-in electric field was not known, due to the nonohmic contacts and the doping inhomogeneities (Wang *et al.*, 1993a). The Schottky barrier, on the other hand, though giving a large Stark shift and a well-known potential discontinuity (about 1.4 V for the Au/ZnSe interface), results in a worse current-voltage characteristic (Kawakami *et al.*, 1993a; Wang *et al.*, 1994). Low-threshold electroptic modulation recently has been obtained in ZnCdSe/ZnSe quantum wells both in photocurrent and luminescence. The bleaching of the oscillator strength, the stark-shift, and the well width dependence of the non-linearity have been quite well described by a variational model for the exciton in electric field (Giugno *et al.*, 1996). Finally, bistable self-electro-optic operation (Wang *et al.*, 1993b) has been demonstrated in ZnCdSe/ZnSe quantum wells.

Many-body nonlinearities, on the other hand, have been extensively studied in recent years, with special attention to III–V heterostructures. In general, the presence of a dense carrier/exciton plasma influences the excitonic states through the screened Coulomb potential and the exchange and correlation interactions which reduce the exciton binding energy and change the self-energy of the electron–hole pair. These many-body processes are treated by means of complicated theoretical models based on the random phase approximation, which allow one to describe the carrier density dependence of the electronic and excitonic states (Klingshirn and Haug, 1981; Schmitt-Rink *et al.*, 1989; Haug and Schmitt-Rink, 1985). Most of these theories have been developed for III–V semiconductors, in which excitons are not strongly bound (typical binding energy of the order of 10 meV). Consistent with the experimental observations, it is found that at carrier densities well below 10^{11} cm^{-2} the excitonic resonances are bleached and the optical absorption is due to interband transitions. In terms of basic thermodynamics, the free-carrier gas causes a rapid ionization of the exciton gas, and the two phases can coexist only at low carrier densities and temperatures (Cingolani and Ploog, 1991).

In II–VI semiconductors, and especially in quantum wells with enhanced excitonic stability, the coexistence of the exciton and free carrier gas occurs on a wider density and temperature range. Due to the large exciton binding energy, carrier densities in excess of 10^{11} cm^{-2} can easily be achieved in the material without relevant excitonic bleaching. This has a tremendous impact on the operation principle of optoelectronic devices based on wide band gap II–VI quantum wells.

Many-body processes like electron–electron, exciton–electron, and exciton–exciton interactions, cause the reduction of the exciton binding energy, due to the renormalization of the electronic and excitonic states, and the saturation of the excitonic absorption (exciton bleaching). The exact calculations of these processes are very difficult. In this section, we will discuss a few relevant perturbative approximations, valid in the limit of intermediate carrier densities, which provide handable analytical formulas for the evaluation of the self-energy corrections of the interacting free-exciton/free carrier gas (Cingolani *et al.*, 1996a).

The macroscopic excitonic bleaching occurring at high carrier densities is explained by the reduction of the overlap of the electron and hole wavefunctions caused by the surrounding interacting carriers (Schmitt-Rink *et al.*, 1989). The exciton oscillator strength depends on the electron carrier density in the well, and is proportional to the squared amplitude of the exciton wavefunction in the volume where the electron and the hole overlap. It is expected that the overlap amplitude of the electron and hole rapidly decreases as the sheet carrier density increases. The bleaching of the oscillator strength f is described by a simple saturation model (Schmitt-Rink *et al.*, 1985)

$$f(n) = f(0) \cdot \frac{1}{1 + \dfrac{n}{N_c}} \tag{15}$$

where n is the total carrier density and $f(0)$ is the unperturbed oscillator strength. N_c is a critical saturation density, i.e., the carrier density at which

$f(n)$ becomes one-half of the value of the unperturbed crystal. This quantity has to be calculated by a full many-body treatment of the various carrier–exciton interactions. A useful analytical approximation valid in the non-degenerate limit and at low temperature, including the effect of phase space filling (PSF) and exchange (EXC), has been presented by Schmitt-Rink *et al.* (1985)

$$\frac{1}{N_c} = \frac{1}{N_c^{\text{PSF}}} + \frac{1}{N_c^{\text{EXC}}} \tag{16}$$

where $N_c^{\text{PSF}} = \dfrac{1}{8\pi a_0^2}$ and $N_c^{\text{EXC}} = \dfrac{1}{9.89\pi a_0^2}$.

Both these terms originate from the Pauli exclusion principle. The PSF describes the blocking of the conduction and valence states available for the formation of new excitons. The short-range exchange interaction represents the energy gained by electrons with same spin orientation in avoiding each other. Long-range Coulomb interactions, i.e., the classical screening among carriers is neglected, because it has been shown to be substantially reduced in quantum well structures with respect to the bulk.

The reduction of the exciton binding energy (E_b) originates from the renormalization of the electron–hole pair energy. At first order we can write a density-dependent exciton binding energy

$$E_b(n_{tot}) = E_b - BGR(n_{eh}) - \pi a_0^2(3.84 n_x + 13.1 n_{eh}) E_b \tag{17}$$

In Eq (17) $n_{tot} = n_{eh} + n_x$ indicates the total density of electron–hole pairs (injected or photogenerated) which consists of free carriers (n_{eh}) and excitons (n_x), distributed according to a temperature- and density-dependent phase diagram. The right hand part of Eq (17) contains the energy corrections due to the different many-body interactions. BGR is the band gap renormalization induced by free carriers and is discussed in more detail below. The last term includes the blue-shift of the exciton resonance induced by the exciton–exciton interaction, which depends only on the exciton density n_x (Schmitt-Rink *et al.*, 1985), and the effect of free carriers on the exciton wavefunction (phase space filling). Actually an additional term accounting for the renormalization of the single particle states induced by the excitons should be included (Cingolani *et al.*, 1996a). For sake of simplicity we will neglect this term in the following, and we will calculate the *BGR* for the total carrier density, no matter whether the electron–hole pairs are bound or not. These analytical approximations are strictly valid in the limit of non-degenerate Fermi gas and for $kT/E_b < 1$. These conditions are reasonably fulfilled in II–VI quantum wells under strong photogeneration rates, by virtue of their large effective masses and binding energy. The band gap renormalization term consists of the exchange (Σ^{exc}) and Coulomb-hole (Σ^{eh}) self-energy corrections. According to Campi and Coriasso (1995)

$$\Sigma^{exc} = \frac{e\hbar^2}{2a_o\mu} \cdot \left[\chi\left(\frac{\chi}{k_F + \chi}\right) + k_F\right] \qquad (18)$$

where $k_F = (2\pi n_{eh})^{1/2}$ is the Fermi wavevector, μ is the electron–hole reduced mass and a_o is the quasi-two-dimensional exciton Bohr radius. For the Coulomb-hole term we have

$$\Sigma^{ch} = \frac{e\hbar^2}{2a_o\mu} \cdot \chi \qquad (19)$$

In Eqs (18) and (19) χ represents the screening wavevector (inverse screening length), given by

$$\chi = \chi_e + \chi_h = \frac{2\pi e^2}{\varepsilon} \cdot \left(\frac{\partial n_{eh}}{\partial E_F^e} + \frac{\partial n_{eh}}{\partial E_F^h}\right) \qquad (20)$$

where $E_F^{e,h}$ is the Fermi energy of electrons and holes. The total band gap renormalization for the neutral electron and hole plasma is thus given by $BGR(n_{eh}) = 2\Sigma^{exc} + 2\Sigma^{ch}$, where the factor 2 accounts for the mutual electron–hole and hole–electron interactions. Similarly, the band gap renormalization for the single component (electron or hole) plasma is given by one-half of the BGR obtained for the neutral electron–hole plasma.

Experimentally, one can investigate these processes either by injecting optically or electrically a dense neutral electron–hole plasma in the quantum well, or by using a single component plasma created by the modulation doping in the barrier (either p- or n-type). In the former case a two-component plasma (TCP) is formed, whereas the latter case allows one to obtain a one-component plasma (OCP) in equilibrium conditions in the well.

2. Excitons and the One-component Electron Plasma

The one-component plasma is formed by introducing donor or acceptor atoms in the middle of the barrier during the growth. In order to keep the Fermi level constant through the whole material, the impurity atoms are

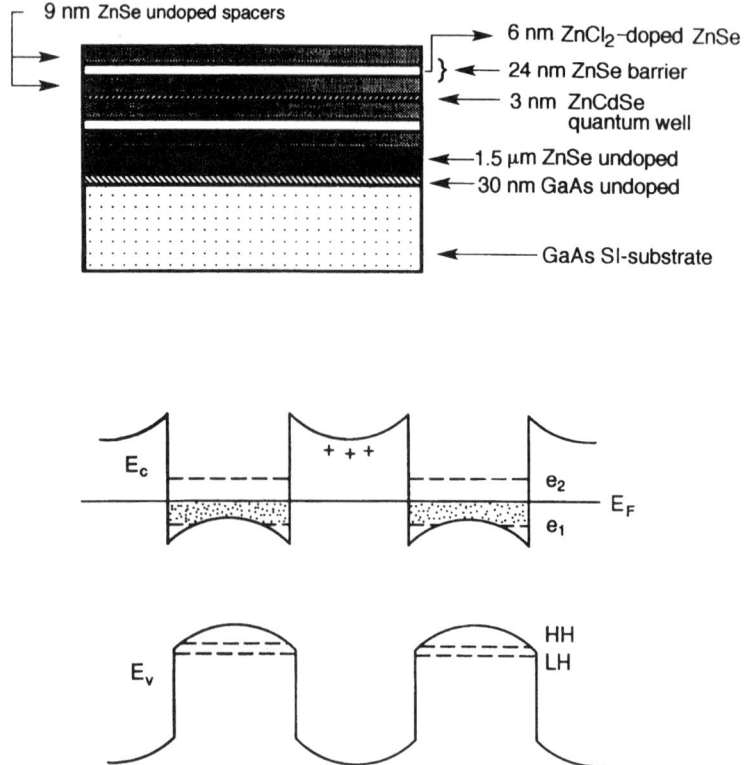

FIG. 27. Layer sequence and band profile of the modulation-doped quantum well heterostructures.

ionized and the free carriers transfer into the well, where the plasma forms (Livescu et al., 1988; Cingolani et al., 1990). The band structure of the undoped material is modified by the electric field created by the charge transfer. The typical structure of the modulation-doped quantum wells (MDQWs) and the corresponding band diagram are shown in Figure 27.

The unperturbed eigenstates of the modulation-doped QWs are modified by the free carriers introduced by the doping. The modified energy levels can be evaluated by solving self-consistently the Poisson and Schrödinger equations. A parabolic bending of the bands, proportional to the sheet carrier density n and to the dielectric constant of the material, is assumed in the calculations. In first-order perturbation theory, the correction to the energy levels induced by the electric field generated by the carriers depends on the band bending and on the electron and hole wavefunctions. This in turn affects the excitonic states. In ZnSe-based quantum wells the electro

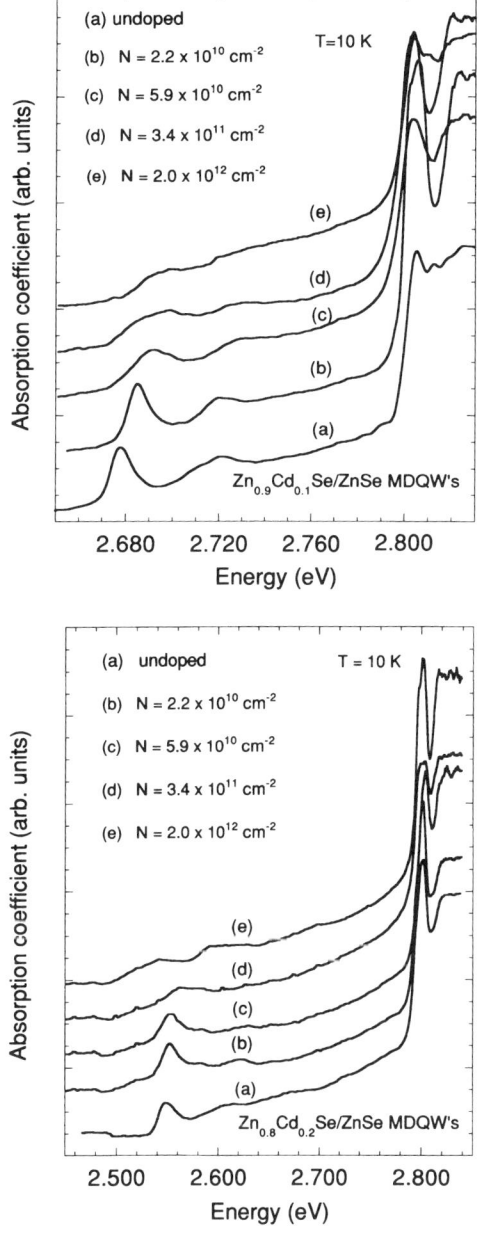

FIG. 28. (a) Absorption spectra of a set of $Zn_{0.9}Cd_{0.1}Se/ZnSe$ modulation-doped multiple quantum wells with different doping concentrations. The well width is $L_w = 3$ nm. (b) The same as in Figure 28a, but for a set of $Zn_{0.8}Cd_{0.2}Se/ZnSe$ modulation doping quantum wells.

static effects on the ground level heavy-hole exciton turn out to be very small, of the order of 1–2 meV at carrier densities of the order of 10^{12} cm^{-2}. We will therefore neglect this correction in the following discussion. Figure 28a shows the low temperature absorption spectra of n-type Zn$_{0.9}$Cd$_{0.1}$Se MDQWs as a function of the sheet carrier density from $2.2 \cdot 10^{12}$ cm^{-2} (Calcagnile et al., 1996). Curve (a) refers to the undoped sample and is shown as reference. This spectrum exhibits the heavy-hole (HH) and light-hole (LH) excitonic transitions, with full width at half maximum (FWHM) of 11 meV and 20 meV, respectively. Moreover, at 2.805 eV the excitonic transition from the undoped ZnSe barrier layer is also observed. The absorption spectra of the modulation-doped samples [curves (b)–(d)] exhibit a clear reduction of the oscillator strength. At the highest density, the spectrum labeled (e) does not show any excitonic feature indicating that the carrier concentration in this sample is above the critical density for the exciton bleaching. Similar results are obtained from the deeper Zn$_{0.8}$Cd$_{0.2}$Se/ZnSe MQWs shown in Figure 28b, though with a stronger excitonic resonance. At the maximum sheet carrier density ($n = 2.0 \cdot 10^{10}$ to $2.0 \cdot 10^{12}$ cm^{-2}) still a weak excitonic peak is observed in the spectrum, as expected from the enhanced exciton binding energy (37 meV for the $x = 0.2$ sample and 30 meV for the $x = 0.1$ sample, respectively).

The effect of carrier concentration on the intensity of the optical transitions can be evaluated by estimating the oscillator strength and the binding energy of the HH exciton. The oscillator strength is directly proportional to the integrated area of the absorption peak (Masumoto et al., 1985):

$$\int \alpha(E)dE = \frac{2\pi^2 e^2 \hbar}{mc\varepsilon^{1/2}} f(0) \tag{21}$$

where e is the electron charge, ε is the dielectric constant, and c the light velocity. Using Eq (21) the HH oscillator strength can be obtained directly from the experimental spectra of Figure 28. As shown in Figure 29 the oscillator strength of the heavy-hole exciton decreases with increasing the sheet carrier density. The experimental results follow the saturation behavior of the exciton oscillator strength predicted by Eq (16). The critical density for exciton bleaching is found to be $N_c = 1.0 \cdot 10^{11}$ cm^{-2} for the shallow MDQWs, and $N_c = 2.0 \cdot 10^{11}$ cm^{-2} for the $x = 0.2$ MDQWs. These values of N_c compare reasonably well with the theoretical critical densities obtained by the perturbative description Eq (15), which amount to $2 \cdot 10^{11}$ cm^{-2} and $3.6 \cdot 10^{11}$ cm^{-2} for $x = 0.1$ and $x = 0.2$, respectively. We should mention that the above saturation model has been obtained for the two-component electron–hole plasma. The straightforward extension to

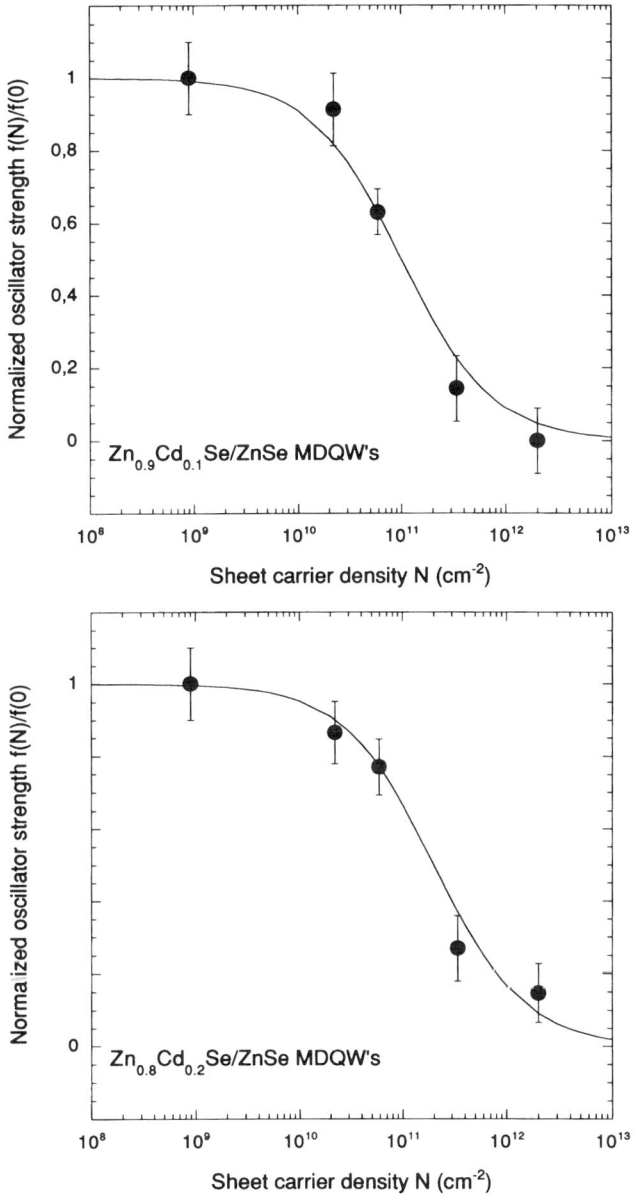

FIG. 29. (a) Carrier density dependence of the normalized integrated absorption of the heavy-hole exciton peak for the $Zn_{0.9}Cd_{0.1}Se/ZnSe$ modulation-doped multiple quantum wells of Figure 28a. The continuous curve is the saturation of the oscillator strength calculated by means of Eq. (16). (b) The same as in Figure 29a but for the set of $Zn_{0.8}Cd_{0.2}Se/ZnSe$ modulation-doped multiple quantum wells studied in Figure 28b.

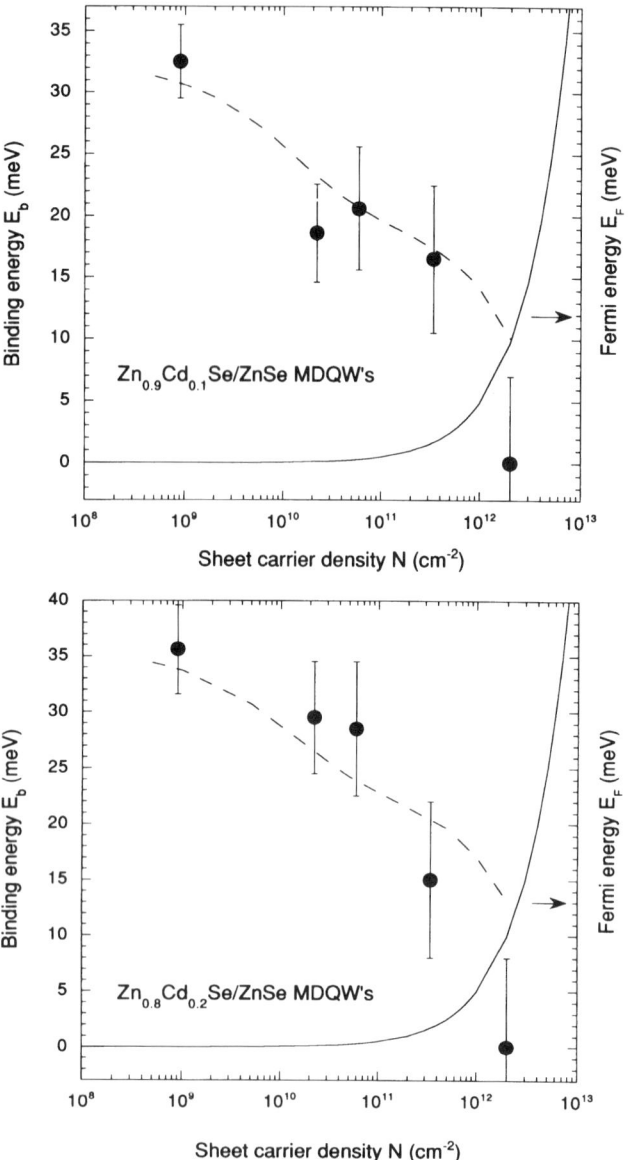

FIG. 30. (a) Carrier density dependence of the heavy-hole exciton binding energy (symbols) obtained from the $Zn_{0.9}Cd_{0.1}Se/ZnSe$ modulation-doped multiple quantum wells of Figure 28a. The dashed curve represents the renormalization of the exciton binding energy calculated by means of Eqs (18)–(20). The continuous curve is the calculated low-temperature quasi-Fermi energy of the electron plasma. The increasing error bars are due to the enhancement of the spectral broadening occurring at high carrier density which affects the line-shape fitting precision. (b) The same as in Figure 30a but for the $Zn_{0.8}Cd_{0.2}Se/ZnSe$ modulation-doped multiple quantum wells of Figure 28b.

the one-component electron-plasma existing in MDQWs can thus be questionable.

The effects of carrier density on the HH exciton binding energy were determined by means of a line-shape analysis (see Section III.1), through the evaluation of the energy difference between the HH exciton energy and its continuum. The experimental results are shown in Figure 30. The curves were calculated assuming that the binding-energy reduction is caused by the band gap renormalization induced by the single-component plasma, neglecting excitonic interactions, i.e. $E'_b(n) = E_b - \text{BGR}$, where $\text{BGR} = \Sigma^{\text{exc}} + \Sigma^{\text{ch}}$ is given by Eqs (18) and (19). The overall agreement is good at low and intermediate densities, where both the theory and the line shape fitting of the spectra are rather precise. At high carrier densities the agreement becomes poor. This is expected since the binding energy obtained from the line shape fitting is rather unprecise due to the broadening of the spectra (see the large error bar). Also, the analytical expressions for the self-energies become unreliable in the high density limit.

3. Excitons and the Electron–Hole Plasma

In this section we discuss the nonlinear properties of excitons in the presence of a dense neutral electron–hole plasma. The basic physics is analogous to the OCP case. However, under stationary conditions, excitons can form and coexist with free carriers leading to a complicated free carrier/exciton phase boundary. The key-parameter is the exciton binding energy modified by the many-body interactions and by the finite temperature of the distribution functions, which governs the dissociation of the exciton phase into an electron–hole plasma phase. Depending on the degree of confinement of the exciton wavefunction, identical injection conditions can create either an exciton-rich or a free-carrier-rich phase. This in turn determines the main stimulated emission mechanism in quantum wells of different size and composition. The most direct way to study these phenomena is the nonlinear pump and probe transmission spectroscopy, in which a strong pump beam creates the electron–hole plasma while a weak tunable beam probes the changes of the excitonic absorption.

In Figure 31, we display the intensity dependence of the pump and probe transmission spectra measured at 10 K in two samples with exciton binding energies of 25 meV (Fig. 31a) and 38 meV (Fig. 31b) (Cingolani *et al.*, 1994e, 1996a). In both cases a clear exciton bleaching is observed. The bleaching intensity amounts to about 1 kW cm^{-2} for the shallow wells and increases up to about 50 kW cm^{-2} in the deeper wells, having considerably stabler exciton. A quantitative interpretation of these experiments is based on the line-shape

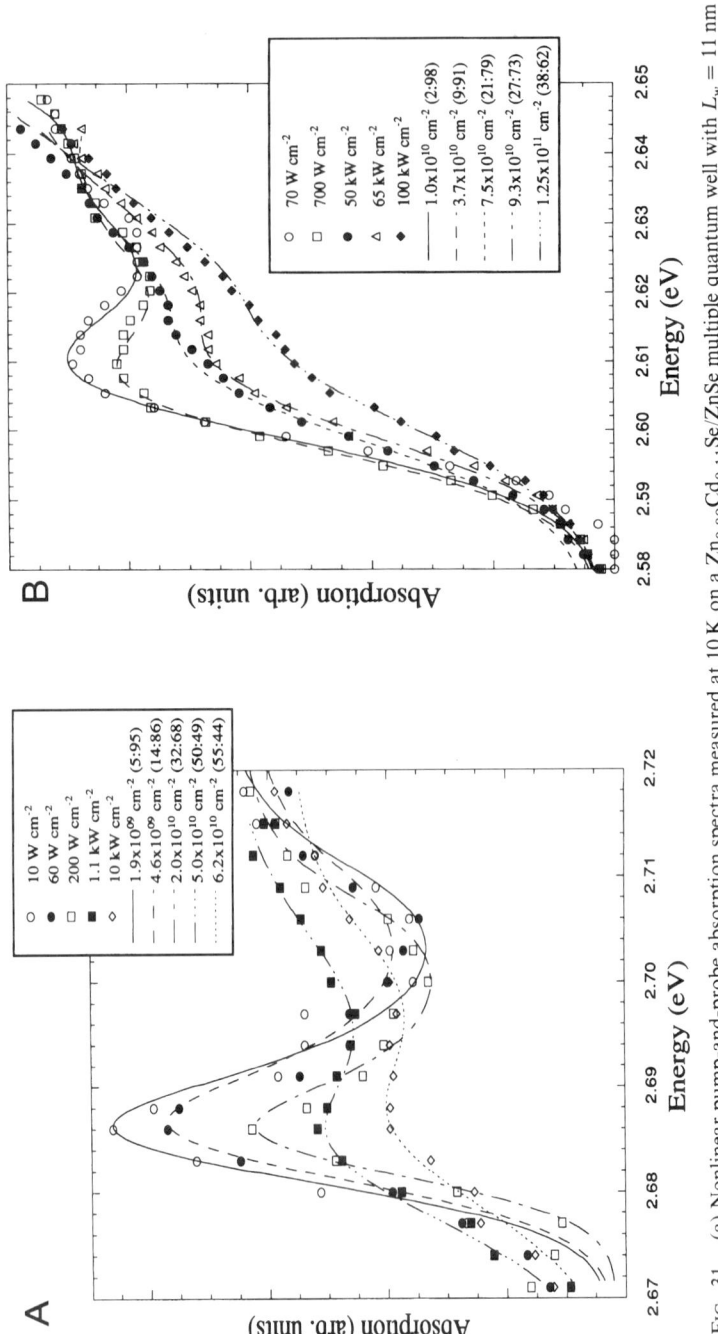

FIG. 31. (a) Nonlinear pump-and-probe absorption spectra measured at 10 K on a $Zn_{0.89}Cd_{0.11}Se/ZnSe$ multiple quantum well with $L_w = 11$ nm under different pump intensities (symbols). The continuous lines are the calculated nonlinear absorption spectra. For sake of clarity the number of experimental points plotted on the figure has been reduced. The inset shows the total carrier densities corresponding to different pump intensities. In parenthesis we report the ratio of free carriers and excitons obtained from the self-consistent calculation of the pump-and-probe spectra for each pumping condition $(n_{eh}:n_x)$. (b) The same as in Figure 31b but for $Zn_{0.77}Cd_{0.23}Se/ZnSe$ multiple quantum wells of well width $L_w = 3$ nm.

analysis of the nonlinear pump-and-probe spectra. Following Section IV.1, the nonlinear absorption profile is given by:

$$\alpha(\hbar\omega, n_{tot}) = \text{const} \cdot \left(\frac{1}{1 + \frac{n_{tot}}{N_c}} \right) \cdot G\{\hbar\omega - [E_g - E'_b(n_{tot})], \Gamma(n_{tot})\}$$

$$+ \text{const} \cdot \Theta[\hbar\omega - E'_g(n_{eh}), \Gamma(n_{tot}), \gamma)] \cdot S(\hbar\omega - E_b) \cdot f_e(n_{eh}, T) - f_h(n_{eh}, T) \quad (22)$$

In Eq (22) n_{tot} is the total density of elementary excitations given by the sum of free-carrier and exciton density, distributed according to the temperature dependent phase diagram of the exciton and electron–hole plasma phase. The first factor of Eq (22) accounts for the saturation of the oscillator strength of the excitonic transition, through the critical saturation density N_c (Eqs (15) and (16)). The exciton density of states is modeled by a Gaussian function G, whose position depends on the renormalized exciton binding energy (Eq (17)). The Θ function in Eq (22) models the steplike density of states, which includes a phenomenological broadening factor γ to account for compositional fluctuations in the ternary well. Coulomb coupling at the continuum, though very weak in these structures, is included by using the Sommerfeld factor $S(\hbar\omega)$. A density-dependent lifetime broadening factor $[\Gamma(n_{tot})]$ is used in the calculations. Finally, T indicates the thermodynamic temperature of the distribution functions (f_e, f_h).

The calculated pump and probe spectra are shown by the continuous lines in Figure 31, showing excellent agreement with the experimental spectra. The exciton bleaching is found to occur at $4 \cdot 10^{10}$ cm^{-2} and $1.5 \cdot 10^{11}$ cm^{-2} in the two samples, consistent with the different exciton binding energies. These values are considerably smaller than those expected from Eq. (16). This result is mainly due to the finite thermodynamic temperature assumed for the electron–hole and exciton populations and to the complete description of the many-body renormalization of the exciton binding energy included in our model. The equilibrium between the two phases is established by a mass-action law. A set of coupled equations for the total density of photogenerated elementary excitations

$$n_{tot} = n_{eh} + n_{exc} \quad (23)$$

and for the relative balance of free carriers and excitons

$$\frac{n_{eh}^2}{n_{exc}} = \frac{m_e m_h k_b T}{2\pi\hbar^2(m_e + m_h)} \cdot \exp\left(\frac{-E'_b(n_{tot})}{k_b T}\right) \quad (24)$$

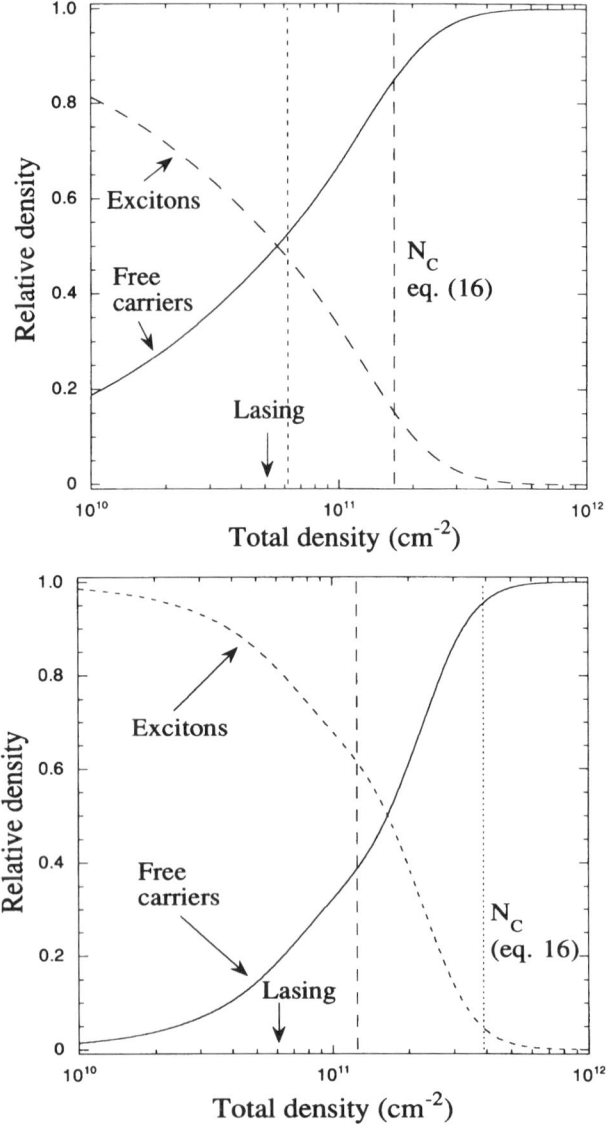

FIG. 32. (a) Phase diagram of the exciton gas and free-carrier gas coexisting in the sample investigated in Figure 31a. The dashed and continuous curves indicate the exciton and free-carrier populations, respectively. The experimental exciton saturation density obtained from the pump and probe line-shape analysis is indicated by the vertical short-dashed line, to be compared with the N_c value calculated by means of Eq (16) (long-dashed vertical line). The carrier density of the lasing threshold (corresponding to $4-6\,\mathrm{kW\,cm^{-2}}$ pump intensity) is indicated by the vertical arrow. Depending on sample characteristics, this falls either in the free-carrier-rich or in the exciton-rich part of the phase diagram. (b) The same as in Figure 32a but for the sample investigated in Figure 31b.

has to be solved self-consistently as a function of the total density of elementary excitations (n_{tot}). All the many-body renormalizations are taken into account according to Eq (17).

The results of the calculations are shown in Figure 32 for the same samples investigated in Figure 31 (Cingolani et al., 1996a). The key parameters are the temperature of the distribution function and the total density n_{tot}, which affect the boundary between the free-carrier and exciton phase in a dramatic way. The carrier temperature adopted in Figure 32 was evaluated from the slope of the high-energy tail of the luminescence spectra recorded at different excitation intensities under the photogeneration conditions of the pump and probe experiments. Typical values ranged between 2 meV and 5 meV in the investigated excitation intensity range, and were found to vary sublinearly with the excitation intensity. The phase diagrams of Figure 32 indicate that the transition from an exciton-rich to a free-carrier-rich phase indeed occurs at n_{tot} values below those expected from the simple approach of Eq. (16), provided the actual carrier temperature and the many-body corrections due to free carriers and excitons are taken into account. In shallow quantum wells the free-carrier phase sets in at injected densities well below $1 \cdot 10^{11}$ cm^{-2}, consistent with the observation of exciton bleaching in the pump and probe spectra. In deeper wells the exciton phase dominates up to higher n-values resulting in a stable exciton gas even under very intense injection rates. This finding has a strong impact on the operation of quantum well lasers based on II–VI materials. In fact, lasing occurs at injected carrier densities which can be either above or below the boundary of the exciton/free-carrier phase transition (resulting in a free-carrier laser or in an excitonic laser), depending on the temperature and on the binding energy of the exciton, i.e. on the compositional and structural parameters of the quantum wells.

V. Role of Excitons in the Lasing of ZnSe-based Quantum Wells

In this section we will discuss the interplay between exciton and free-carrier recombination in the stimulated emission mechanism of ZnCdSe/ZnSe and ZnSe/ZnSSe MQWs. It is well known that the operation principle of a laser is based on a three-level system in which the electronic population of the intermediate level is inverted by the fast relaxation of electrons injected in the topmost level. In semiconductors the three-level system involves the electron and hole quasi-Fermi levels and the renormalized band gap from which stimulated emission occurs. Such process involves free carriers distributed in the conduction and valence bands. In II–VI wide band gap quantum wells, the strong Coulomb correlation may lead to stable excitons even in the presence of a dense free-carrier background. Lasing can occur through the stimulated recombination of excitons as well, provided an intermediate level exists to form the three-level system, namely a localized exciton state or the final state of a scattering process (exciton–exciton, exciton–electron, exciton–phonon).

These processes can be distinguished spectroscopically by the characteristic energy of their luminescence. In general, energy and momentum conservation and Boltzman distribution for the excitons are assumed in the scattering process (Klingshirn and Haugh, 1981). For the free exciton–exciton scattering the resulting emission should occur at energy:

$$\hbar\omega_{\text{exc-exc}} \simeq E_{HH} - E_b - \frac{\hbar^2 k^2}{2\mu} \tag{25}$$

where E_{HH} is the free exciton (heavy-hole) energy, and μ is the reduced exciton mass. If the scattering involves a free exciton and a localized exciton the resulting photon energy can be obtained by replacing the exciton binding energy with the localization energy of the trapped exciton, which is ionized after the scattering. Therefore, the exciton–exciton collision should result in a characteristic emission band which is red-shifted with respect to the free exciton emission by an amount equal to the exciton binding energy (neglecting the kinetic energy).

The exciton–electron scattering should occur at energy

$$\hbar\omega_{\text{exc-el}} = E_{HH} - \frac{\hbar^2 k^2}{2m_e} \tag{26}$$

As the kinetic energy term can usually be neglected, this recombination mechanism mostly contributes to the broadening of the excitonic line. Eventually, a spectral shift from the excitonic band can be observed with increasing the excitation intensity or the temperature. In particular, an anomalous temperature dependence of the emission is expected, which follows the simple relation (Klingshirn and Haug, 1981):

$$\hbar\omega(T)_{\text{exc-el}} \simeq E_{HH}(T) + \sigma\left(\frac{m_e}{m_{HH}}\right) K_b T \tag{27}$$

where σ is a weak function of the effective mass ratio.

The energy conservation in the exciton–phonon scattering leads to the well-known relation

$$\hbar\omega_{\text{exc-LO}} = E_{HH} - \hbar\omega_{\text{LO}} \tag{28}$$

In ZnSe and related ternary alloys, the LO phonon energy amounts to about 31 meV, resulting in a strongly red-shifted emission band.

We should mention that biexcitons also can play a role in the lasing processes. The formation of biexcitons in II–VI quantum wells has been investigated by few groups (Fu et al., 1988; Kuroda et al., 1992a). The reduced binding energy of the biexciton with respect to the exciton, and the competition of inelastic scattering processes make the biexciton particle rather unstable, so that biexciton lasing is very unlikely to be dominant in ZnSe MQWs at high temperatures. Recent observations of biexciton recombination and lasing have been reported by Wang and Simmons (1995), Kreller et al. (1995), Yamada et al. (1995), and Kozlov et al. (1996).

Finally, we recall that the free-carrier recombination is treated by the usual Fermi gas model. The electron and hole population in quasi-stationary conditions is modeled by Fermi-Dirac distribution functions ($f_{e,h}$), taking into account the effective carrier temperature, the band gap renormalization, and the density-dependent quasi-Fermi level ($F_{e,h}$) (Lasher and Stern, 1964). Actually, many-body effects and partial removal of the k-conservation are included in the PL line-shape according to Section IV.1:

$$I(\hbar\omega) \simeq \int_{\infty}^{\infty} D_{\text{joint}}(E) f_e(E) \cdot [1 - f_h(-E)] L(\hbar\omega - E'_g, \Gamma) \cdot dE \qquad (29)$$

where D_{joint} is the joint density of states and E'_g is the renormalized band gap edge. Under these conditions, the PL is expected to raise around the renormalized band gap edge, with considerable spectral broadening in the low-energy tail, and a bandwidth somewhat proportional to the total quasi fermi level of the electron–hole plasma.

The unambiguous identification of one of the above processes needs a systematic investigation, either in terms of spectral positions or in terms of temperature and intensity dependence. In particular, the free-carrier recombination is expected to red-shift with the carrier density, proportionally to the band gap shrinkage, whereas the other processes exhibit a spectral shift due to the increasing exchange of kinetic energy in the scattering. These spectral features are often difficult to be distinguished, and are strongly affected by the broadening of the emission bands occurring with increasing the excitation power. In order to facilitate the comparison of the experimental spectra with Eqs (25)–(28), we plot in Figure 33 the well width dependence and the intensity dependence of the stimulated emission band measured in a set of $ZnS_{0.18}Se_{0.82}/ZnSe$ superlattices (Cingolani et al., 1996b). The exciton energy determined by photoluminescence excitation or absorption spectroscopy is used as the origin of the energy axis, to determine the relative energy position of the photoluminescence line below and above the stimulation threshold. Two main conclusions can be drawn

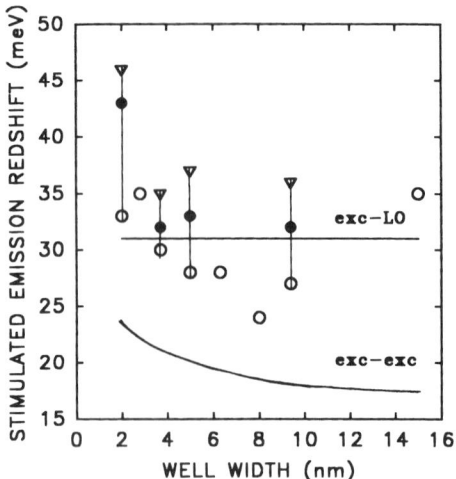

FIG. 33. Well width dependence of the stimulated emission red shift measured in a set of symmetric $ZnS_{0.18}Se_{0.82}/ZnSe$ superlattices at 10 K. The empty circles were measured just at the threshold for the stimulated emission I_o. The full circles and the triangles correspond to excitation intensities equal to $5 I_o$ and $10 I_o$ respectively. The curves represent the expected red shift of the recombination due to inelastic exciton–phonon (exc-LO) and exciton–exciton (exc-exc) scattering processes.

from these results: first, the red shift of the stimulated emission around threshold ranges between 20 and 35 meV irrespective of the well width (empty dots in Fig. 33). The measured values compare neither with the characteristic exciton binding energy (E_b, from Fig. (13)) nor with the LO phonon energy (LO line) in the superlattices. This rules out the free exciton–exciton [Eq. (25)] or exciton–phonon scattering [Eq (28)] as dominant lasing mechanisms in ZnSe-based MQWs.

Second, the splitting is found to increase considerably with increasing the excitation intensity. This is shown by the different symbols plotted in Figure 33, representing the red shift of the stimulated emission for excitation intensities close to the stimulation threshold I_0 (empty dots), at $5 I_0$ (full dots) and at $10 I_0$ (triangles), as measured in few selected samples. This might explain why, accidentally, particular combinations of sample size and excitation intensity can result in characteristic stimulated emission energies which are apparently consistent with Eqs (25)–(28), thus leading to wrong attributions (Dabbico et al., 1992). We should mention that the absolute position of the stimulated emission is found to depend not only on the temperature and excitation intensity, but also on the physical size of the small cavities cleaved for the high excitation intensity studies.

The exciton electron scattering process is somewhat more difficult to identify. The main emission band should have a clear onset of the low-energy tail of the exciton band, and a broad low-energy tail due to the kinetic energy distribution of the scattered electrons recombining after the collision. However, the stimulated emission of the samples investigated in Figure 33 is always observed with a considerable red shift even at the stimulation threshold. This rules out the exciton–electron scattering as dominant recombination mechanism in our samples. Similar results are found for the ZnCdSe/ZnSe material system. Based on these considerations, we are forced to deal with the two remaining mechanisms: namely, scattering involving localized excitons and free-carrier recombination. Both processes are likely to be dominant in the ZnSe-based MQWs. The presence of localized excitons is demonstrated by a number of optical experiments, namely photocurrent and PLE spectroscopy, in both ZnSe/ZnSSe (Kuroda et al., 1992b) and ZnCdSe/ZnSe (Ding et al., 1990b; Cingolani et al., 1995). The possibility of optical gain at the low-energy tail of a strongly inhomogeneous excitonic density of states has been recently discussed in ZnCdSe/ZnSe heterostructures with rather high Cd content (Ding et al., 1992; Ding et al., 1993). On the other hand, electron–hole plasma recombination can be obtained provided the photogeneration rate is strong enough to overcome the Mott threshold for exciton screening.

The spectroscopic identification of lasing from localized excitons or free carriers is very difficult. A systematic comparison of the pump and probe spectra displayed in Figure 31 and the stimulated emission spectra collected from ZnCdSe MQWs of different composition and size reveals that stimulated emission threshold (of the order of $4-6\,\text{kW cm}^{-2}$) can be either larger or smaller than the exciton bleaching intensity, depending on the exciton stability of the investigated sample. Therefore, excitonic lasing or free-carrier lasing can be obtained depending on the phase diagram of the exciton and free-carrier gas in thermal equilibrium. Two limiting cases can be clearly identified: (a) the deep and narrow quantum wells ($E_b > 32\,\text{meV}$) exhibit stimulated emission at power densities such that distinguishable excitonic features can still be observed in the pump and probe spectra; (b) lasing in the shallow and thick quantum wells ($E_b < 25\,\text{meV}$) occurs at power densities well above the bleaching intensity. In the wide range of intermediate size and composition, the bleaching intensity and the stimulated emission threshold are comparable, preventing a straightforward correlation between the two processes.

An unambiguous identification of these two different recombination mechanisms can be obtained by measuring the diamagnetic shift of the stimulated emission in high magnetic field. The basic idea, which was already successfully applied to GaAs quantum wells (Cingolani et al., 1991), relies on the different magnetoluminescence shift exhibited by excitonic and free-carrier-related features in the optical spectra. The excitonic recombination in magnetic field (B) exhibits a diamagnetic shift quadratic in the field:

$$\Delta E \simeq \frac{\hbar^4 \varepsilon^2}{e^2 \mu^3} \cdot B^2 \qquad (30)$$

where μ is the reduced exciton mass. By using the correct in-plane heavy-hole mass, a very small diamagnetic shift, of the order of 2–4 μeV/T^2 is expected for $Zn_{0.9}Cd_{0.1}Se$. Conversely, the free-carrier recombination is characterized by the well-known Landau shift, linear in the magnetic field:

$$\Delta E \simeq \frac{1}{2} \cdot \left[\frac{\hbar e}{m_e} + \frac{\hbar e}{m_h} \right] \cdot B \qquad (31)$$

In this case a stronger spectral shift of about 0.65 meV/T is found.

A summary of the experimental data obtained from the ZnCdSe/ZnSe MQWs is displayed in Figure 34 (Cingolani et al., 1994e). Clearly two classes of results can be identified: excitonic shift for the deepest quantum wells, with binding energies as high as 35 meV, and Landau shift for the other samples with binding energies smaller than 30 meV. These data demonstrate that both excitonic or free-carrier lasing can occur in these quantum wells.

Shallow quantum wells exhibit a free-carrier behavior. The stimulated emission always occurs at carrier densities corresponding to bleached exciton features in the pump and probe spectra, though smaller than expected from Eq. (16), corresponding to the free-carrier-rich side of the phase diagram. The excitonic stimulated emission occurs in deep quantum wells at carrier densities falling in the exciton-rich side of the phase diagram, as expected from the strong stability of the elementary excitation. In these samples the excitonic recombination is observed up to the strongest injection densities, when the excitonic features have been bleached in the pump and probe spectra. This is probably due to rather high Cd content of these samples causing appreciable compositional fluctuations ($\pm 1\%$) leading to exciton localization. This phenomenon is expected to raise considerably the screening threshold of the exciton, and is not accounted for in the calculation of the exciton/free-carrier gas phase diagram.

Similar results are obtained from the ZnSe/ZnSSe superlattices. In Figure 35 we display a summary of the magnetophotoluminescence experiments performed on a representative ZnSSe/ZnSe sample (5 nm/5 nm superlattice) (Cingolani et al., 1996b). The symbols represent the energy positions of the

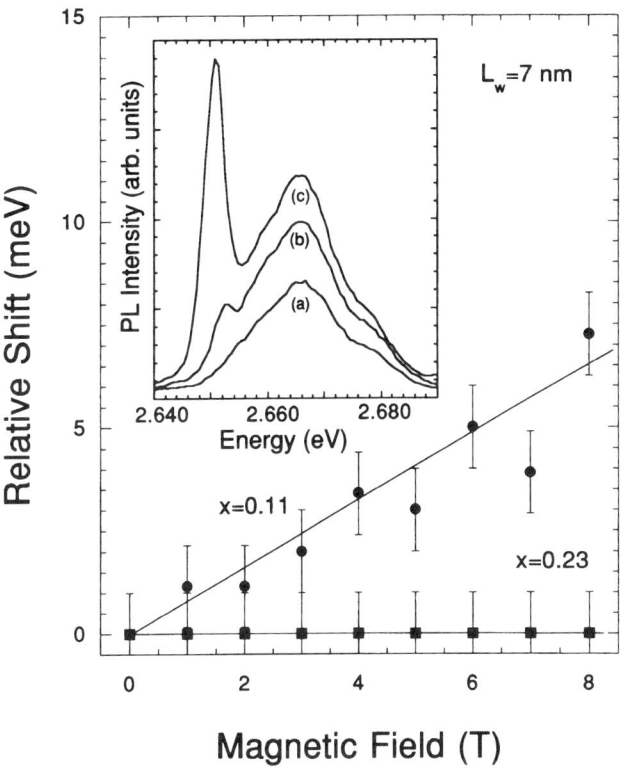

FIG. 34. Shift of the stimulated emission in magnetic field for a deep ($z = 0.23$) and a shallow ($x = 0.11$) $Zn_{(1-x)}Cd_xSe/ZnSe$ multiple quantum well heterostructure at 4 K. The well width is $L_w = 7$ nm. *Inset:* Stimulated emission spectra around the threshold. (a) $1\,kW\,cm^{-2}$, (b) $4\,kW\,cm^{-2}$, and (c) $10\,kW\,cm^{-2}$.

various features displayed in Figure 14. The heavy-hole exciton measured by PLE (HH, empty squares), the strong localized exciton (LE, empty dots), and the first phonon replica (PR_1, empty triangles) measured under low-power continuous-wave excitation exhibit an almost constant energy position up to magnetic field of 8 T. This is consistent with the small excitonic diamagnetic shift predicted by Eq. (30) (continuous lines). Actually an even smaller diamagnetic shift should be observed for localized excitons. However, this is below the resolution limit of the equipment used in this experiment. The spontaneous emission band measured under strong excitation, just below threshold (labeled SP in Fig. 35) also exhibits no detectable shift with increasing the magnetic field (downwards empty triangles). This is a clear signature of excitonic recombination, even though the photo-

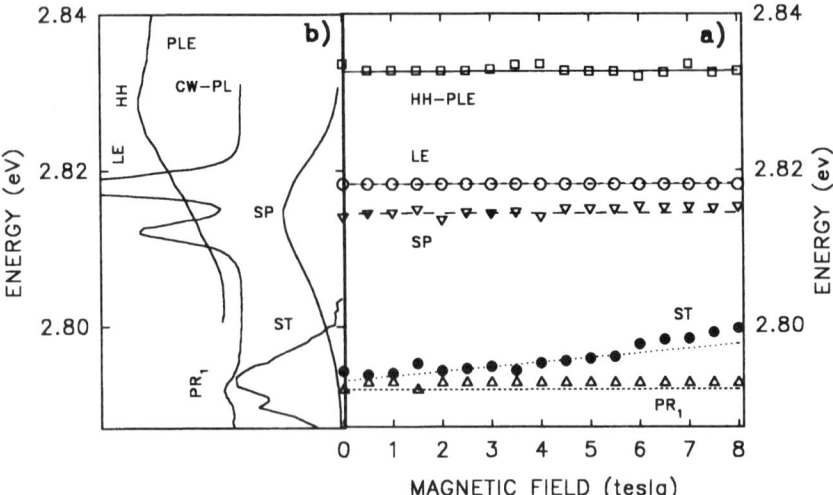

FIG. 35. (a) Magnetoluminescence shift of the main emission lines of a symmetric $ZnS_{0.18}Se_{0.82}/ZnSe$ superlattice at 4 K. Empty symbols indicate excitonic processes: the heavy-hole exciton band in photoluminescence excitation (HH-PLE, empty squares), the localized exciton continuous wave luminescence (LE, empty circles) and its phonon replica (PR and upwards empty triangles), and the spontaneous emission obtained under strong pulsed excitation just below threshold (SP, downwards triangles). The full symbols indicate free-carrier processes associated with the stimulated emission (ST, full circles).

generated carrier density has been raised in the 10^{11} cm^{-2} range. On the contrary, the stimulated emission band (ST and full dots in Fig. 35) is found to shift linearly with the magnetic field, consistent with Eq (31) (dotted line), clearly indicating that lasing occurs through electron–hole recombination in our ZnSSe/ZnSe superlattices. We should mention that the stimulated emission band in Figure 35 has been measured under strong optical pumping of intensity far above the stimulation threshold ($I \sim 10\, I_0$). The energy position is therefore strongly red-shifted with respect to the HH peaks shown in Figure 14 (in this case it is even coincident with the phonon replica PR_1, resulting in possibly erroneous attributions). However, the intensity-dependent measurements confirmed the Landau shift of the stimulated emission band at any injection level above the threshold (with the zero-field spectral position of the lasing line in the range 2.79–2.81 eV). Only at excitation intensities very close to I_0, the stimulated emission line disappeared and the spontaneous emission band (SP) showed up a clear diamagnetic behavior (empty downwards triangles).

VI. Conclusions

In conclusion, we have overviewed the main linear and nonlinear optical properties of excitons in ZnSe-based quantum wells. The fundamental importance of strongly bound excitons in terms of polariton properties, time dynamics, and oscillator strength has been addressed through the discussion of recent spectroscopic experiments. The role of excitons in the operation of optoelectronic devices operating in the blue-green region has been emphasized in the discusion of the nonlinear optical properties. The unique combination of large exciton binding energy and reduced screening and phonon coupling occurring in quantum wells permits the observation of novel phenomena connected with the coexistence of the exciton gas and the free-carrier gas at high density. Besides their fundamental importance, these phenomena are found to play an important role in the physics of blue-green lasers and modulators.

ACKNOWLEDGMENTS

This work is the outcome of a collaboration between different groups. The author wishes to thank the group of optoelectronics at the Material Science Department of the University of Lecce: L. Calcagnile, G. Coli', M. DeVittorio, A. Cola, D. Greco, P. V. Giugno, P. Prete, and R. Rinaldi for the collaborative work on the spectroscopy and transport experiments. M. Lomascolo, M. DiDio, and L. Tapfer of the CNRSM laboratory (Brindisi, Italy) are gratefully acknowledged for their collaboration on the time-resolved luminescence experiments and structural analysis. Special thanks to A. Franciosi, L. Sorba, and L. Vanzetti at the TASC laboratory (Trieste, Italy) for their continuous collaboration and for the MBE growth of the high-quality ZnCdSe/ZnSe heterostructures. The author is also indebted to I. Suemune (Hokkaido University, Japan) for supplying the MOCVD-grown ZnSe/ZnSSe superlattices.

REFERENCES

Alonso, R. G., Parks, C., Ramdas, A., Luo, H., Samarth, N., Furdyna, J. K., and Ram-Mohan, L. R. (1992). *Phys. Rev.* **B45**, 1181.

Andreani, L. C., D'Andrea, A., and Del Sole, R. (1992). *Physics Lett.* **A168**, 451.

Baranovskii, S. D., Doerr, U., Thomas, P., Naumov, A., and Gebhardt, W. (1993). *Phys. Rev.* **B48**, 17149.

Bastard, G., Mendez, E. E., Chang, L. L., and Esaki, L. (1982). *Phys. Rev.* **B26**, 1974.

Bastard, G. (1991). *Wave Mechanics Applied to Semiconductor Heterostructures*. Editions de Physique, Les Ulis, Paris.

Bertho, D., and Jouanin, C. (1993). *Phys. Rev.* **B47**, 2184.
Bratina, G., Nicolini, R., Sorba, L., Vanzetti, L., Mula, G., Yu, X., and Franciosi, A. (1993). *J. Cryst. Growth* **127**, 387.
Calcagnile, L., Lomascolo, M., Rinaldi, R., Prete, P., Cingolani, R., Vanzetti, L., Bonanni, A., Bassani, F., Sorba, L., and Franciosi, A. 22nd International Conference on the Physics of Semiconductors August 15–19, 1994, Vancouver, Canada.
Calcagnile, L., Rinaldi, R., Prete, P., Stevens, C. J., Cingolani, R., Vanzetti, L., Sorba, L., and Franciosi, A. (1996). *Phys. Rev.* **B54**, 17248.
Campi, D., and Coriasso, C. (1995). *Phys. Rev.* **B51**, 7985.
Chemla, D. S., Miller, D. A. B., Smith, P. W., Gossard, A. C., and Wiegmann, W. (1984). *IEEE J. Quantum El.* **QE-20**, 265.
Chung, H. Y. A., Uhle, N., and Tschudi, T. (1993). *Appl. Phys. Rev.* **63**, 1378.
Cingolani, R., Stolz, W., Ploog, K. (1990). *Phys. Rev.* **B40**, 2950.
Cingolani, R., LaRocca, G. C., Kalt, H., Ploog, K., Potemsky, M., and Maan, J. C. (1991). *Phys. Rev.* **B43**, 9662.
Cingolani, R., and Ploog, K. (1991). *Adv. Phys.* **40**, 535.
Cingolani, R., Rinaldi, R., Calcagnile, L., Prete, P., Sciacovelli, P., Tapfer, L., Vanzetti, L., Mula, G., Bassani, F., Sorba, L., and Franciosi, A. (1994a). *Phys. Rev.* **B49**, 16769.
Cingolani, R. Lomascolo, M., Lovergine, N., Dabbicco, M., Ferrara, M., and Suemune, I. (1994b). *Appl. Phys. Rev.* **64**, 2439.
Cingolani, R., Di Dio, M., Lomascolo, M., Rinaldi, R., Prete, P., Vasanelli, L., Vanzetti, L., Bassani, F., Bonnani, A., Sorba, L., and Franciosi, A. (1994c). *Phys. Rev.* **B50**, 12179.
Cingolani, R., de Vittorio, M., DiDio, M., Lomascolo, M., Cola, A., Vasanelli, L., Vanzetti, L., Sorba, L., and Franciosi, A. (1994d). *Superlatt. Microstr.* **16**, 363.
Cingolani, R., Calcagnile, L., Franciosi, A., Sorba, L., and Vanzetti, L. (1994e). *SPIE II–VI Blue Green Laser Diodes* **112**, 2346.
Cingolani, R., Prete, P. Greco, D., Giugno, P. V., Lomascolo, M., Rinaldi, R., Calcagnile, L. Vanzetti, L., Sorba, L., and Franciosi, A. (1995). *Phys. Rev.* **B51**, 5176.
Cingolani, R., Calcagnile, L., Colí, G., Rinaldi, R., Lomascolo, M., Didio, M., Franciosi, A., Vanzetti, L., La Rocca, G. C., and Campi, D. (1996a). *J. Opt. Soc. Am.* **B13**, 1268.
Cingolani, R., Colí, G., Calcagnile, L., Convertino, A. L., Lomascolo, M., Didio, M., and Suemune, I. (1996). *Phys. Rev.* **B54**, 15 December.
Citrin, D. S. (1994). *Phys. Rev.* **B50**, 5497.
Dabbicco, M., Cingolani, R., Ferrara, M., Suemune, I., and Kuroda, Y. (1992). *Appl. Phys. Lett.* **72**, 4969.
D'Andrea, A., and Del Sole, R. (1990). *Phys. Rev.* **B41**, 1413.
Ding, J., Jeon, H., Nurmikko, A. V., Luo, H., Samarth, N., and Furdyna, J. K. (1990a). *Appl. Phys. Lett.* **57**, 2756.
Ding, J., Pelekanos, N., Nurmikko, A. V., Luo, H., Samarth, N., and Furdyna, J. K. (1990b). *Appl. Phys. Lett.* **57**, 2885.
Ding, J., Jeon, H., Ishihara, T. Hagerott, M., Nurmikko, A. V., Luo, H., Samarth, N., and Furdyna, J. K. (1992). *Phys. Rev. Lett.* **69**, 1707.
Ding, J., Hagerott, M., Ishihara, T., Jeon, H., and Nurmikko, A. V. (1993). *Phys. Rev.* **B47**, 10528.
Fu, Q., Lee, D., Mysyrowicz, A., Nurmikko, A. V., Gunshor, R. L., and Kolodziejski, L. A. (1988). *Phys. Rev.* **B37**, 8791.
Fujii, Y., Suemune, I., and Fujimoto, M. (1994). *Jpn. A. Appl. Phys.* **33**, 840.
Galbraith, I. (1992). *Phys. Rev.* **B45**, 6950.
Givgno, P. V., de Vittorio, M., Rinaldi, R., Cingolani, R., Quaranta, F., Vanzetti, L., Sorba, L., and Franciosi, A. (1996). *Phys. Rev.* **B54**, 16934.
Gil, B., Cloitre, T., DiBlasio, M., Bigenwald, P., Aigouy, L., Briot, N., Briot, O., Bouchara, D., Aulombard, R. L., and Calas, J. (1994). *Phys. Rev.* **B50**, 18231.
Gorczyca, I., and Christensen, N. E. (1993). *Phys. Rev.* **B48**, 17202.
Gowar, J. (1984). *Optical Communication Systems.* Prentice Hall, New York. Chapter 8.

Grahan, H. T., Fisher, A., and Ploog, K. (1992). *Appl. Phys. Lett.* **61**, 2211.
Greco, D., Cingolani, R., D'Andrea, A., Tommasini, N., Vanzetti, L., and Franciosi, A. (1996). *Phys. Rev.* **B54**, 16998.
Gurioli, M., Martinez-Pastor, J., Colocci, M., Deparis, C., Chastaingt, B., and Massies, J. (1992). *Phys. Rev.* **B46**, 6922.
Gutowsky, J., Diessel, A., Neukirch, U., Weckendrup, D., Beher, T., Jobst, B., and Hommel, D. (1995). *Phys. Stat. Sol.* **B187**, 423.
Haug, H., and Schmitt-Rink, S. (1985). *J. Opt. Soc. Am.* **B2**, 1135.
Hohnoki, S., Katayama, S., and Hasegawa, A. (1994). *Solid State Commun.* **89**, 41.
Hopfield, J. J., and Thomas, D. G. (1963). *Phys. Rev.* **132**, 563.
Hwang, S. J., Sham, W., Song, J. J., Zhu, Z. Q., and Tao, T. (1994). *Appl. Phys. Lett.* **64**, 2267.
Kawakami, Y., Wang, S. Y., Simpson, J., Hauksson, I., Adams, S. J. A., Stewart, H., Cavenett, B. C., and Prior, K. A. (1993a). *Physica* **B185**, 508.
Kawakami, Y., Hauksson, I., Stewart, H., Simpson, J., Galbraith, I., Prior, K. A., and Cavenett, B. C. (1993b). *Phys. Rev. B* **B48**, 11994.
Klingshirn, C., and Haug, H. (1981). *Phys. Report* **70**, 316.
Kozlov, V., Kelkar, P., Nurmikko, A. V., Chu, C.-C., Grillo, D. L., Han, J., Hua, C. G., and Gunshor, R. L. (1996). *Phys. Rev.* **B53**, 10837.
Kreller, F., Lowisch, M., Puls, J., and Henneberger, F. (1995). *Phys. Rev. Lett.* **75**, 2420.
Kuroda, Y., Suemune, I., Fujimoto, M., and Fuji, A. (1992a). *J. Appl. Phys.* **72**, 3029.
Kuroda, Y., Suemune, I., Fuji, A., and Fujimoto, M. (1992b). *Appl. Phys. Lett.* **61**, 1182.
Kuroda, T. Inoue, K., Suemune, I., and Minami, F. (1994). *Proc. 22nd Int. Conf. Physics of Semiconductors*, Vol. II. Vancouver, 1388.
Landolt-Bornstein. (1982). *Numerical Data and Functional Relationship in Science and Technology*, ed. O. Madelung. Springer, Berlin. Gp. III, Vol. 17a.
Lasher, G., and Stern, F. (1964). *Phys. Rev.* **A133**, 553.
Lawaetz, P. (1971). *Phys. Rev.* **B4**, 3460.
Leavitt, R. P., and Bradshaw, J. L. (1991). *Appl. Phys. Lett.* **59**, 2433.
Lee, J., Koteles, E., and Vassel, M. O. (1986). *Phys. Rev.* **B33**, 5512.
Liaci, F., Bigenwald, P., Briot, O., Gil, B., Briot, N., Cloitre, T., and Aulombard, P. L. (1995). *Phys. Rev.* **B51**, 4699.
Lippens, P. E., and Lanoo, M. (1989). *Phys. Rev.* **B39**, 10935.
Livescu, G., Miller, D. A. B., Chemla, D. S., Ramaswamy, M., Chang, T. Y., Sauer, N., Gossard, A. C., and English, J. H. (1988). *IEEE J. Quant. Electron.* **QE-24**, 1677.
Lomascolo, M., Didio, M., Greco, D., Calcagnile, L., Cingolani, R., Vanzetti, L., Sorba, L., and Franciosi, A. (1996). *Appl. Phys. Lett.* **69**, 1145.
Lomascolo, M., Li, G. H., Syassen, K., Cingolani, R., and Suemune, I. (1994) *Phys. Rev.* **B50**, 14635.
Lozykowsky, H. J., and Shastri, V. K. (1992). *J. Appl. Phys.* **69**, 3235.
Marquadt, E., Opitz, B., Scholl, M., and Heuken, M. (1994). *J. Appl. Phys.* **75**, 8022.
Marshal, T., Colak, S., and Cammack, D. (1989). *J. Appl. Phys.* **66**, 1753.
Masumoto, Y., Matsuura, M., Tarucha, S., and Okamoto, H. (1985). *Phys. Rev.* **B32**, 4275.
Michler, P., Hangleiter, A., Moser, M., Geiger, M., and Scholz, F. (1992). *Phys. Rev.* **B46**, 7280.
Miller, D. A. B., Chemla, D. S., Damen, T. C., Gossard, A. C., Wiegmann, W., Wood, T. H., and Burrus, C. A. (1985). *Phys. Rev.* **B32**, 1043.
Mohammed, K., Olego, D. J., Newbury, P., Cammack, D. A., Dalby, R., and Cornellisen, H. (1987). *Appl. Phys. Lett.* **50**, 1820.
Moss, T. S., Burrell, G. J., and Ellis, B. (1973). *Semiconductor Optoelectronics.* Butterworth and Co., London. Chapter 5.
Neukirch, U. Weckendrup, D., Gutowski, J., Hommel, D., and Landwher, G. (1994). *J. Cryst. Growth* **138**, 861.
Pelekanos, N. T., Haas, H., Magnea, N., Mariette, H., and Asiela, A. W. (1992a). *Appl. Phys. Lett.* **61**, 3154.

Pelekanos, N. T., Ding, J., Hagerott, M., Nurmikko, A. V., Luo, H., Samarth, N., and Furdyna, J. K. (1992b). *Phys. Rev.* **B45**, 6037.
Pellegrini, V., Atanasov, R., Tredicucci, A., Beltram, F., Amzulini, C., Sorba, L., Vanzetti, L., and Franciosi, A. (1995). *Phys. Rev.* **B51**, 5176.
Pollak, F. H., and Cardona, M. (1968). *Phys. Rev.* **172**, 816.
Quiroga, L., Rodriguez, F. J., Camacho, A., and Tejedor, C. (1990). *Phys. Rev.* **B42**, 11198.
Ren, S.-F., Gu, Z.-Q., and Chung, Y.-C. (1994). *Phys. Rev.* **B49**, 7569.
Rudin, S., Reinecke, T. L., and Segal, B. (1990). *Phys. Rev.* **B42**, 11218.
Schmitt-Rink, S., Chemla, D. S., and Miller, D. A. B. (1985). *Phys. Rev.* **B32**, 6601.
Schmitt-Rink, S., Chemla, D. S., and Miller, D. A. B. (1989). *Advances in Physics* **38**, 89.
Shahzad, K., Olego, D. J., and Van de Walle, C. G. (1988). *Phys. Rev.* **B38**, 1417.
Shan, W., Hwang, S. J., Hays, J. M., Song, J. J., Zhu, Z. Q., and Yao, T. (1993). *J. Appl. Phys.* **74**, 5699.
Shen, A., Wang, H., Wang, Z., and Lu, S. (1992). *Appl. Phys. Lett.* **60**, 2640.
Singh, R. K., and Singh, S. (1987). *Phys. Stat. Sol.* **B140**, 407.
Stevens, C. J., Taylor, R. A., Ryan, J. F., Cingolani, R., Dabbicco, M., Ferrara, M., and Suemune, I. (1994). *Semicond. Sci. Technol.* **9**, 762.
Suemune, I. (1993). *Appl. Phys. Lett.* **63**, 2612.
Suemune, I, and Yamanishi, M. (1993) *New Functionality Materials, Volume A: Optical and Quantum-structural Properties of Semiconductors*, ed T. Tsuruta, M. Doyama, M. Seno, and S. Fujita. Elsevier Science Publ., Amsterdam.
Suemune, I., Fujii, Y., and Fujimoto, M. (1994). *J. Cryst. Growth* **138**, 750.
Sugawara, M. (1992). *J. Appl. Phys.* **71**, 277.
Tamargo, M. C., Brasil, M. J. S. P., Nahory, R. E., Martin, R. J., Waever, A. L., and Gilchrist, H. L. (1991). *Semicond. Sci. Technol.* **6**, A8.
Tapfer, L., and Ploog, K. (1986). *Phys. Rev.* **B33**, 5565.
Thomas, R. J., Chandrasekhar, H. R., Chandrasekhar, M., Samarth, N., Luo, H., and Furdyna, J. (1992). *Phys. Rev.* **B45**, 9505.
Tokizaki, T., Sakai, H., Nakamura, A. Manabe, Y., Hayashi, S., and Mitsuyu, T. (1994). *Proc. 22nd Int. Conf. Physics Semiconductors*, Vol. II. Vancouver, 1532.
Tuffigo, H., Cox, R. T., Magnea, N., Merle d'Aub≤igné, Y., Million, A. (1988). *Phys. Rev.* **B37**, 4310.
Tsutsumi, T., Yen, J. Y., Suoma, I., and Oka, Y. (1994). *Superlatt. Microstr.* **16**, 000.
Ves, S., Schwartz, U., Christensen, N. E., Syassen, K., and Cardona, M. (1990). *Phys. Rev.* **B42**, 9113.
Wang, L. and Simmons, J. (1995). *Appl. Phys. Lett.* **67**, 1450.
Wang, S. Y., Simpson, J., Stewart, H., Adams, S. J. A., Hauksson, I., Kawakami, Y., Taghizadeh, M. R., Prior, K. A., and Cavenett, B. C. (1993a). *Physica* **B185**, 508.
Wang, S. Y., Horsburgh, G., Thompson, P., Hauksson, I., Mullins, J. T., Prior, K. A., and Cavenett, B. C. (1993b). *Appl. Phys. Lett.* **63**, 857.
Wang, S. Y., Thompson, P., Horsburgh, G., Mullins, J. T., Livingstone, M., Galbraith, I., Prior, K. A., and Cavenett, B. C. (1994). *Proc. 22nd Int. Conference Physics of Semiconductors*, Vol. 2. 1536.
Wu, J. W., and Nurmikko, A. V. (1988). *Phys. Rev.*, **B38**, 1504.
Yamada, Y., Masumoto, Y., Taguchi, T., and Takemura, K. (1991). *Phys. Rev.* **B44**, 1801.
Yamada, Y., Mishina, T., Masumoto, Y., Kawakami, Y., Suda, J., and Fujita, S. (1995). *Phys. Rev.* **B52**, R2289.
Yang, F., Henderson, B., and O'Donnel, K. P. (1993). *Solid State Commun.* **88**, 687.
Yao, F., Fujimoto, M., Chang, S. K., and Tanino, H. (1991). *J. Cryst. Growth* **111**, 823.
Yokogawa, T., Ishikawa, T., Merz, J. L., and Taguchi, T. Y. (1994). *J. Appl. Phys.* **75**, 2189.
Yokuda, Y., Abe, Y., and Tsukada, N. (1994). *J. Appl. Phys.* **75**, 1620.
Young, P. M., Runge, E., Ziegler, M., and Ehrenreich, H. (1994). *Phys. Rev.* **B49**, 7424.
Zhu, W., Yoshihara, H., Takebayashi, K., and Yao, T. (1993). *Appl. Phys. Lett.* **63**, 1678.

CHAPTER 6

II-VI Diode Lasers: A Current View of Device Performance and Issues

A. V. Nurmikko

BROWN UNIVERSITY
PROVIDENCE, RHODE ISLAND

A. Ishibashi

SONY CORPORATION RESEARCH CENTRE
YOKOHAMA, JAPAN

I. INTRODUCTION 227
II. DESIGNS CONSIDERATIONS 228
   1. Electronic Confinement: Bandoffsets and Quantum Wells 228
   2. Electrical Contacts and Vertical Transport 236
III. DIODE LASER PERFORMANCE AND CHARACTERISTICS 239
   1. Evolution of Diode Laser Design 240
   2. Diode Laser Characteristics 242
   3. Diode Laser Degradation and Reliability 252
IV. PHYSICS OF GAIN AND STIMULATED EMISSION IN ZnSe-BASED QUANTUM
    WELL LASERS 256
   1. Excitonic Molecules and Lasing in ZnSe Quantum Wells 258
   2. Gain Spectroscopy of Blue-Green Diode Lasers at Room Temperature 262
V. SUMMARY 267
    REFERENCES 268

I. Introduction

As illustrated by many examples in this volume, the widegap II–VI semiconductor materials, their heterostructures in particular, have come of age in the past few years. A large amount of work, concentrating on the basic electronic properties of quantum wells and superlattices, has been performed, summarized for example in the recent proceedings of the International Conference on II–VI Compounds (Proceedings 1994, 1996). With parallel progress in important areas such as doping, defect control, and light-emitting device design, there are at this writing a number of indicators that suggest the possibility of technologically viable II–VI lasers in the near future. In this chapter we attempt to provide the reader a view of the progress in the diode lasers, both in terms of the successes and present key challenges. A review article such as this has the danger of "attempting to predict history," given the intensity of research efforts that may provide key breakthroughs on the scale of a year, or discover major obstacles, e.g., in the device processing and manufacturability, subjects that have so far been given little attention.

Work on II–VI lasers is not immune from other contemporary developments, of course. In particular, the remarkable advances that have been made in the GaN-based light-emitting diodes have directed intense attention to the widegap semiconductors and greatly increased the number of researchers in the overall field. While the first rudimentary demonstrations of the InGaN/GaN blue diode laser have been just announced, it is very difficult to anticipate the rate of progress in this case, given, for example, the idiosyncrasies of material microstructure that characterize the present GaN epitaxy. In a broader context, the entry of green and blue diode lasers into optical recording and storage technologies is very much dictated by commercial competition and establishment of recording formats and standards, following in the footsteps of the recent announcements by major international consortia for a common new digital videodisk format in conjunction with a red (670 nm) diode laser. In this sense the "windows of opportunity" for the short wavelength semiconductor lasers are strongly dependent on the dynamics of the marketplace.

This chapter is organized as follows. We first give an overview of the design considerations for a II–VI quantum well emitter from the viewpoint of basic heterojunction physics in Section II. This is followed by a description of the current diode laser devices and their performance characteristics in Section 3. We then discuss the presently all-important topic of defects and device degradation in Section 4, while attempting to make extrapolations to the future. In Section 5 early work on vertical cavity designs is reviewed. The physically very interesting question of the microscopics of gain in the II–VI lasers is the focus of section 6 which complements the chapter by R. Cingolani in this volume. The possibilities that emerge from the strong electron–hole correlations in microresonator light-emitting structures are introduced in Section 7.

II. Design Considerations

1. ELECTRONIC CONFINEMENT: BANDOFFSETS AND QUANTUM WELLS

Since the question of bandoffsets and electronic confinement in II–VI heterostructures is also considered elsewhere in this volume, we examine the

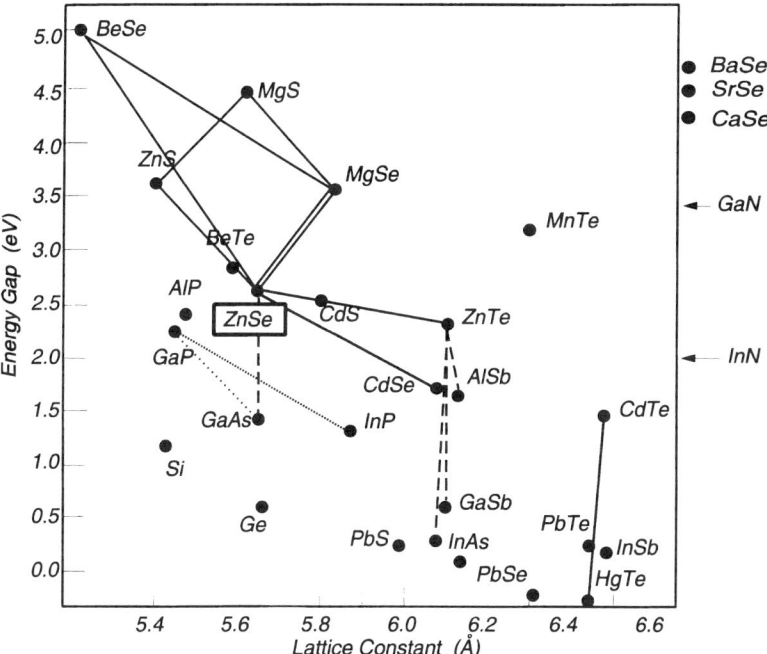

FIG. 1. Energy bandgap vs. lattice constant for common II–VI, III–V, IV, and IV–VI semiconductors in the cubic phase. The quadrangles indicate two important quaternary systems that originate from the ZnSe "hub". Note that there is some uncertainty in the values for MgSe, MgS, and BeSe in the cubic phase. Bandgaps for the widegap III–V noncubic GaN and InN are also shown as well as those estimated for the hypothetical cubic BaSe, SrSe, and CaSe.

issue here mainly from the direct standpoint of the diode lasers. Earlier experience with GaAs and related semiconductor lasers has clearly shown the benefit of the two-dimensional density of states in reducing the threshold current density in a quantum well (QW) laser, provided that requisite optical design demands are met in fabricating the entire device. As discussed below in Sections 6 and 7, as well as in the chapter by R. Cingolani, the presence of a QW active medium assumes additional importance in the widegap II–VI heterostructures due to electron–hole Coulomb enhancement of the interband transition strength.

Figure 1 shows a design map for heteroepitaxy of II–VI compounds in relation to some cubic III–V and IV–VI semiconductors, by plotting the bandgap (at room temperature) against the lattice constant. Both lattice matching and quantum well prospects can be roughly appreciated from this map, although little insight into bandoffsets for the QWs (i.e., partitioning of the bandgap difference between the conduction and valence bands) is gained in the figure. In comparison with the first diode laser demonstrations in 1991 by the 3M group (Haase et al., 1991), followed by the Brown/Purdue team (Jeon et al., 1991), the map reflects the subsequent evolution and new design possibilities in two important ways. First, researchers at Sony Research Laboratories introduced the column IIa element Mg in 1993 (Okuyama et al., 1991) to facilitate the epitaxy of the ZnMgSSe quaternary material which greatly impacted on the subsequent evolution of the diode lasers, as illustrated below. Second, very recently, the group at the University at Würzburg has succeeded in introducing Be as an ingredient to Te- and Se-based widegap II–VI compounds, including the quaternary ZnBeMgSe (Waag et al., in press), further widening the range of design options for blue and green diode lasers. The bandgap value for BeSe is based on a rough theoretical estimate, as are the bandgaps for the "hypothetical" BaSe and CaSe which have not so far been synthesized epitaxially. For reference, the figure also includes the bandgaps for the (noncubic) GaN and InN, respectively.

While the principle motivation for considering group IIa cations (such as Mg and Be) to supplement the traditional group IIb cations (Zn and Cd) in the synthesis of widegap ternary and quaternary materials is the desire to obtain large bandgaps with respect to the binary endpoints ZnTe and ZnSe in lattice-matched heterostructures, other physical attributes must also be considered. Issues of growth (specifically molecular beam epitaxy), doping (especially p-type), and stability of the crystal structure against defects (including strained layer epitaxy) are critical to device applications. Some useful insight can be obtained from simple ideas of consideration of the bond energy from the viewpoint of apportioning the total energy between covalent and polar components. Table I is a selective extraction of data from Harrison's textbook (1980) in which these quantities, obtained from a tight binding model for tetrahedrally bonded (sp^3-hybridized) compounds, are compared in terms of the cation–anion bond length d and the 'hybrid' covalent and polar energies. The total bond energy, apart from an additive average bond energy, is proportional to $(V_2^2 + V_3^2)^{1/2}$. The dependence of the actual bandgap is more complicated due to the inaccuracies of the tight binding approach in determining conduction band energies, but can be very roughly viewed as increasing with the bond energy. We learn from the table, for example, that when compared with GaAs, ZnSe has a comparable covalent energy but is clearly more polar. The larger polarity (roughly connected also to a chemist's definition of ionicity of a bond) implies increasing interaction of electronic and lattice states (hence difficulties with amphoteric doping), and a reduction in the elastic constants (Young's modulus and shear strength) as well as in thermal conductivity. These

TABLE I

Compound	Bond Length (Å)	Covalent Energy (eV)	Polar Energy (eV)
GaAs	2.45	5.55	3.80
CdTe	2.81	4.22	3.13
ZnTe	2.64	4.78	3.04
ZnSe	2.45	5.55	3.80
ZnS	2.34	6.08	4.13
MgTe	2.76	4.37	3.39
BeTe	2.40	5.78	2.79
BeSe	2.20	6.88	3.54
GaN	1.94	8.85	3.92
BeO	1.65	12.23	6.37

implications are only qualitative; the issue of doping is, for example, a much more complex issue at the microscopic level. Nonetheless, we anticipate that the incorporation of Mg into ZnSe and ZnTe would increase the mechanical strength, a trend that should be further enhanced with the inclusion of Be. The benefits of the latter choice have been recently also pointed out by Verie (1995), based on a starting point similar to that of Table I. Until recently, however, the known high degree of toxicity of Be has not encouraged research with these II–VI compounds.

For comparison, Table I also includes values for GaN and BeO. The striking robustness of GaN against high density defect microstructure in the present light-emitting diodes is rooted in the large degree of covalency of the tetrahedral bond. As an extreme example of a very widegap II–VI covalent material, BeO presents an extreme case as a possibility for future consideration.

At present, although some research and development efforts are under way for testing the concept of homoepitaxy on ZnSe-based substrates, GaAs remains the substrate of choice. In order to reach a level of material quality required even for a very brief demonstration of cw diode laser operation, the (molecular beam) II–VI epitaxy is performed predominantly on GaAs homoepitaxial buffer layers. Exact lattice matching can be achieved with ZnS_xSe_{1-x} ($x \approx 0.07$). An including the choice of InGaAs for lattice matching to ZnSe (Kolodziejski et al., 1994). However, the growth of ZnSe on InGaP buffer layers by gas source epitaxy has also been demonstrated quite recently (Ho et al., 1995), thereby widening the potential scope for lattice-matched choices for the II–VI superstructures. Generally, we can expect much more exploration into the use of both III–V and II–VI (and perhaps other) semiconductor mixed crystal substrates in facilitating the exploration of an increasingly wider range of short visible and near UV light emitters. We return to the choice of the GaAs substrate specifically in connection with defects and device degradation later in this chapter.

To date, approximately 20 wide bandgap II–VI heterostructures (by the authors' count) have been investigated in terms of their physical properties. Among the first measurement objectives, following successful epitaxy, has been the determination of the conduction and valence band offsets, usually by optical spectroscopy. Thus, for example $CdTe/Cd_{1-x}Mn_xTe$ and $ZnSe/Zn_{1-x}Mn_xSe$ were early determined to be weakly of type I (i.e., with electrons strongly confined within the QW but holes only weakly so), whereas the ZnTe/ZnSe band alignment is of type II (electrons and holes confined in ZnSe and ZnTe, respectively). Similar early conclusion has been reached with the BeTe/ZnSe heterostructure (Waag et al., in press). The fact that the valence band maximum of ZnTe (and probably of BeTe) is approximately 1 eV higher in energy than that for ZnSe has implications for the design of electrical cotacts to p-type ZnSe and its alloys, as referred to in the next section and discussed elsewhere in this volume.

In terms of the blue-green pn-junction II–VI light emitters which are discussed in this chapter, the most common choice for the quantum well has been $Zn_{1-x}Cd_xSe$, typically with Cd concentration of $x \approx 0.15$–0.25. The epitaxy of this system was first introduced at the University of Notre Dame (Samarth et al., 1990). For the concentration range in question, ZnCdSe is lattice-mismatched with respect to ZnSe barrier layers by a value on the order of 1% (somewhat larger for ZnS_ySe_{1-y} with $y \leqslant 0.07$), restricting the critical QW layer thickness, as found empirically, to about 100–200 Å in a pseudomorphic heterostructure. For the ZnCdSe/ZnSe junction, a common cation/anion design rule is roughly applicable in defining the bandoffsets: most of the bandoffset occurs in the conduction band as the cation Cd lowers the conduction bandedge energy so as to define the QW layer. The useful type I nature of this QW system was recognized early and verified in magneto-optical experiments in which the diluted magnetic semiconductor (Zn, Mn)Se was used as the barrier layer material (Walecki et al., 1990). The biaxial lattice mismatch strain leads to a uniaxial component which enhances the valence band confinement by splitting the heavy–light hole degeneracy so that the effective mass in the growth direction ($m_{h\perp}$) is heavy-hole-like, while the in-plane mass ($m_{h\parallel}$) and its dispersion resembles that of the light hole near $k \approx 0$ (see inset of Fig. 2). Consequently, the in-plane density of states is reduced from the bulk value for ZnCdSe, and aids in reducing the threshold inversion density required for gain. This conclusion is immediate in the conventional degenerate electron–hole plasma model of diode laser gain, and is broadly applicable also in the regime where strong electron–hole Coulomb correlations define the gain, as in the widegap

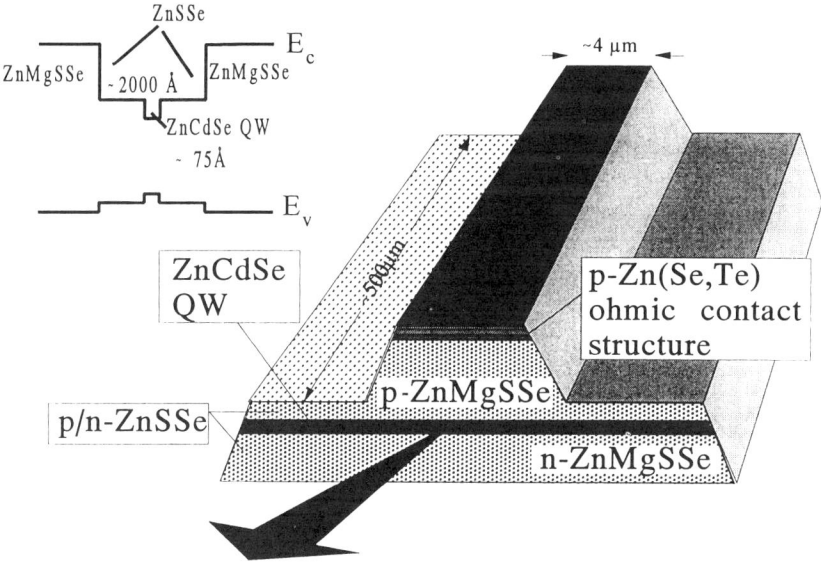

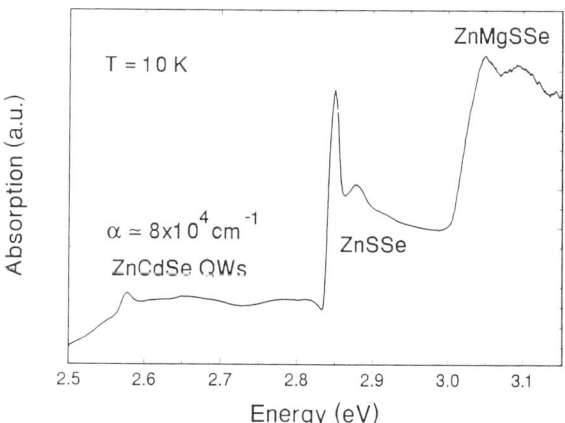

FIG. 2. (top): Schematic of a ZnCdSe/ZnMgSSe SCH-SQW index-guided ridge laser structure; (bottom): Absorbance of a typical laser structure (through the layered structures) at $T = 10$ K, showing the relationship between the excitonic energy bandgaps of the constituent materials.

II-VI systems. Precise hole effective mass values in ZnSe-based alloys are not accurately known. Indirectly, however, magneto-optical spectroscopy of the so-called 1S and 2S exciton states in the ZnCdSe/ZnSe QW has indicated the following in-plane effective mass values: $m_e = 0.17 m_0$ and $m_{h\parallel} \approx 0.2 m_0$ (e.g., Pelekanos et al., 1992).

Optical spectroscopy of the pseudomorphic (Zn,Cd)Se/ZnSe QWs has also yielded approximate values for the bandoffsets from the comparison of observed interband transition energies (from absorption or luminescence), in general agreement by different researchers. This aspect is discussed in detail in the chapter by R. Cingolani in this volume. For example, given a Cd concentration $x \approx 0.24$, representative values are: $\Delta E_c \approx 180$ meV and $\Delta E_v \approx 60$ meV, with the valence bandoffset mainly induced by the strain component (Ho et al., 1995). Replacing the ZnSe barriers by $ZnS_y Se_{1-y}$ pulls the valence band maximum down in energy, hence increasing the QW confinement. For sulfur concentration $y \approx 0.07$, the valence band offset deepens to about 100 meV (Pelekanos et al., 1992), while lattice match with GaAs substrate is obtained (pseudomorphic structure remaining from the growth temperature down to cryogenic temperatures). The energy separation between the lowest heavy-hole subbands in the quantum well ΔE_{h1-h2} is on the order of kT at room temperature (the $n = 1$ heavy-light hole separation is split by the strain by at least 50 meV). This implies a finite population of the second hole subband in a laser structure; not an optimal circumstance. The situation has an impact both on the gain and leakage currents expected in a laser diode and hence further "bandgap engineering" is necessary for optimizing the carrier (and optical) confinement. Accordingly, the first ZnCdSe/ZnSSe II-VI diode lasers typically operated at threshold current densities in excess of 1 kA/cm^2 at room temperature and no cw lasing could be reached. The ZnSe/ZnMgSSe QW offers a larger degree of confinement (ΔE_g up to 400 meV in the composition range which is within relatively easy access by molecular beam epitaxy [MBE] growth); this system has been used as a basis for demonstrating diode laser operation deeper into the blue by researchers at Sony (Okuyama et al., 1993) and the Brown/Purdue team (Grillo et al., 1994).

The introduction of quaternaries into the II-VI heterostructures has made it possible to realize pseudomorphic heterostructures in which the combined requirements for electronic and optical confinement could be better optimized for a diode laser. As already mentioned, the group at Sony Laboratories first showed how the ternary $Zn_{1-x}Mg_x Se$ could be grown epitaxially and used as a QW barrier material in a ZnSe QW diode laser. As demonstrated amply in the development, e.g., of the phosphide-based III-V semiconductor lasers, the choice of a quaternary compound allows much flexibility in the joint control of the bandgap and the lattice constant.

With the inclusion of $Zn_{1-x}Mg_xS_ySe_{1-y}$, a pseudomorphic separate confinement heterostructure (SCH) for a blue-green II–VI diode laser became possible. In 1994, the group at Philips Laboratories (Briarcliff Manor, NY) obtained room temperature threshold current densities with short electrical pulse injection ($\tau_p \approx 50$ nsec) below 500 A/cm^2 from an SCH device (Gaines et al., 1993), which at the time was the lowest of any reported in the II–VI lasers. This basic structure, with some modifications including a particular electrical contacting scheme, was also employed at Sony Laboratories and in the authors' laboratories to demonstrate brief room temperature cw operation (Nakayama et al., 1993; Salokatve et al., 1993); such structures were also realized by the group at 3M (St. Paul, MN) (Haase et al., 1993). Further discussion of the design and performance of these lasers is deferred to section 3. Figure 2 (top) shows a schematic for an index-guided SCH laser structure, the prototype laser geometry at this writing, which also incorporates a graded Zn(Se,Te) pseudoalloy contact discussed below. Figure 2 (bottom) displays the relationship of the bandgap energies between the ZnCdSe Qw, ZnSSe optical waveguide layer, and the ZnMgSSe outer barrier layer, respectively, as measured in optical absorption of a pn-junction device structure at $T = 10$K (Salokatre et al., 1993). At this temperature both the thin QW layer and the ZnSSe layer show a well-

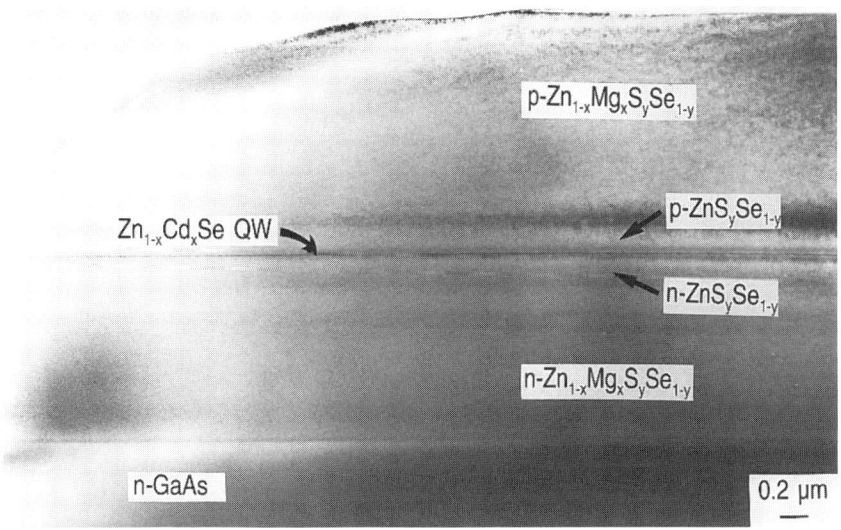

FIG. 3. Cross-sectional transmission electron microscopy image of a pseudomorphic ZnCdSe SCH-QW diode laser structure demonstrating the absence of misfit dislocations (from Hua, 19??).

defined excitonic absorption peak at energy $h\omega = E_g - E_x$, where E_x is the exciton binding energy. (The excitonic bandedge of the quaternary is also sharp, but the $\sim 2\,\mu m$ thick layers attenuate the optical probe so heavily that the peak itself cannot be seen in the figure.)

An illustration of the present quaternary material quality for the SCH-SQW pseudomorphic diode laser structures, Figure 3 shows the details of the multilayer structure in a cross-sectional bright field transmission electron microscope image, including the ZnCdSe QW and the ZnSSe optical waveguide layers. On this spatial scale, the absence of threading/misfit dislocations and stacking faults is also qualitatively consistent with the pseudomorphic condition (Hua, 19??).

2. Electrical Contacts and Vertical Transport

The critical subject of doping and transport in II–VI heterostructures is discussed in detail in this volume in the chapter by J. Han and R. Gunshor. In an attempt to keep the present chapter self-contained, we also review here the question of electrical contacts and vertical transport, specifically from the viewpoint of the blue-green diode laser devices. While considerable progress has been made both at the fundamental level of applying advanced theoretical concepts to p-doping in ZnSe and experimental characterization of doping from the binary compounds to ternaries and quaternaries, at a microscopic level there is much room for an improved understanding, specifically in terms of the maximum possible hole concentration (and mobility). Important questions related to transport are related, for example, to optimizing the injection efficiency in diode laser heterostructures so as to reduce the requisite threshold current densities further. Lateral injection schemes are only in infant stages of investigation for possible new diode laser designs such as vertical cavity emitters.

After the first demonstrations of the II–VI blue-green lasers, it was clear that the lack of proper electrical contacting scheme was responsible for the very large required operating voltages ($\sim 20V$). Today, voltages as low as 4V have been demonstrated by using a graded bandgap scheme already schematically included in Figure 2(a).

To obtain a low resistance contact between a metal and a wide bandgap semiconductor might at first sight present a fundamental dilemma of electrically bridging two vastly dissimilar materials in terms of their electronic band structure. In the case of ZnSe and its alloys, the relative ease in achieving heavy n-type doping (beyond $10^{19}\,cm^{-3}$), together with the position of its surface Fermi level, however, presents a relatively low barrier for electron injection with many common metals (such as indium). Furthermore, in the typical diode laser structures (Fig. 2, top), the electron injection occurs from n-GaAs into the n-type ZnSe-based material so that the problem is transformed to a question about the heterovalent heterointerface. Some potential energy discontinuity is present, perhaps up to 300 meV for GaAs/ZnMgSSe), and, together with interface traps, contributes to a finite heterojunction impedance (hence a voltage drop). While this is not at present the major hurdle, issues about the influence of the microstructure at the heterointerface, i.e. interface states, remain open and are the subject of ongoing research, especially in connection with device degradation as discussed in Section 4. The main challenge exists at the p-ZnSe/metal (or p-alloy/metal contact), due to the fact that the valence band maximum, E_{vm}, in ZnSe (with respect to the vacuum level) is larger than the work function for all common metals so that $E_{vm} < E_f$, where E_F is the Fermi level. Here we highlight the implementation of a graded p-ZnTe/ZnSe heterostructure that greatly facilitates an ohmic contact approach, independently developed by two groups (Fan et al., 1992; Hiei et al., 1993). We also note that, at this writing, it is still undetermined whether this contact scheme is sufficiently robust and offers low enough contact resistance to withstand the requirements of a long lived, *practical* (cw) blue-green diode laser device.

A "bandengineered" contact scheme became available following the discovery that a very high degree of p-doping within the MBE growth of ZnTe could be obtained by nitrogen doping (Fan et al., 1993), with free-hole concentrations approaching 10^{19} cm^{-3} readily achieved. It was also shown that e.g. palladium (including its multilayers with platinum and gold) form a low-resistance contact to such highly doped p-ZnTe epilayers (Ozawa et al., 1994). The Pd/Pt/Au contact has yielded values for the specific contact resistance as low as $5 \times 10^{-6}\,\Omega\,\text{cm}^2$. Hence one is naturally led to consider the use of ZnTe as an intermediate "electrical buffer" for contacting to p-ZnSe. However, the large valence band offset ($\Delta E_v \approx 1\,\text{eV}$) between ZnTe and ZnSe layers forms a barrier to the hole injection at a p-ZnTe/p-ZnSe interface (ZnTe/ZnSe is a type II QW). A ready solution for removing the expected potential energy barrier "spike" at such an abrupt interface due to the electrostatic charge redistribution is to introduce a graded bandgap p-ZnSe$_{1-y}$Te$_y$ layer in which the Te concentration varies continuously from $y = 0$ to 1 across such a contact layer. Due to the practical difficulty of controlling the Te concentration in an MBE grown Zn(Se, Te) alloy (selenium and tellurium compete for surface incorporation), a "pseudograded" p-type, strained ZnTe/ZnSe ultrathin layer structure (20 Å per cell) was designed and grown, with the ZnTe and ZnSe layer thicknesses in each cell varying to approximate a graded bandgap material. The physical action of such a discrete structure can be viewed from the standpoint of a quasi-continuous graded potential (the hole wavelength larger than the supercell

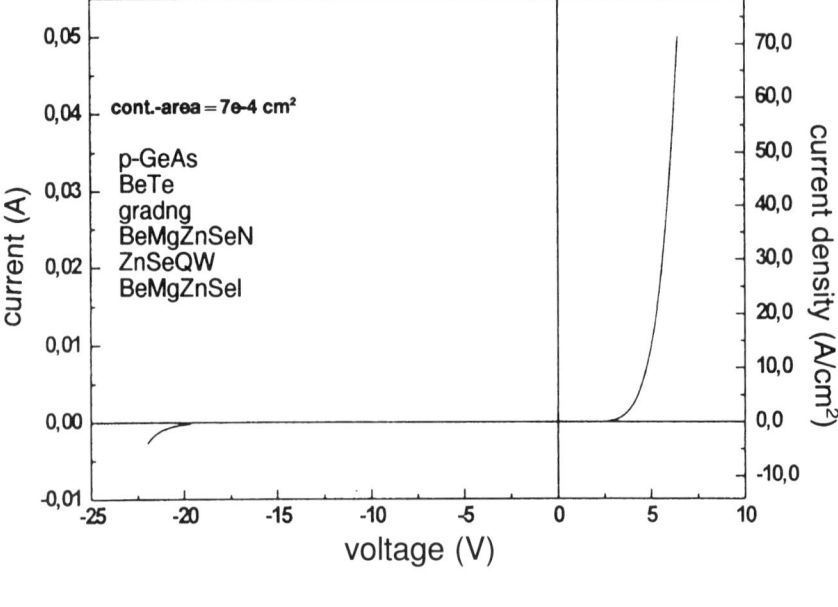
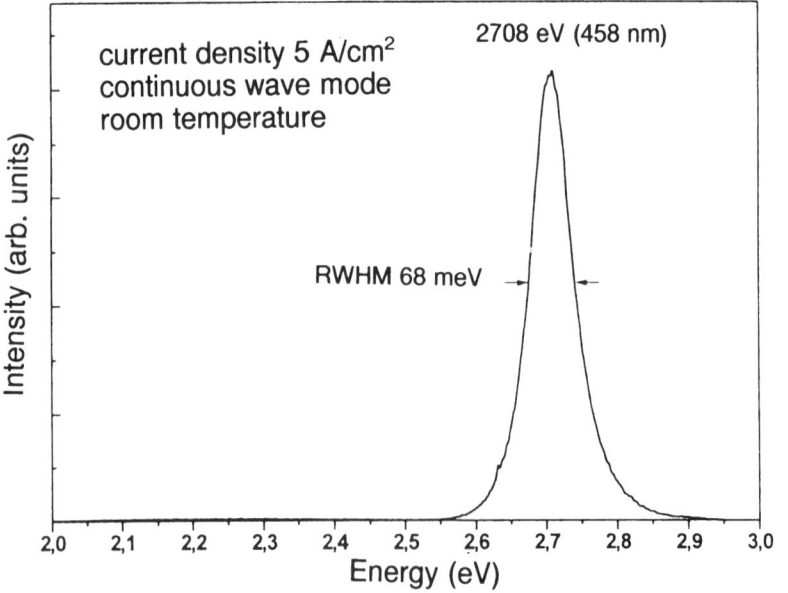

FIG. 4. I–V characteristic of a BeMgZnSe/ZnSe light-emitting diode grown on p-type GaAs. The corresponding emission spectrum at 5 A/cm² is shown in the bottom (from Waag et al., in press).

size) (Fan et al., 1992) or by evoking resonant tunneling in the suerlattice (Hiei et al., 1993). Although the overall "superlattice" thickness (≈ 340 Å) exceeds substantially the critical thickness for the 7% lattice mismatch between ZnSe and ZnTe, electron microscopy shows that the dislocations in the contact layer do not propagate into the adjacent quaternary in a SCH-SQW diode laser structure.

As a consequence of the Zn(Se, Te) graded gap scheme, the specific contact resistance for contacting to p-ZnSe epitaxial layers has been determined to be typically about $3 \times 10^4 \,\Omega\,\text{cm}^2$. This value is certainly acceptable for LEDs but is somewhat high for laser diode devices. One might well expect an improvement in performance of this contact scheme as the growth is modified to more closely approach a continuous alloy grading and as p-doping levels in the selenide layers are increased.

Very recently, an earlier proposal to employ Be-containing II–VI materials as a means to synthesize a low resistance contact to p-ZnSe (Mensz, 1994) has been realized in practice by the group at University of Würzburg (Waag et al., in press). In contrast to the Zn(Se, Te) pseudoalloy, the BeTe/ZnSe system can be lattice-matched to GaAs, and hence MBE grown on p-GaAs to facilitate hole injection from the substrate. As with ZnTe, BeTe can be doped to yield high hole densities and its valence band maximum is at a comparable energy. Waag and co-workers have incorporated a (Be, Zn)(Se, Te) pseudoalloy into blue LED structures whose characteristics are shown in Figure 4 (Waag et al., in press). The active portion of the heterostructure consisted of a ZnSe QW layer with quaternary (Zn, Mg, Be)Se barriers, grown on p-GaAs and the graded contact. The upper portion of this figure shows the I–V characteristics of a device (contact area $7 \times 10^{-4} \,\text{cm}^{-4}$) while the bottom trace shows the electroluminescent spectrum at room temperature. While comparable current densities can be achieved at lower voltages in the more established structures such as that of Figure 2(a), the early results shown in Figure 4 represent a very attractive alternative, given in part by arguments about the increased covalency in the Be-compounds already referred to above. We return to the subject below in the context of device degradation.

III. Diode Laser Performance and Characteristics

In this section we focus on the device concepts and engineering of II–VI blue-green diode lasers, which at this writing (late 1995) are under basic research and development in a number of laboratories worldwide. It is very important for the reader to recognize that from a technological perspective, the device subject is still immature, following the first demonstrations of the diode laser at a cryogenic temperature in 1991 (Haase et al., 1991; Jeon et al., 1991). The diode lasers are subject to a wide spectrum of vigorous studies that range from their basic physics of operation, to device engineering, to degradation studies. Given the pace of progress and the technical accomplishments in the past 3 years, there is cause for optimism about further advances that can be expected with the II–VI lasers on their journey toward future optoelectronic applications. The present laser devices are still subject to rapid degradation and failure (by the standards of a mature technology), and continued improvements in the ability to control defects, both in as-grown heterostructure materials as well as in subsequent device processing, are examples of key issues in this context. From another viewpoint, the ability to approach material challenges by designing heterostructures for tailored physical properties within an increasingly wide choice for the constituent materials, coupled with the very attractive optical properties inherent to II–VI semiconductors, offers considerable room for innovative device physics and engineering in the future.

1. EVOLUTION OF DIODE LASER DESIGN

We have already introduced in the above (Fig. 2(a)) the most common heterostructure arrangement that has formed the testbed device structure for the II–VI edge emitting diode laser in a number of laboratories. This separate confinement quantum well heterostructure typically operates near 500 nm wavelength and longer (to 530 nm), depending chiefly on the choice of the Cd-concentration in the active ZnCdSe QW. Much of the evolution toward this design has been shaped by concurrent developments in II–VI epitaxy, with the aim toward the implementation of the pseudomorphic SCH–QW configuration. The advantages of the joint electronic/optical confinement in an SCH structure are, of course, very well known in III–VI semiconductor lasers, but limitations of lattice matching constraints prevented its realization in the ZnSe-based emitters until the introduction of the (Zn, Mg)(S, Se) quaternary material. Other pseudomorphic configurations such as the graded index design, which have been hampered by strain-induced defects in II–VI lasers, are only now becoming possible. We also note that a high misfit dislocation density (beyond 10^7 per cm^2) was present in the strain-relaxed structures that were first demonstrated as diode lasers in 1991. On the other hand, ZnCdSe/ZnSSe pseudomorphic configurations possessed a poor optical confinement factor (the optical intensity overlap with the confined electronic states of the QW of $\Gamma \sim 10^{-2}$ and lower), and were subject to large leakage currents. In either case, it was

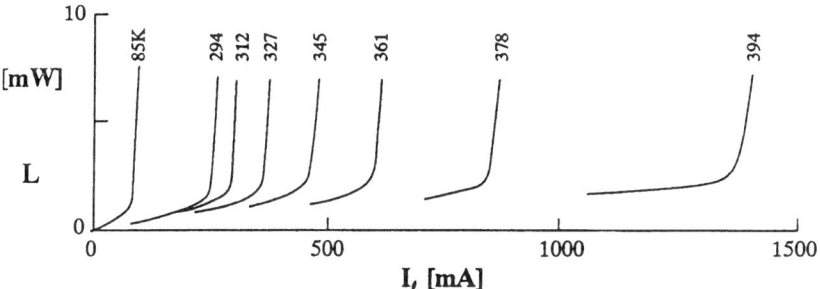

FIG. 5. Output characteristics of a gain-guided ZnCdSe/ZnSSe/ZnMgSSe SCH-SQW diode laser, operated under low duty cycle pulsed injection (5×10^{-4}; $\tau \approx 50$ nsec) (from Gaines et al., 1993).

difficult to reach room temperature operation, although with reflective end facet coatings, pulsed threshold current densities of about $1\,\text{kA/cm}^2$ were obtained (Walker et al., 1993; Jeon et al., 1993). Even at low duty cycles ($\sim 10^3$), however, such devices operated only a few minutes. On the other hand, cw operation at T = 77K was relatively easily accomplished, producing output powers on the order of a mW.

A laser design composed of a ZnSe QW with (Zn, Mg)Se barrier layers was introduced by the group at Sony Laboratories in 1992 to show wavelength shortening effects by the choice of the wide bandgap ternary material for (Okuyama et al., 1993); device operation was still limited to cryogenic temperatures, probably due to both inadequate optical and electronic confinement. Indicating marked improvement and the direct impact of the SCH configuration, Figure 5 shows the output characteristics obtained by the group at Philips in 1993 from a gain-guided ZnCdSe/ZnSSe/ZnMgSSe SCH-SQW diode laser, operated under low duty cycle, very short pulsed injection ($\tau \approx 50$ nsec) but reaching well beyond room temperature (Gaines et al., 1993). In these devices, with the Cd concentration typically $x \approx 0.15$ to 0.20, sulfur concentration $y = 0.07$, and the Mg concentration $x' \approx 0.10$ to 0.20, the room temperature current injection density was measured as approximately $500\,\text{A/cm}^2$ for devices of about 1 mm in length, an excellent figure of merit. Differential quantum efficiency of about 17% was also obtained. From the measured values for the index of refraction for ZnMgSSe, a typical value for the electronic/optical confinement factor is about $\Gamma \approx 0.03$ for a quantum well thicknesses of $L_w \approx 75\,\text{Å}$ and $\sim 2000\,\text{Å}$ for the ZnSSe optical guiding layer. Very similar diode laser structures were soon fabricated by the Brown/Purdue team, but now incorporating the graded gap ZnSeTe contact layer (Grillo et al., 1993), as well as by the groups at 3M and Sony. The 3M group reported a very low threshold current for their (pulsed) operation of 2.5 mA in narrow (2 μm mesa width) index-guided devices (Haase et al., 1993).

2. DIODE LASER CHARACTERISTICS

The benchmark development with the ZnCdSe/ZnSSe/ZnMgSSe SCH/SQW lasers, namely the observation of operation under *continuous* electrical injection, with the device lasting about 1 sec came from Sony Laboratories in 1993 (Nakayama et al., 1993). A relatively high voltage (~ 14 V) and current, however, was required in these gain-guided devices. Also in 1993, the Brown/Purdue team achieved the room temperature cw operation in ridge waveguide device (~ 4.2 μm wide) into which the Zn(Se, Te) graded bandgap ohmic contacts were incorporated (Salokatve et al., 1993). With reflective facet coatings in these ~ 500 μm long devices, laser threshold currents in the 5–7 mA and voltages of about 5.8 V were obtained. Still, the device lifetime was short, approximately 30 sec. As an example of subsequent progress, Figure 6 shows the optical and electrical characteristics from a recent device study from Sony Laboratories in very similar structures, for which the room temperature pulsed and cw threshold voltages were 4.5 V and 4.7 V, respectively (Itoh et al., 1994). The Philips/3M team has reported

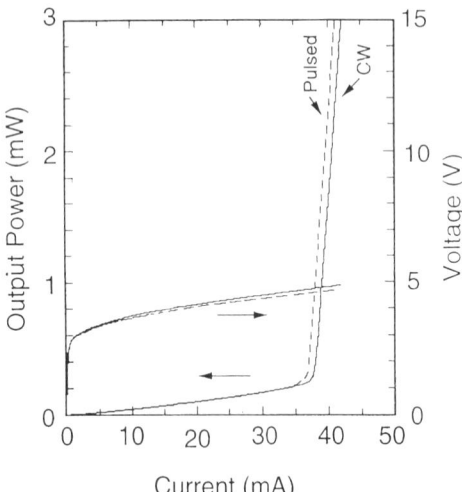

FIG. 6. I–V, L–I, characteristics of a ZnMgSSe-based QW diode laser under CW room temperature operation. The performance under pulsed injection is shown for comparison (after Itoh et al., 1994).

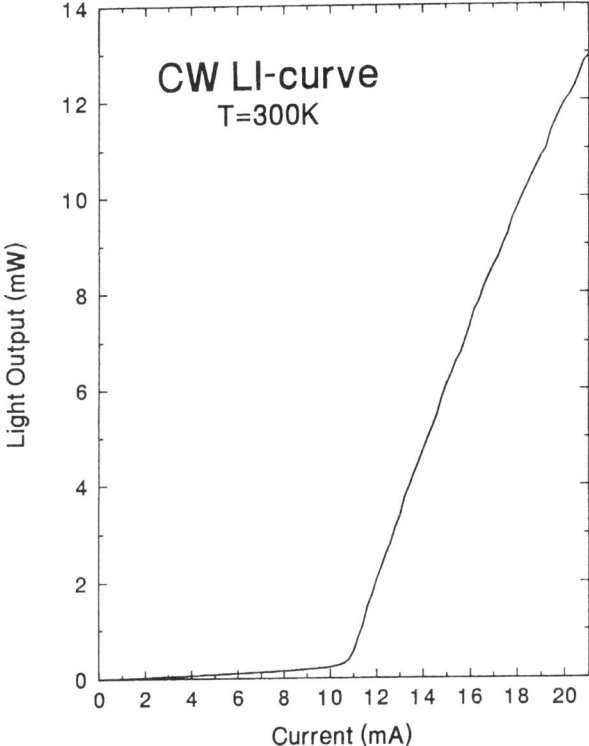

FIG. 7. Laser output characteristics of a CW ridge waveguide device under higher injection exhibiting a differential quantum efficiency of approximately 60% (from Salokatve, 19??).

even lower voltages in their recent devices, on the order of 4 V (Marshall et al., 1994), again on very comparable designs.

The output characteristics of the ZnCdSe/ZnSSe/ZnMgSSe SCH/SQW lasers have been studied to show single transverse mode characteristics have been obtained in the index-guided geometry for both single and multiple quantum well devices. Continuous wave powers up to several tens of mW have been reported by the leading groups and, although under such high injection the device lifetimes are still very short, such power levels demonstrate the capability at least in principle for the II–VI lasers to meet the requirements of rewriteable optical disk technology. Figure 7 shows the cw output from a laser structure with one end facet coated, demonstrating a differential quantum efficiency of about 60%. Furthermore, the best cw threshold current densities that are being reported today are in the range of 300–400 A/cm^2 for SCH devices without facet coatings. In general, an important observation about the II–VI diode lasers is that their key performance characteristics are very competitive with those of the red and infrared III–V lasers, demonstrating that the widegap semiconductors are, at least at a fundamental level, excellent choices for sources of stimulated emission.

As a specific example of such characteristics, results have appeared (Kozlov et al., 1994) that employ the Hakki-Paoli method (Hakki et al., 1975) for the measurement of the gain spectra in the SCH diode lasers. In this method, the gain is retrieved by analyzing the ratio r between the maxima and minima in the Fabry-Perot oscillations in the emission spectrum. This ratio can be calculated at any wavelength by first averaging the intensity of each two adjacent maxima as $I^+ = (I_i + I_{i+1})/2$. By measuring the intensity of the associated minimum I^-, one obtains for the particular resonance $r_i = I^+/I^-$. The corresponding net modal gain is then obtained from

$$\alpha = \gamma - \Gamma g = \frac{1}{L} \ln\left[\frac{\sqrt{r}+1}{\sqrt{r}-1}\right] + \frac{1}{L} \ln \sqrt{R_1 R_2} \qquad (1)$$

where g is the QW gain coefficient, Γ is the optical confinement factor, γ is the internal cavity loss, L is the length of the laser, R_1 and R_2 are the reflection coefficients of the cleaved, uncoated resonator facets ($R_1 = R_2 \approx 0.21$).

Figure 8 shows the gain/loss spectra (α vs. λ) determined for a range of different injection current levels (Kozlov et al., 1994). The threshold current for the particular device was $I = 28$ mA. The transparency current is reached at about 4 mA ($J < 100$ A/cm^2) and the peak gain reaches the value of $\alpha \approx -38$ cm^{-1} at laser threshold. From ellipsometric measurements of the index of refraction for the constituent layers in the SCH heterostructure, we calculate for the confinement factor a value in the range of $\Gamma \approx 0.02$–0.03. Hence the peak QW gain at threshold approaches the value $g_{QW} \approx 1500$ cm^{-1} and beyond. The transparency condition ($\alpha = 0$) is reached at a current of approximately I ≈ 5 mA. A current independent loss ($\gamma_{sc} \approx 6$ cm^{-1}) is attributed to scattering losses in the ridge waveguide structure. Kondo et al. (1994) have reported even lower losses (2–4 cm^{-1}), confirming that the threshold current density is dominated by the mirror losses.

In conjunction with the efforts to optimize the cw laser design for the SCH structures discussed above, efforts have also been underway to shorten the wavelength of the emission further into the blue regime. For example, the Sony group reported cw operation at room temperature at $\lambda = 489.9$ nm of a structure in which somewhat larger bandgap energies

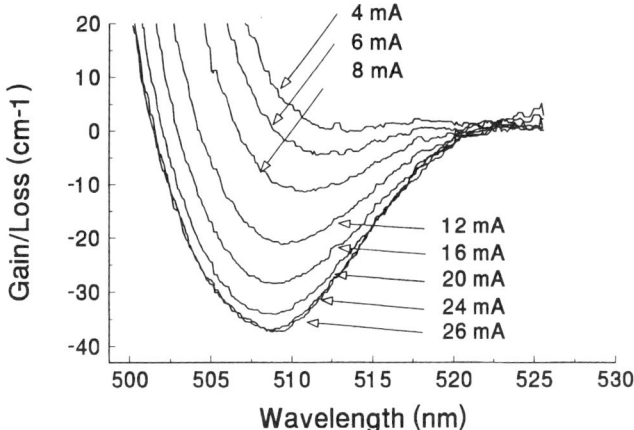

FIG. 8. The gain/loss spectra as a function of injection current retrieved from the Hakki-Paoli method for a ZnCdSe/ZnSSe/ZnMgSSe SCH-SQW ridge waveguide laser (from Kozlov et al., 1994).

were adopted for the active and the cladding layers (Nakayama et al., 1993). The threshold current density and operating voltage were $1.5 \, \text{kA/cm}^2$ and $6.3 \, \text{V}$, respectively. In recent exploratory work on ZnCdSe/$\text{Zn}_{1-x'}\text{Mg}_{x'}\text{S}_{y'}\text{Se}_{1-y'}$/$\text{Zn}_{1-x''}\text{Mg}_{x''}\text{S}_{y''}\text{Se}_{1-y''}$ and ZnSe/$\text{Zn}_{1-x'}\text{Mg}_{x'}\text{S}_{y'}\text{Se}_{1-y'}$/$\text{Zn}_{1-x''}\text{Mg}_{x''}\text{S}_{y''}\text{Se}_{1-y''}$ heterostructures at Sony Laboratories and the authors' group, the emission wavelengths under pulsed excitation at room temperature have been shortened from about 480 nm (Okuyama et al., 1994) to 470 (Ohata et al., 1994) to 460 nm (Grillo et al., 1994) by judiciously increasing the Mg concentration in the barrier layers. As an example, Figure 9 shows the spectrum from a blue ZnSe QW laser both below and above the lasing threshold ($J_{\text{th}} \approx 2 \, \text{kA/cm}^2$ per QW) (Okuyama et al., 1994). A portion of the spectrum form an InGaN/GaN LED is included for comparison. Two particular problems await a solution in attempts to reach into the cw regime of operation with these devices: (1) the rather rapid drop of the free-hole concentration in p-ZnMgSSe with Mg-concentration (typically into the range of $<10^{16} \, \text{cm}^{-3}$ for $E_g \approx 3.0 \, \text{eV}$), and (2) the appearance of ordered alloy structure in the quaternary along specific crystallographic directions. Cross-sectional transmission electron microscopy (TEM) revealed the presence of one-dimensional quasi-periodic compositional modulation along the [110] direction in the ZnMgSSe outer cladding layers of the laser structure (Okuyama et al., 1994). The microscopic issues that underlie the reduction in the level of p-doping are discussed by J. Han and R. Gunshor elsewhere in this volume; the effects of spontaneous ordering of the alloy are not yet clear on the diode laser performance. By and large, these devices are

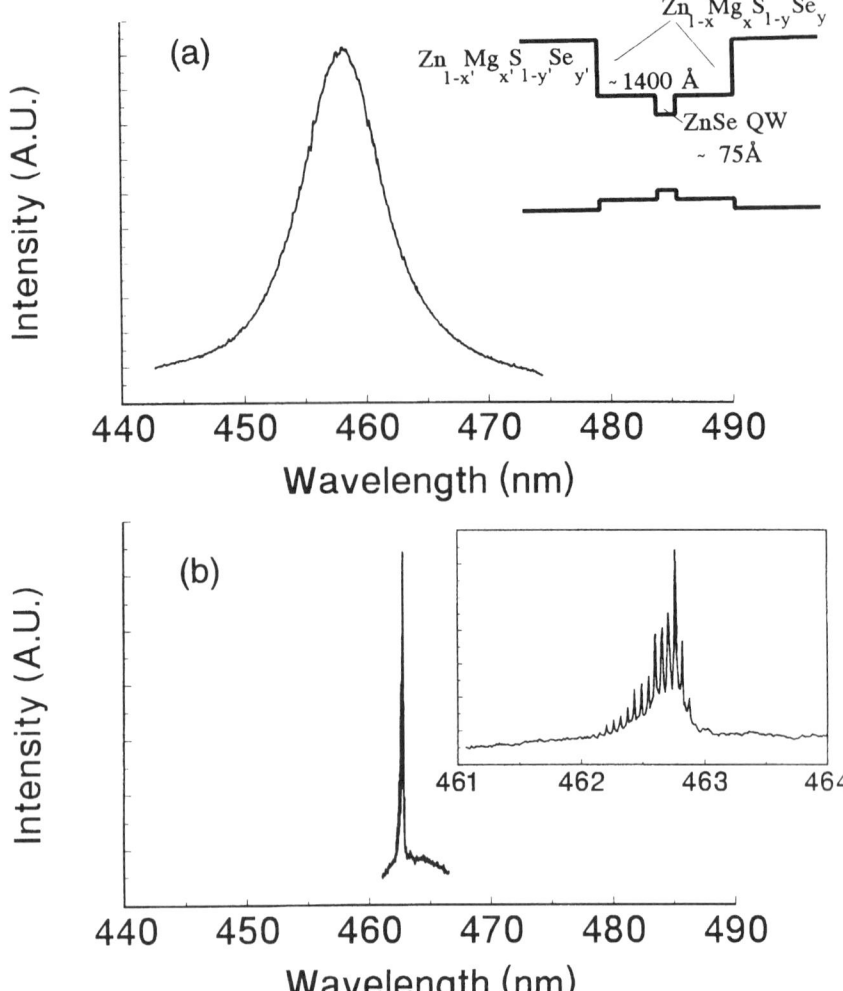

FIG. 9. (a) Top emission spectrum from a ZnSe/ZnMgSSe/ZnMgSSe SCH heterostructure, in the LED mode, at room temperature, and comparison with a commercial InGaN/GaN LED. The inset shows a bandstructure schematic of the active portion. (b) Laser (edge) emission spectrum at slightly above the threshold at $T = 300$ K; the inset displays longitudinal mode structure (from Grillo et al., 1994).

presently test structures designed for feasibility studies, whose evaluation will pave the way for shortening the emission wavelengths deeper into the blue at lower threshold current densities by appropriate adjustment of the compositional ratios and the choice of substrate.

For optimizing the performance of blue and green II–VI diode lasers in terms of a low threshold current density, the details of vertical transport are presently being researched further, specifically from the viewpoint of carrier capture and leakage in the various heterostructure configurations. Complicating the issue is a degree of uncertainty about the bandoffset values, carrier lifetimes, dopant profiles, etc., which require more systematic experimental and theoretical studies. Furthermore, the cw devices have only become available rather recently for such investigations. The issue of leakage was considered early in the II–VI diode lasers by Suemune (1992) and Kuramoto et al. (1993). Subsequently, Mensz (1994) performed fairly comprehensive numerical simulations of the SCH device structures to show, for example, how the smaller bandgap difference in the blue laser devices, based on ZnSe/ZnMgSSe/ZnMgSSe heterostructures, would lead to a significant increase in the threshold current density due to insufficient carrier confinement and the low conductivity of the p-ZnMgSSe cladding layers. Itoh et al. (1994) have argued on empirical experimental grounds that a minimum bandgap difference of $\Delta E_g \approx 0.35\,\text{eV}$ is required for adequate suppression of the leakage and carrier overflow.

Recently, Buijs et al. (1995) have investigated the issue of carrier confinement and leakage in the ZnCdSe/ZnSSe/ZnMgSSe SCH diode lasers by measuring the threshold current density J_{th} and differential quantum efficiency as a function of device length and temperature. Figure 10 shows results from such temperature dependence, which can be connected to empirical models used earlier in the III–V lasers (????, 1993). In the absence of leakage, conventional electron–hole plasma model for gain and transport predicts that the laser quantum efficiency is inversely proportional to the device length. For sufficiently short devices, however, the mirror output loss coefficient is proportionately larger, necessitating higher injection. The increased injection can lead to finite leakage which is observed in the figure. Assuming an empirical exponential temperature dependence for J_{th}, model calculation can then be performed for the (temperature-dependent) leakage current. A main conclusion of the analysis is that, due to the low conductivity of the p-type ZnMgSe, drift current contributes significantly to the leakage (even dominating the diffusive contribution), a fact that rapidly reduces the performance characteristics of a ZnCdSe/ZnSSe/ZnMgSSe SCH diode laser at elevated temperatures.

In addition to the gain and index-guided SCH laser devices described above, we wish to mention recent work on edge emitting devices in which other design approaches and designs have been adapted. Kawasumi et al. (1995) have fabricated an index-guided diode laser in which the ZnCdSe/ZnSSe/ZnMgSSe SCH-QW heterostructure has been grown on a structured n-GaAs substrate having a p-GaAs current blocking epilayer. The blocking

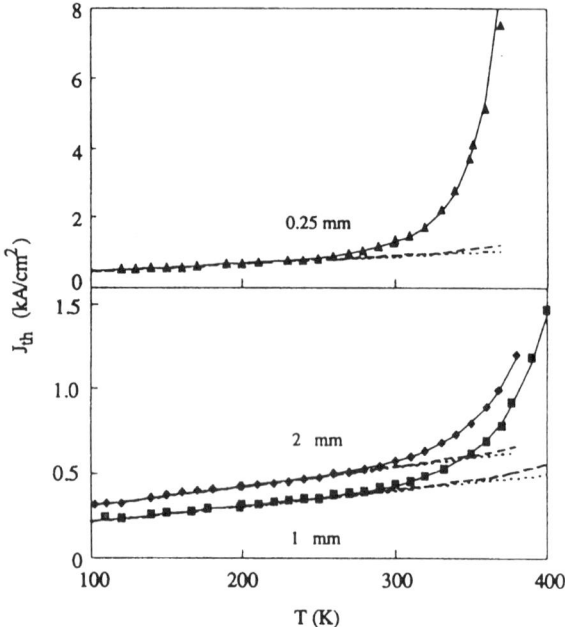

FIG. 10. Threshold current density vs. temperature for ZnCdSe/ZnSSe/ZnMgSSe SCH diode lasers of different length L (measured under pulsed injection). Triangles: $L = 0.25$ mm; squares: $L = 1$ mm; diamond: $L = 2$ mm. Solid curve represent a calculation which accounts for drift and diffusion; broken line for diffusion only, and dotted line a linear fit to low temperature data (from Buijs et al., 1995).

concept is based on the relatively low electron mobility of the n-type II–VI material when compared with that in p-GaAs. Figure 11 shows the structure (inset) together with the optical output characteristics. Fabrication and design approaches such as this are attractive in that they minimize the possible damage which can accompany the dry etching methods that are normally used to define the II–VI mesa in the conventional index-guided lasers. Elsewhere, following optically pumped demonstrations of lasing in distributed feedback ZnCdSe QW structures (Ishihara et al., 1992; Eisert et al., 1996), operation of an electrically injected device in first order Bragg diffraction has been demonstrated very recently (Bacher et al., 1996). In the optical pumping experiments by Bacher et al., both gain and index modulation were accomplished by nanofabrication techniques with grating periods below 90 nm. The gain modulation was accomplished by selective ion implantation with Ga^+ ions through the active QW layer of a ZnSe/ZnSSe/ZnMgSSe SCH structure, while wet etching of the optical cladding layer was

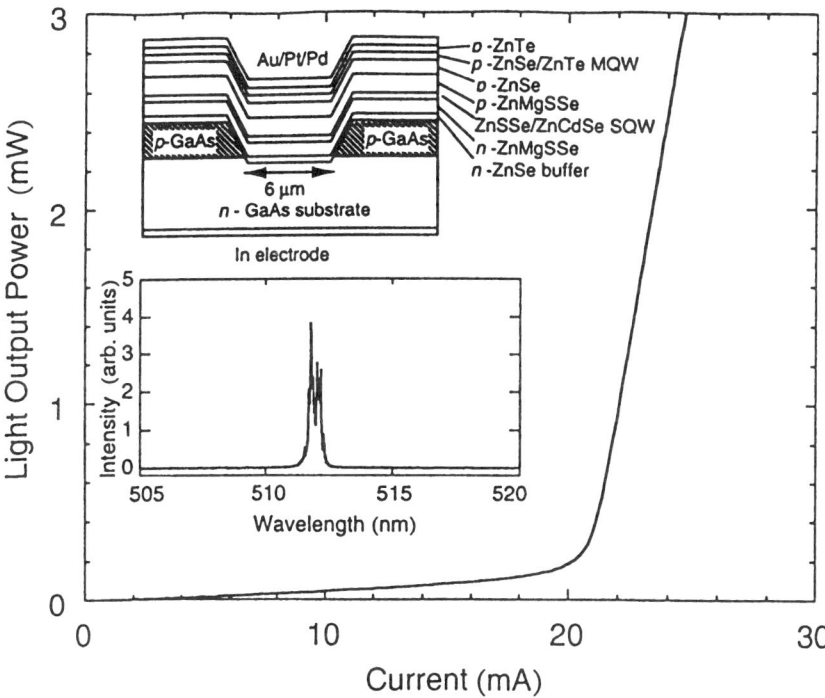

FIG. 11. The L–I characteristic of an index-guided diode laser grown on a structured GaAs substrate. Top inset shows the structure while bottom inset shows the lasing spectrum (from Kawasumi et al., 1995).

used to produce the index-modulating grating structure. Figure 12 shows the spectral output characteristics of such DFB lasers at room temperature (Eisert et al., 1996).

For future applications, there will be a role for blue-green diode lasers that operate in geometries other than the edge emitters described above. At this writing preliminary steps are being taken in the laboratory to explore the device challenges for II–VI surface emitting heterostructures. This exploration has included the demonstration of an optically pumped "whispering gallery" (Hovinen et al., 1993) and vertical cavity lasers (Jeon et al., 1995, 1995a). We review briefly the progress in the latter area.

A number of fabrication and design questions can be addressed by the demonstration of an optically pumped VCSEL as a first step. Such VCSEL operation has been demonstrated in the 480–500 nm range, including room temperature, in ZnCdSe/ZnSSe/Zn, MgS, Se pseudomorphic separate confinement heterostructures (SCH), containing three active QW layers and

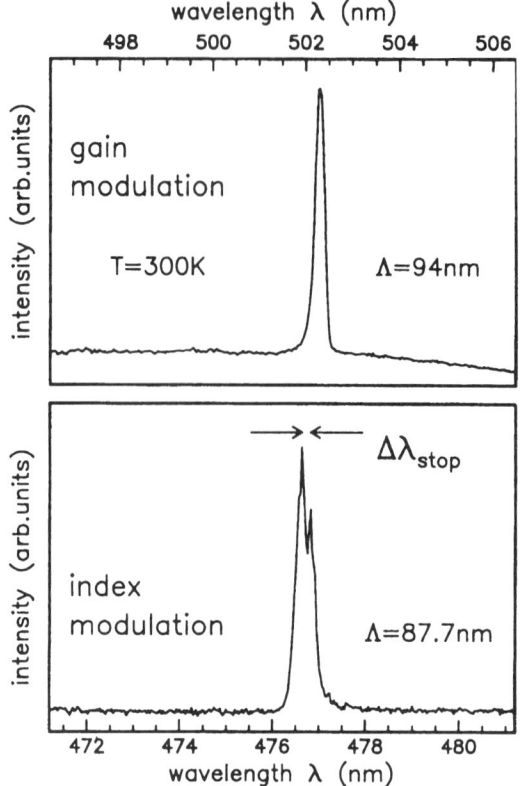

FIG. 12. Optical output characteristic of a gain-modulated and index-modulated DFB optically pumped ZnCdSe SCH QW laser operating in first order diffraction (from Bacher et al., 1996).

equipped with dielectric high reflectivity mirrors sandwiching an approximately 4λ thick II–VI heterostructure. The rather small index difference between a binary compound and its ternary and quaternary alloys in widegap semiconductors (at moderate composition levels) places serious challenges for the *in situ* epitaxy of distributed Bragg reflectors (DBR) for a blue-green VCSEL, typically requiring up to 100 layer pairs of materials whose composition control is quite nontrivial. On the other hand, it is possible to fabricate the VCSEL planar resonator with dielectric mirrors added *ex situ*; in particular, reactive ion-beam sputter deposition of low-loss multilayer SiO_2/HfO_2 stacks has provided high Q resonators for ZnSe-based heterostructures $(Q > 10^3)$. The required mirror performance can be estimated from gain spectral measurements for edge emitters (Kozlov

6 II-VI DIODE LASERS

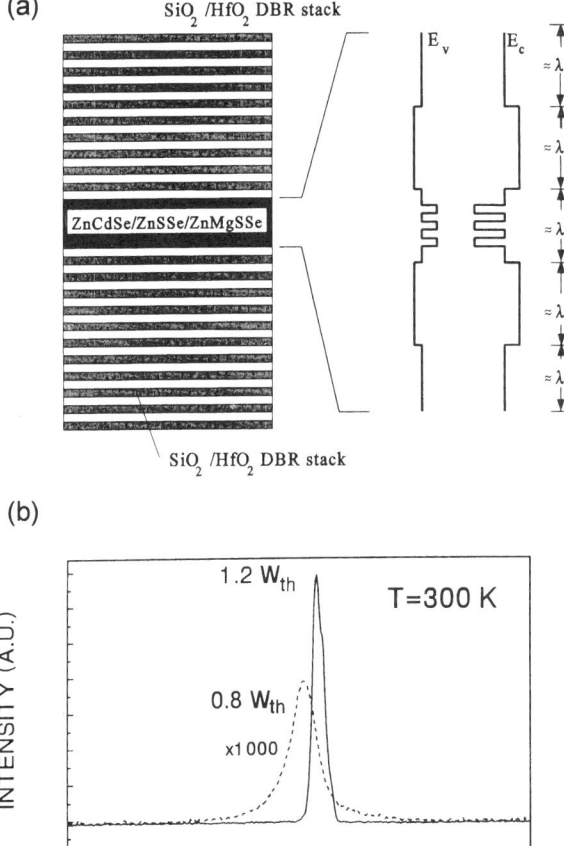

FIG. 13. (a) The schematic of a vertical cavity laser structure utilizing dielectric mirrors; (b) spontaneous emission (dashed lines) and stimulated emission (solid lines) spectra of an optically pumped device indicating the transition to lasing at room temperature (from Jeon *et al.*, 1995).

et al., 1994) which show room temperature QW threshold gain $g_{QW} \approx 1.5 \times 10^3 \text{cm}^{-1}$ at a current density of $I_{th} \approx 500 \text{A/cm}^2$. This relatively large gain parametrizes the VCSEL requirements so that a reflectance of roughly $R \approx 0.997$ is required in the standard plane parallel resonator geometry.

A schematic for an optically pumped VCSEL test structure is shown in Figure 13, composed of three ZnCdSe QWs ($L_w = 75$ Å) defining the active region, cladded by ZnSSe ($L \approx 2 \times 690$ Å) and ZnMgSSe ($L \approx 2 \times 1830$ Å) inner and outer confinement layers. The spectral characteristics, recorded in the vertical direction of emission near and above the threshold at $T = 300$ K and under pulsed excitation are also shown in Figure 13 (Jeon et al., 1995, 1995a). Far field measurements confirmed the spatial coherence expected of VCSEL emission above threshold; typical beam divergence is $\theta \approx 4°$. In this case, while the QW $n = 1$ exciton and cavity resonances nearly coincided at $T \approx 50$ K, their overlap is quite nonoptimal at $T = 300$ K. Nontheless, in part due to the homogeneous broadening of the gain profile, laser emission was obtained. A conversion efficiency (emitted photons/absorbed pump photons) of approximately 20% was measured, with average output powers up to 1 mW. These results, as well as the extensive analysis by Honda et al. (1995) on blue-green II–VI VCSELs show that, supported by the large gain that can be achieved in the ZnSe-based QW, an electrically injected blue-green VCSEL is quite feasible. A major practical challenge in realizing such a device in practice concerns the actual scheme for injection which must take into account the demand for a graded bandgap contact and the generally low p-type conductivity in the present edge emitter SCH diode lasers. In case of the ZnSeTe contact, optical absorption losses are severe, though this can, in principle, be solved with the larger bandgap BeTe/ZnSe contacts. The low p-conductivity implies large ohmic losses in a VCSEL in which II–VI multilayer stacks are used as the DFB mirrors (the feasibility of such mirrors has been recently demonstrated (Uusimaa et al., 1995); on the other hand, the choice for dielectric mirrors scheme requires lateral electrical injection which is also constrained by the low hole concentration and mobility.

While all the devices described so far in this section (and chapter) have been produced by MBE growth techniques, it is important to note that progress with MOCVD synthesis is also being made. As noted in the chapter by Sz. Fujita and S. Fujita in this volume, there have been serious impediments to the p-type doping of ZnSe in the MOCVD growth that have inhibited progress in the development of blue-green light emitting structures. Fujita et al., (1994) have reported progress in the doping and demonstrated LED devices. More recently, the group at Sony has succeeded in fabricating a MOCVD-grown ZnSe-based edge emitting diode laser, operating at $T = 77$ K, using ethyliodide for the n-type doping and di-isopropylamine for p-type doping (Toda et al., 1995; Yu et al., 1995). Elsewhere, the use of homo-epitaxy in the MBE growth on ZnSe substrates has led to the demonstration of a diode laser at cryogenic temperatures (Waters et al., 1990).

3. DIODE LASER DEGRADATION AND RELIABILITY

A major engineering challenge in the further development of the present blue-green II–VI light emitters toward technologically viable sources has been raised by concerns about the device degradation and their lifetime. The topic is under intensive research in all the leading laboratories and much progress is being made. The subject of degradation and reliability is covered in detail in the chapter by S. Guha and J. Petruzzello in this volume; hence we restrict our remarks here to general observations.

To date, the best external efficiencies reached in the edge emitting ZnCdSe cw diode lasers (differential quantum efficiency of 0.6–0.7, and wall-plug conversion efficiency up to 20%) indicate that much has been accomplished in optimizing the heterostructure designs in terms of their optical/electronic confinement, improved p-type doping, elegant contact schemes, etc. Yet, since the degradation issues are intimately related to the current density in the active device, further improvement in the device design continues to be very important in order to reduce the threshold densities toward the feasible goal of $100\,A/cm^2$. At the same time, we recall how the mechanical properties, of the relatively polar ZnSe suggest, for example, the likelihood of more vigorous dislocation dynamics than in III–V materials. Since both point and extended defects were found to be the main reason for the failure of the early GaAs diode lasers in the late 1970s, this fundamental fact, coupled with the low MBE growth temperature of the ZnSe-based semiconductors, raises both intrinsic and extrinsic challenges to the II–VI blue-green lasers in terms of controlling and balancing the defect-induced device degradation processes. On the other hand, we have already noted in Section II, how the inclusion of Mg, and most recently Be, into the ternary and quaternary II–VI heterostructures increases the covalent part of the chemical bond, thereby offering a means to "harden" the laser material against defect generation and especially subsequent propagation.

Because the availability of room temperature cw diode lasers for systematic degradation studies is still a rather recent development, and because the material quality in terms of preexisting defect has been rapidly improving, it is not possible to offer a unified single quantitative model for the II–VI device degradation and failure at this time. It is clear, however, that at this point the degradation is due to defects in as-grown material and not due to intrinsic properties of the material. For example, there have been earlier arguments that the bond energy in ZnSe is critically weak, when compared to the bandgap (photon energy). Based on available evidence to date, however, this seems not to be the case.

In general, the following consensus has emerged from several laboratories where the defects and laser degradation have been extensively studied. The main structural defects in the as-grown heterostructure materials (until today) have been due to preexisting stacking faults and their associated threading dislocation features which lead to further defect propagation and multiplication in the active QW layer during laser operation. The entire degradation is very likely to include a participating role of point defects in the QW (that act as nonradiative recombination centers and provide a local electronic "power supply," e.g., for further dislocation climb). Furthermore, the ZnCdSe QW in the present prototype laser devices is under large biaxial compressive strain. While such strain has been shown (somewhat counterintuitively) to reduce the dark line formation in InGaAs/AlGaAs QW lasers (Waters *et al.*, 1990), it is far from clear whether such a scenario is plausible in case of the ZnCdSe case. In fact, given the fact that the crystal structure of the binary end point CdSe is a wurtzite one, there exists a possibility that the strained ZnCdSe layer might be less stable that the adjacent ternary and quaternary cladding layers.

The stacking fault defects in as-grown material originate at or very near the GaAs buffer layer/ZnSe heterovalent heterointerface. When viewed by spatially resolved luminescence (real time) imaging, for example, dark spots (emanating from the stacking faults) are initially observed in the active QW layer, dark mobile defects emanate from the spots and propagate in the $\langle 100 \rangle$ directions, leaving a dark line defect in their wake, and eventually the dark spots and lines grow and coalesce into macroscopically degraded regions which correspond to a dense, intricate network of dislocation structure. As described by Guha and Petruzzello in their chapter, several groups have made important and complementary contributions to the understanding and origin of the nonluminescent dark defects which grow and propagate in an II–VI LED or a diode laser, leading to decreased optical output and impairing the electrical characteristics until a catastrophe failure occurs. We mention here that Guha and coworkers (1993, 1994) first employed spatially resolved electroluminescence and transmission electron microscope (TEM) techniques to study the formation and propagation of the system of dislocations that are associated with the degradated LED device microstructure; they also provided detailed evidence about the as-grown defects that nucleate these processes. Hua and co-workers (1994) have employed similar techniques to study in detail the microstructure of degraded SCH-QW diode lasers to identify dislocation networks created in the QW layer and the origin of their nucleation from the as-grown defects as stacking at or near the II–VI/GaAs interface. Kuo and collaborators (1994) have performed complementary work, also by electroluminescence and TEM methods, to highlight the interplay of the stacking faults and dislocations in the degradation process in ZnCdSe heterostructure. Hovinen and coworkers (1995, 1995a) have identified correlations between the electrical-optical degradation of diode lasers and their defect microstructure, and introduced a means of simulating the electrical injection by precisely wavelength tuned optical excitation, a technique that has permitted further insight into the defect dynamics. Other contemporary investigations have focused on the details of the defect structure.

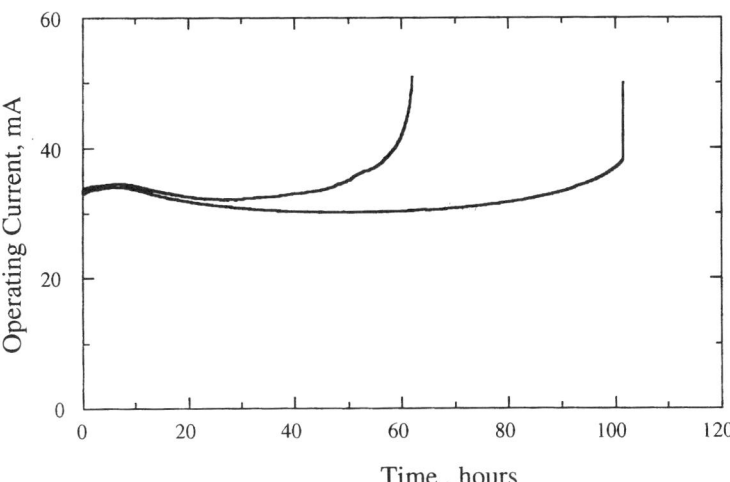

FIG. 14. Aging test for a cw ZnCdSe QW diode laser under constant output optical power of 1 mW illustrating a lifetime in excess of 100 h (from Taniguchi et al., in press).

The identification of the stacking faults at the II–VI/GaAs interface as the limiting factor for II–VI laser lifetime has motivated research where details of the nucleation of growth are being carefully examined under a variety of growth conditions. The use of migration enhanced epitaxy [x] and careful control of the specific surface reconstructions of the GaAs growth surface (either c(4 × 4) or 2 × 4) has led to significant progress within the past year in the reduction of the preexisting extended defects in the heterostructures. With the defect densities being reduced from the range of 10^5–10^6 cm^{-2} to 10^3–10^4 cm^{-2}, the cw diode laser lifetime has increased dramatically. As vivid evidence of recent rate of progress in the control and reduction of defects, Figure 14 shows blue-green diode laser lifetime test data from Sony Laboratories for two devices, with a lifetime reaching up to 100 h under constant optical output power of 1 mW (Taniguchi et al., 1996).

Although quantitative modeling of the relationship between the preexisting defect density and the device lifetime is somewhat uncertain due to the complexity and many unknowns in the problem, we may roughly extrapolate that the estimated defect density should approach the level of 10^2 cm^2, in order for the device lifetime to reach technologically meaningful values. As another example of very recent progress, Han and Gunshor (*submitted*) describe in their article in this volume a result of an etch pit study of a ZnSSe epitaxial layer where the extended defect densities remain well below 10^4 cm^{-2} over a full 3-inch GaAs substrate wafer. Test LED devices fabricated from material of comparable quality show lifetimes well in excess of 100 h for a current density of 100 A/cm^2.

With such improvements in the control of extended defects and device lifetime, it is clear that further attention must be given to other factors that may emerge as being critical in determining the degradation processes as the device operation is extended on an increasingly longer timescale. In terms of as-grown material, point defects are an important aspect which is being presently addressed. The question of generation of unwanted defects during device processing is also becoming increasingly crucial, and is also connected with the eventual manufacturability of the II–VI lasers. Finally, heterostructure engineering schemes by which the propagation of defects and degradation is spatially controlled (e.g., by introduction of selective dislocation filters) is a subject still in its infancy and will undoubtedly be pursued further.

IV. Physics of Gain and Stimulated Emission in ZnSe-based Quantum Well Lasers

In this section we discuss spectroscopic work which focuses on the microscopic mechanism of gain in ZnSe-based quantum well (QW) lasers, under optical and electrical injection, respectively. The discussion parallels, in part, that developed by R. Cingolani in another chapter of this volume where the subject of exciton physics in II–VI heterostructures is covered extensively.

In contrast to the case in e.g., III–V semiconductors, the lasing process, especially in optically pumped ZnCdSe/ZnSe QW lasers at cryogenic temperatures, can be entirely an excitonic, or even a biexcitonic process. That is, the Coulomb interaction between electrons and holes even in the density range required for optical gain and laser action preserves its bound state character. Figure 15 shows an empirical map that summarizes a phenomenological view of the different manifestations of electron–hole correlations in ZnSe-based QWs over a range of temperatures and densities. The key overall experimental observation is that laser action, including the room temperature diode laser operation, takes place in a density regime which is below that associated with the "Mott transition" to an electron–hole plasma ($n_s a_B^2 \sim 1$, where n_s is the sheet density and a_B the exciton Bohr radius).

The empirical map of Figure 15 is supported by spectroscopic results, especially those obtained through optical pump/probe studies, by a number of groups at cryogenic temperatures. The experiments range from steady-

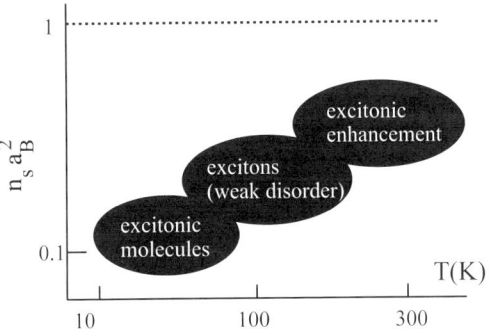

FIG. 15. A schematic illustrating the different facets of electron–hole Coulomb interaction in optically and electrically pumped ZnSe-based lasers. The pair density that corresponds to the approximate laser threshold remains in each case well below the phase filling limit that corresponds to a "Mott transition" to an electron–hole plasma.

state spectroscopy to resonant pumping experiments by femtosecond pulsed techniques. Perhaps more strikingly for this strong quasi two-dimensional exciton system, recent studies on room temperature ZnCdSe/ZnSSe/ZnMgSSe diode lasers show that pairwise electron–hole Coulomb correlation remains quite pronounced in this case as well (Ding et al., 1994). The microscopic theoretical approaches use the so-called semiconductor Bloch equations (Haug and Koch, 1993) as a starting point in computer-intensive calculations that can present qualitative features in broad agreement with experiment in the room temperature case, as shown below. Overall, the subject matter of many-electron interactions in lower dimensional widegap semiconductors presents many open questions and we suggest that interested readers should follow the technical literature. We also underline that while exciton-like contribution to gain in the widegap II–VI lasers is a characteristic feature of such widegap systems, this by no means excludes the possibility of a more electron–hole plasma (EHP-like) driven gain mechanism. At sufficiently high injection levels an exciton gas will eventually make a transition to the EHP state, as the Coulomb correlations are screened out; hence the microscopics of gain depend importantly on the particular experimental conditions. The key point to remember is that if laser operation at the lowest possible electron–hole density (i.e., current density) is sought for, the Coulomb enhancement effects can be of significant benefit for device applications.

The major quantitative distinction between III–V and the II–VI QW lasers is the intrinsic strength of Coulomb interaction between charged particles. The exciton binding energy in a ZnCdSe QW can reach up to

40 meV, hence exceeding the LO-phonon energy and kT at room temperature. Such conditions, which cannot be found in GaAs-based QWs, make the electron–hole Coulomb pairwise correlation robust against both many-body and thermal (LO-phonon intermediated) dissociation processes. The corresponding enhancement in the interband oscillator strength increases the radiative recombination rate significantly, as illustrated below. As a consequence, the resonance at or near which the inversion condition and maximum in the gain spectrum is achieved is the $n = 1$ heavy hole (HH) exciton transition in a ZnCdSe or ZnSe QW. We illustrate the point by choosing two regions from the "map" of Figure 15, namely that of the case of excitonic molecules at low temperatures and the diode laser at room temperature, respectively. For a broader discussion of the general case of excitonic lasing, the reader is referred to the chapter by R. Cingolani. We also note that ideas about excitonic lasing were explored in bulk II–VI semiconductors quite some time ago, under very intense pulsed excitation at low temperatures (Klingshirn and Haug, 1981). The theoretical work was focused on ideal 3D crystals without disorder effects, with main emphasis of population redistribution by exciton–exciton and exciton–free particle scattering in the polariton picture.

1. EXCITONIC MOLECULES AND LASING IN ZnSe QUANTUM WELLS

Aided in part by continued improvement in the heterostructure quality, spectroscopic studies have within the past year yielded fascinating new indications that excitonic molecules, as opposed to excitons, can dominate the gain and stimulated emission at cryogenic temperatures. Whether such phenomena have any impact on practical room temperature devices is unclear at this writing; on the other hand, with advanced nanofabrication techniques it is possible to envision lower dimensional structures in which the binding energy of an excitonic molecule (up to 20% of the exciton binding energy in the ZnSe-based QWs) can be enhanced to the point of being stable at elevated temperatures. In studies of excitonic gain in the ZnCdSe QWs in 1995, Henneberger, Kreller, and co-workers first raised the possibility that allowing for spin degeneracy in this (weakly) disordered alloy introduces the possibility of biexciton (excitonic molecule) contribution to lasing and showed convincingly that indeed is the case in a series of important experiments. The point has been also emphatically made through recent experiments by Kozlov *et al.* (1996). In the latter case, the degree of inhomogeneous broadening at the $n = 1$ HH exciton resonance has been substantially reduced by choosing the binary ZnSe as the QW material, so that both the exciton and the excitonic molecule emerge as spectroscopically isolated and clearly defined resonances in luminescence, nonlinear spectra, as well as in the gain measurements. Such a battery of experiments were applied to a single ZnSe QW by Kozlov *et al.* possessing a sharp ($\Delta E = 0.6$ meV) spectrally isolated exciton resonance, to show the additional presence of narrow excitonic molecular transition, characterized by a 5.5 meV binding energy. Spectrally resolved, transient four-wave mixing (FWM) experiments with fsec pulses were performed at $T = 10$ K to study the dynamics of biexciton–exciton interactions in a density regime in which gain and stimulated emission occur in an edge emitting laser device geometry (≈ 1–4×10^{11} cm^{-2}). Pronounced temporal quantum beats between the exciton and biexciton transitions were observed on a subpsec timescale, with a period matching the biexcitonic binding energy. Furthermore, the clear polarization dependence of the FWM signal implies the presence of strong exchange interactions in the exciton–biexciton system that were also revealed by the blue shift of the excitonic resonance at higher electron–hole pair densities.

To study the stimulated emission in the single ZnSe QW system, the heterostructures were cleaved into 300–500 μm long cavities and subjected to resonant optical excitation at the HH exciton resonance, hence creating a cold excitonic system. Figure 16 shows how the lasing emission emerged from the low-energy portion of the ≈ 1 meV wide biexcitonic transition (labeled $|B\rangle$, and corresponding to emission of a photon in a biexciton–exciton transition), indicating the direct role of excitonic molecules in the process of an optical gain formation in these conditions. Note how the laser emission is distinct from the exciton resonance at $|X\rangle$. The experiment was configured in such a way that in addition to edge emission, spontaneous emission through the top of the layered structure, as well as scattered pump radiation were directed into the spectrometer for spectral reference. Detailed optical spectra obtained from time-resolved "pump and probe" type experiments at low temperatures clearly show how the QW gain forms at the biexcitonic resonance on a psec timescale, with peak values up to $g \approx 7 \times 10^4$ cm^{-1}. On a subpsec timescale, coherent effects that couple the exciton and the biexciton polarization states were clearly in evidence and provide one means of creating gain at the molecular transition. Figure 17 shows one illustration of such transient gain spectroscopy at a density corresponding to that required for laser action (Kozlov, submitted). The right panel shows the partial bleaching of the HH exciton resonance (reduced absorption coefficient, $\Delta\alpha < 0$), whereas real gain appears at the transition involving the biexcitons. On a subpsec timescale there are strong spectral interference effects as the exciton and molecular states are coupled coherently (quantum mechanically), and the gain emerges as dephasing of this coherence occurs. These experiments hence also isolate the formation

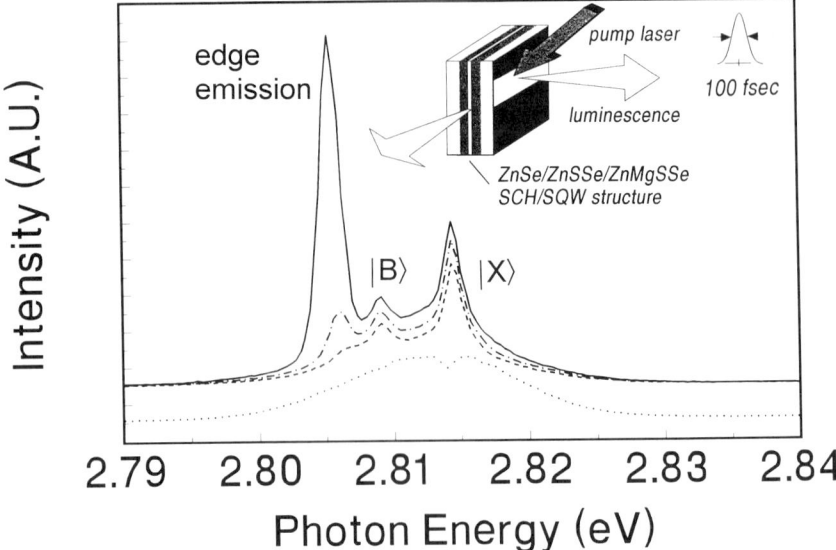

FIG. 16. Superposition of edge emission, top emission (contributions labeled as $|B\rangle$ and $|X\rangle$), and scattered pump radiation as laser threshold is approached in an optically pumped ZnSe SQW device at $T = 10$ K. The dotted spectrum at the bottom displays the transmitted pump spectrum indicating the absorption at the heavy-hole exciton $|X\rangle$ resonance. Inset shows experimental geometry (after Kozlov et al., 1996).

process of biexcitons from a gas of excitons, where inelastic scattering processes are involved as intermediate steps, with a typical time constant of 2–5 psec in the density range considered here. When circular polarization is used, the observed gain is very weak, with a much longer biexciton formation time ($\gg 10$ psec), consistent with requirements for spin-flip scattering processes are imposed by the optical selection rules.

On the theoretical front, there has been renewed interest in the optical properties of dense electron–hole systems at a fundamental level, spurred in part by suggestions from experiment that excitons play a role in determining gain characteristics. Flatte and co-workers (1995) have proposed a concept where a BCS-like condensed exciton state provides a lasing transition below the lowest exciton energy state. Such a coherent state has been calculated to exist several tens of meV below the absorptive resonance for the ZnSe/ZnCdSe case, separated by a well-defined gap. Elsewhere, Zhu et al. (1995) have suggested that excitonic condensates may form in specifically configured quantum wells in which a double-layer electron-hole system is calculated to exhibit a pairing gap and excitation spectrum in analog to a

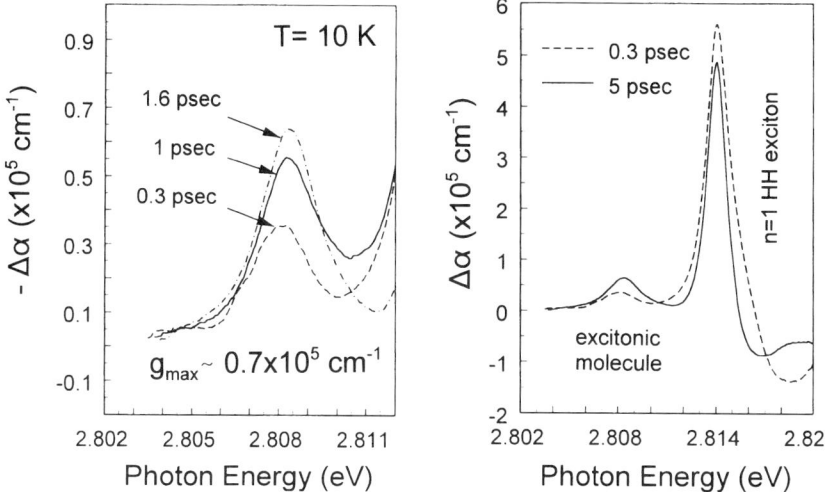

FIG. 17. Results from transient pump-probe spectroscopy for a SQW ZnSe heterostructure, showing the changes in the absorption coefficient at the $n = 1$ HH exciton and the excitonic molecule at 0.3 psec and 5 psec, respectively (right panel). The left panel shows details of the actual gain spectrum at the biexciton (after Kozlov et al., submitted).

BCS-like system. While it is yet unclear whether these theoretical concepts have a bearing on the operation of practical devices such as a diode laser, they are giving impetus to further basic physics experiments in the widegap II–VI heterostructures whose intrinsic electronic and optical properties appear to be very well suited for the study of fundamental properties of dense electron–hole systems in lower dimensions.

2. Gain Spectroscopy of Blue-Green Diode Lasers at Room Temperature

Here we focus on the high temperature range of the schematic map of Figure 15. Recently, as already mentioned, first results have appeared (Kozlov et al., 1994) that employ the Hakki-Paoli method for the measurement of the gain spectra in the SCH diode lasers. While the Hakki-Paoli method and related techniques are useful for characterizing the diode laser gain, the spectral range that they offer is much too limited for studying the microscopic mechanism of the gain. This is particularly relevant in the ZnSe-based systems given the interest in possible excitonic features at room temperature. It is necessary to measure the optical constants (absorption and gain) over a spectral range which is wide enough to cover the entire $n = 1$ QW transition (including the HH and LH exciton resonances). The approach that the Brown/Purdue team has found useful makes use of the correlation between spontaneous emission and stimulated emission spectra, first employed by Henry *et al.* to obtain the gain spectra for a GaAs double heterostructure laser (Flatte *et al.*, 1995). It has been used to study the gain/absorption spectra of GaAs QW diode lasers as well (Zhu *et al.*, 1995). The formulation of gain in the approach by Henry *et al.* is independent of the details of the gain mechanism or the nature of the electronic states that participate in the radiative process, and draws from the fundamental connection between gain and absorption by detailed balance arguments. This leads to the following explicit relationship between the gain and experimental spontaneous emission spectra, $S(E)$, where the separation between the quasi-Fermi levels, ΔE_F, is also experimentally obtained from the spontaneous emission spectrum at threshold conditions from the known position of laser emission:

$$g(E, \Delta E_F) = C * \frac{S(E)}{E^2} \left\{ 1 - \exp\left(\frac{E - \Delta E_F}{kT}\right) \right\} \tag{2}$$

In this expression C includes fundamental constants and experimental amplitude calibration factors. In the limit where the homogeneous broadening is negligible, the expression is exact. However, when substantial broadening is present, the quantitative use of this formulation for extracting the value for the gain coefficient can lead to errors. Spectral lineshapes are much more immune to the effects of broadening, however.

In applying this approach to the SCH ZnCdSe/ZnSSe/ZnMgSSe gain-guided diode lasers, 20 μm wide stripes were fabricated with a top transparent electrode of indium–tin oxide. Spectroscopy was performed by recording both the spontaneous emission through the top electrode and the stimulated emission through the cleaved end facets, as shown schematically in the inset of Figure 18 (Ding *et al.*, 1994). The room temperature pulsed threshold current density for this device, housing three 75 Å thick ZnCdSe QWs was $J = 1.4 \text{kA/cm}^2$, i.e. approximately 470 A/cm² per QW. Figure 18 shows the results for gain spectra at room temperature under different levels of current injection. On the low energy side where gain eventually develops, the spectra are consistent with those obtained using the Hakki-Paoli method. A substantial broadening of the resonance is observed. At low current level, the $n = 1$ HH and LH exciton resonances are quite evident at photon energies of $\hbar\omega = 2.52 \text{eV}$ and 2.625eV, respectively. While partly saturated, the $n = 1$ HH resonance is clearly distinguisable at laser threshold

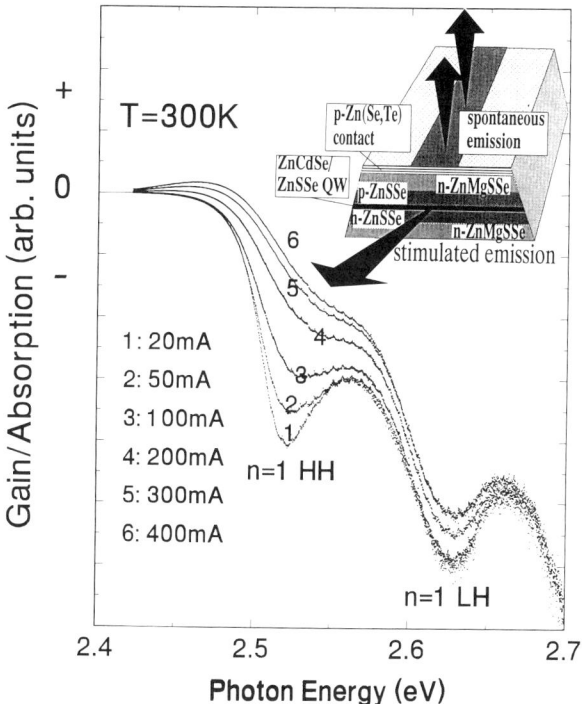

FIG. 18. The gain/absorption spectra of a gain-guided ZnCdSe/ZnSSe/ZnMgSSe SCH diode laser at room temperature as a function of injection current, including the $n = 1$ QW exciton resonance region. The inset shows the experimental schematic (after Ding et al., 1994).

(<400 mA). Note that the free e–h pair continuum edge for the $n = 1$ HH transition is merged into the LH exciton transition. The key spectroscopic result is that lasing occurs approximately 90–100 meV below the free e–h pair states (the latter corresponding to the 2D density of states step in the elementary model of a QW).

At room temperature, the many-body effects, as well as the additional presence of a background free electron–hole population make the microscopic description of the exciton-like effects complicated. However, that electron–hole Coulomb pairwise effects remain important in the ZnCdSe QW diode lasers is strongly suggested by the data in Figure 18. Furthermore, the powerful theoretical formulation through the semiconductor Bloch equations allows a numerical calculation to be performed, under the assumption that free carrier screening only is taken into account (i.e., bound–bound state interactions are excluded). Figure 19 shows the result of a calculation by Ell and Haug (1989) for ZnSe in the 2D limit, using the

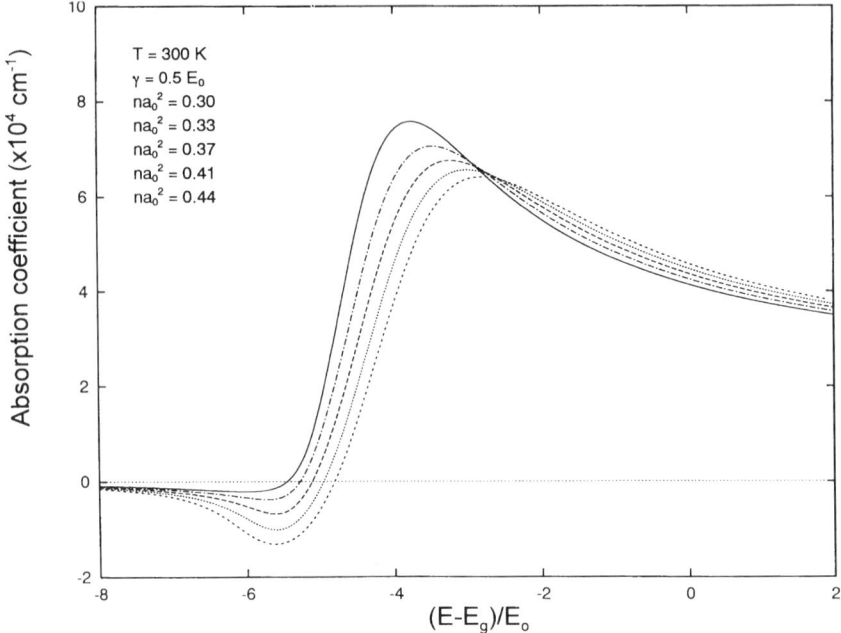

FIG. 19. A calculation for the absorption (positive values on vertical axis) and gain based on semiconductor Bloch equations for ZnSe in the 2D limit at room temperature. The electron–hole pair density is indicated as the product na_0^2, and a phenomenological damping equal to half of the exciton binding energy is used (after Ell and Haug, 1989).

typical exciton effective mass parameters ($m_e = 0.16\,m_0$, $m_{hh} = 0.60\,m_0$, $\varepsilon = 9.2$. A phenomenological damping coefficient $\gamma = 0.5\,E_0$ has been used, where E_0 is the (3D) exciton binding energy. Clearly, the calculation produces the main features observed in the experiment: namely that a gain maximum and an absorption maximum are distinct and adjacent to each other, straddling the $n = 1$ HH exciton resonance region. The theory does, however, suggest stronger blueshift with increasing excitation (expressed as the product of sheet density and Bohr radius squared) of the absorption maximum than is observed experimentally. Overall, however, both theory and experiment clearly demonstrate the importance of "excitonic enhancement" of gain in the blue–green, II–VI diode lasers under normal operating conditions. We add that these results are in sharp contrast to the observations by Kesler and Harder, where the same experimental techniques have been applied to GaAs-based QW diode lasers. By the time lasing threshold is reached in the GaAs lasers, exciton absorption peaks, which are observable at low current injection levels, totally disappear and a step-like gain profile consistent with an electron–hole plasma emerges.

The electron–hole pairwise Coulomb correlation will result in an enhancement in the electron–hole wave function overlap and hence impacts directly on the optical transition matrix element which determines the luminescence decay time. In addition, details of the temporal behavior of the luminescence also depend on the recombination kinetics, monomolecular or bimolecular. It has been shown recently both experimentally and theoretically that the Coulomb correlation between electrons and holes results in an enhancement in radiative recombination rate at room temperature in the InGaAs QW system (Michler et al., 1993). The authors' group has performed a picosecond time resolved luminescence experiment on the ZnCdSe/ZnSSe/ZnMgSSe QW structure (the same material employed in the gain spectroscopy in Figure 36) to study the behavior of transient photoluminescence and the associated carrier dynamics. In this case, the absence of a cleaved cavity ensured the absence of stimulated emission. Figure 20 shows the spectrally integrated transient luminescence intensity at $t = 0$, immediately after the arrival of the 5 psec excitation pulse [labeled as PL($t = 0$)], as a function of the excitation intensity on a logarithm–logarithm plot for three different temperatures (Ding et al., 1994). The maximum excitation level is $130\,\mu\text{J/cm}^2$. The dashed and solid lines represent linear and quadratic excitation intensity dependence of PL($t = 0$), respectively. The data show that the recombination mechanism is much closer to that of excitons or correlated pairs (linear dependence) than to that of totally uncorrelated electron–hole pairs (quadratic dependence). Further evidence of the Coulomb correlation comes from the measured rate of luminescence

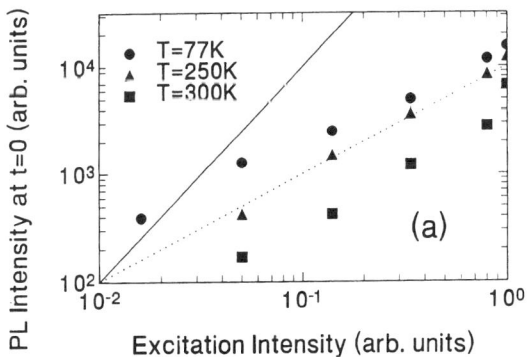

FIG. 20. Spectrally integrated photoluminescence intensity at the $n = 1$ HH exciton resonance at $t = 0$ in a transient PL experiment on a ZnCdSe SCH lase device. The solid and dashed lines depict quadratic and linear dependence, respectively (from Ding et al., 1994).

decay. The experimentally observed luminescence decay time at room temperature is about 0.9 nsec, at least a factor of 5 shorter than a value which is estimated based on the band-to-band transition model (Lasher and Stern, 1964). Such an enhancement in the radiative recombination rate, as pointed out by Michler et al. (1993), is consistent with the role of the electron–hole pairwise correlation in our ZnSe based QWs. Both the gain spectra of a diode laser and the measurements of electron–hole pair radiative lifetime at high densities point to an enhanced interband oscillator strength, very much in keeping with the predictions by theory on the role of excitons. The broader problem of many-body interactions in a practical device setting presents thus a unique opportunity for experimental and theoretical study of a quasi-2D electron–hole system in the blue-green diode lasers; and perhaps even more so as the experiments will develop further into quantum wire and quantum dot regimes. We mention that a range of device science problems of both fundamental and practical interest can emerge from the short vertical cavity structures (or other nanostructured optical resonator configurations) in which the optical field–matter interaction is enhanced by the reduction of the number of cavity modes to a very low number. Initial results in the study of such interactions in ZnCdSe and ZnSe QW heterostructures in high-Q vertical cavities has recently yielded very large normal mode (Rabi) splittings at the $n = 1$ HH exciton resonance, values approaching 20 meV (Kelkar et al., 1995). Such an exceptional coupling strength to the electromagnetic modes, in combination with the excitonic gain, offers the potential of low threshold density blue-green semiconductor light emitters, in which the spontaneous emission is usefully channelled into the "lasing mode" at very low pump/injection levels.

V. Summary

In this review we have covered a range of topics drawn from research connected to the new blue-green II–VI light emitters. The field is moving at a brisk pace but even so, it is difficult to forecast a timetable for future progress, especially as to when and how these compact sources might reach a technical maturity suitable for applications in optical storage, displays, and so on. From a basic research point of view, there exists a broad spectrum of fascinating fundamental problems in II–VI semiconductor nanostructures, ranging from the omnipresent issue of p-type doping to excitonic lasers in quantum microcavity resonators. The extension of the wavelength range into the near ultraviolet range calls for further exploration of the II–VI materials spectrum which, as a glance at the periodic table will show, remains only partially explored. Epitaxial growth techniques are yet to be applied on patterned substrates or on cleaved edges to facilitate growth in more than one dimension of lower dimensional optoelectronic heterostructures. With future advances in II–VI materials processing, photonic bandgap structures fabricated from ZnSe and related materials may also offer broad appeal as a testbed for novel directions in guided wave optics and applications toward integrated optoelectronic circuits.

At the device level, the main issue at this writing continues to be the control of growth-related defects, although excellent progress is being made in this area. Any active semiconductor device in which a current density on the order of 1 kA/cm^2 needs to be sustained for tens of thousands of hours requires a high degree of crystalline perfection and purity; a requirement especially underscored in the polar II–VI semiconductors. At the same time, opportunities for much innovation exist in the study, control, and synthesis of heterointerfaces and other "critical" points for diode lasers and LEDs, to tailor the electronic transport both in vertical and in-plane directions. As a practical matter, device fabrication techniques need further attention, taking into account the rather low temperature processing requirements (typically $T < 250°\text{C}$) that must also conform to the mechanical idiosyncrasies of II–VI semiconductors. In drawing analogies, it is useful to recall that a large number of manpower years were expended in the late 1970s to solve device degradation, lifetime, and reliability problems in the III–V semiconductor lasers which today can be found on the ocean floor as part of long-haul fiber cables or in millions of compact disk players. In the case of the widegap II–VI light emitters, the pace of evolution in the past 3 years has been striking, given the smaller size and overall effort of the field. There is increasing confidence within the community that the II–VI blue–green lasers will have a real impact on optoelectronics applications at short visible wavelengths.

Acknowledgments

The authors are grateful to members of their groups, present and former, who have made major contributions to that portion of the work which has been carried out in their laboratories as described in this article.

The research at Brown University was supported by the Defense Advanced Projects Agency, under the University Research Initiative program (Grant 218-25015XX) and by the National Science Foundation (DMR-9112329; ECS95-08401).

References

Bacher, G., Eisert, D., Mais, N., Reithmaier, J. P., Forchel, A., Jobst, B., Hommel, D., and Landwehr, G. (1996). Symposium on Blue Lasers and LEDs, Chiba, Japan.
Buijis, M., Shahzad, K., Flamholtz, S., Haberem, K., and Gaines, J. (1995). *Appl. Phys. Lett.* **67**, 1987.
Ding, J., Hagerott, M., Kelkar, P. Nurmikko, A. V., Grillo, D. C., He, L., Han, J., and Gunshor, R. L. (1994). *Phys. Rev.* **50**, 5787.
Eisert, D., Bacher, G., Mais, N., Reithmaier, J. P., Forchel, A., Jobst, B., Hommel, D., and Landwehr, G. (1996). *Appl. Phys. Lett.* **68**, 599.
Ell, C., and Haug, H., private communication; the calculation is based in part on the formalism found in C. Ell and H. Haug. (1989). *J. Opt. Soc. Am.* **Bb**, xxxx.
Fan, Y., Han, J., He, L., Saraie, J., Gunshor, R. L., Hagerott, M., Jeon, H., Nurmikko, A. V. (1992). *Appl. Phys. Lett.* **61**, 3160.
Fan, Y., Han, J., Saraie, J., Gunshor, R. L., Hagerott, M. M., and Nurmikko, A. V. (1993). *Appl. Phys. Lett.* **63**, 1812.
Flatte, M. E., Runge, E., and Ehrenreich, H. (1995). *Appl. Phys. Lett.* **66**, xxxx.
Fujita, S. (1994). *Proceedings of the 6th International Conference on II–VI Compounds*, Newport, RI, *J. Cryst. Growth* **138**.
Gaines, J. M., Drenten, R. R., Haberen, K. W., Marshall, T., Mensz, P., and Petruzzello, J. (1993). *Appl. Phys. Lett.* **62**, pp. 2462–2464.
Grillo, D. C., Fan, Y., Han, J., Li, H., Gunshor, R. L., Hagerott, M., Jeon, H., Salokatve, A., Hua, G., and Otsuka, N. (1993). *Appl. Phys. Lett.* **63**, 2723.
Grillo, D. C., Han, J., Ringle, M., Hua, G., Gunshor, R. L., Kelkar, P., Kozlov, V., Jeon, H., and Nurmikko, A. V. (1994). *Electron. Lett.* **30**, 2131.
Guha, S., DePuydt, J., Haase, M. A., Qiu, J., and Cheng, H. (1993). *Appl. Phys. Lett.* **63**, 3107.
Guha, S. Cheng, H., Haase, M. A., DePuydt, J., Qiu, J., Wu, B. J., and Hofler, G. E. (1994). *Appl. Phys. Lett.* **63**, 2723.
Haase, M. A., Baude, J. F., Hagedorn, M. S., Qiu, J., DePuydt, J., Cheng, H., Guha, S., Hofler, G. E., and Wu, B. J. (1993). *Appl. Phys. Lett.* **63**, 2315.
Haase, M, Qiu, J., DePuydt, J., and Cheng, H. (1991). *Appl. Phys. Lett.* **59**, 1272–1274.
Hakki, B. W., and Paoli, T. L. (1975). *J. Appl. Phys.* **44**, 4113 (1973); ibid **46**, 1299.
Han, J., Chu, C.-C., Gunshor, R. L., and Nurmikko, A. V., submitted to *Appl. Phys. Lett.*
Hangleiter, A. (1993). *Phys. Rev.* **B48**, 9146.
Harrison, W. A. (1980). In *Electronic Structure and the Properties of Solids*, ed. Freeman & Co., San Francisco. Chapter 7.
Haug, H., and Koch, S. W. (1993). *Quantum Theory of the Optical and Electronic Properties of Semiconductors.* World Scientific, ????.
Hiei, F., Ikeda, M., Ozawa, M., Miyajima, T., Ishibashi, A., and Akimoto, K. (1993). *Electr. Lett.* **29**, 878.
Honda, T., Katsube, A., Sagaguchi, T., Koyama, F., and Iga, K. (1995). *Jpn. J. Appl. Phys.* **34**, 3527.
Honda, T., Sagaguchi, T., Koyama, F., Iga, K., Yanashina, K., Munekata, H., and Kukimoto, H. (1995). *J. Appl. Phys.* **78**, 4784.
Ho, E., Fisher, P. A., House, J. L., Petrich, G. S., Kolodziejski, L. A., Walker, J., and Johnson, N. M. (1995). *Appl. Phys. Lett.* **66**, xxxx.
Hovinen, M., Ding, J., Nurmikko, A. V., Grillo, D. C., Fan, Y., Han, J., Li, H., and Gunshor, R. L. (1993). *Appl. Phys. Lett.* **63**, 3128.
Hovinen, M., Ding, J., Nurmikko, A. V., Hua, G. C., Grillo, D. C., He, L., Han, J., and Gunshor,

R. L. (1995a). *Appl. Phys. Lett.* **65**, xxxx.
Hovinen, M., Ding, J., Salokatve, A., Nurmikko, A. V., Hua, G. C., Grillo, D. C., He, L., Han, J., Ringle, M., and Gunshor, R. L. (1995). *J. Appl. Phys.* **77**, xxxx.
Hua, C.-G. private communication.
Hua, G. C., Otsuka, N., Grillo, D. C., Fan, Y., Han, J., Ringle, M. D., Gunshor, R. L., Hovinen, M., and Nurmikko, A. V. (1994). *Appl. Phys. Lett.* **65**, 1331.
Ishihara, T., Brunthaler, G., Walecki, W., Hagerott, M., Nurmikko, A. V., Samarth, N., Luo, H., and Furdyna, J. (1992). *Appl. Phys. Lett.* **60**, 2460.
Itoh, S., Nakayama, N., Matsumoto, S., Nagai, N., Nakano, K., Ozawa, M., Okuyama, H., Tomiya, S., Ohata, T., Ikeda, M., Ishibashi, A., and Mori, Y. (1994). *Jpn. J. Appl. Phys.* **33**, L938.
Itoh, S., Nakayama, N., Ohata, T., Ozawa, M., Okuyama, H., Matsumoto, S., Nakano, K., Ikeda, M., Ishibashi, A., and Mori, Y. (1994). *Jpn. J. Appl. Phys.* **33**, L639.
Jeon, H., Ding, J., Patterson, W., Nurmikko, A. V., Xie, W., Grillo, D. C., Kobayashi, M., and Gunshor, R. L. (1991). *Appl. Phys. Lett.* **90**, 3619-3621.
Jeon, H., Hagerott, M., Ding, J., Nurmikko, A. V., Grillo, D. C., Xie, W., Kobayashi, M., and Gunshor, R. L. (1992). *Optics Lett.* **18**, 125-127.
Jeon, H., Kozlov, V., Kelkar, P., Nurmikko, A. V., Grillo, D. C., Han, J., Ringle, M., and Gunshor, R. L. (1995). *Electr. Lett.* **31**, 106.
Jeon, H., Kozlov, V. Kelkar, P., Nurmikko, A. V., Chu, C.-C., Han, J., Hua, G. C., Ringle, M., and Gunshor, R. L. (1995a). *Appl. Phys. Lett.* **67**, xxxx.
Kawasumi, T., Nakayama, N., Ishibashi, A., and Mori, Y. (1995). *Electr. Lett.* **31**, 1667.
Kelkar, P., Kozlov, V., Jeon, H., Nurmikko, A. V., Chu, C.-C., Han, J., Hua, G. C., Ringle, M., and Gunshor, R. L. (Aug. 15, 1995). *Phys. Rev.* **B60**, xxxx.
Kesler, H. INFS. TO COME!!!
Klingshirn, C., Haug, H. (1981). *Phys. Reports* **70**, 315-398.
Kolodziejski, L., Lu, K., Ho, E., Coronado, C. A., Fisher, P. A., and Petrich, G. S. (1994). *J. Cryst. Growth* **138**, 1.
Kondo, K., Ukita, M., Yoshida, H., Kishita, Y., Okuyama, H., Itoh, S., Ohata, T., Nakano, K., and Ishibashi, A. (1994). *Electr. Lett.* **30**, 568.
Kozlov, V., Kelkar, P., Nurmikko, A. V., Grillo, D. C., Han, J., and Gunshor, R. L. (April 1996). *Phys. Rev.* **Bxx**, xxxx.
Kozlov, V., Kelkar, P., Nurmikko, A. V., Grillo, D. C., Han, J., and Gunshor, R. L., submitted to *Phys. Rev.*
Kozlov, V., Salokatve, A., Nurmikko, A. V., Grillo, D., He, L., Han, J., Fan, Y., Ringle, M., and Gunshor, R. *Appl. Phys. Lett.* **65**, 1684.
Kreller, F., Lowisch, M., Puls, J., and Henneberger, F. (1995). *Phys. Rev. Lett.* **75**, 2420.
Kuo, L. H., Salamanca-Riba, L., Wu, B. J., DePuydt, J. M., Haugen, G. M., Cheng, H., Guha, S., and Haase, M. A. (1994). *Appl. Phys. Lett.* **65**, 1230.
Kuramoto, M., Chong, T. C., Kikuchi, A., and Kishino, K. (1993). *Electr. Lett.* **29**, 1260.
Lasher, G., and Stern, F. (1964). *Phys. Rev.* **133**, A533.
Marshall, Y., Gaines, J., Petruzello, J., Drenten, R., Mensz, P., and Habaren, K. (1994). 7th Annual Meeting of IEEE Lasers and Electro-Optics Society, VS4.1, Boston.
Mensz, P. (1994). *Appl. Phys. Lett.* **64**, 2148.
Mensz, P. M. (1994). *Appl. Phys. Lett.* **65**, 2627.
Michler, P., Hangleiter, A., Moritz, A., Harle, V., and Scholz, F. (1993). *Phys. Rev.* **47**, 1671-1674.
Nakayama, N., Itoh, S., Nakano, K., Okuyama, H., Ozawa, M., Ishibashi, A., Ikeda, M., and Mori, Y. (1993). *Electr. Lett.* **29**, 1488.
Nakayama, N., Itoh, S., Nakano, K., Okuyama, H., Ozawa, M., Ishibashi, A., Ikeda, M., and Mori, Y. (1993a). *Electr. Lett.* **29**, 2194.

Nakayama, N., Itoh, S., Okuyama, H., Ozawa, M., Ohata, T., Nakano, K., Ozawa, M., Ikeda, M., Ishibashi, A., and Mori, Y. (1993). *Electr. Lett.* **29**, 2194.
Ohata, T., Itoh, S., Nakayama, N., Matsumoto, S., Nakano, K., Ozawa, M., Okuyama, H., Tomiya, S., Ikeda, M., and Ishibashi, A. (1994). *Proceedings of Int. Workshop on ZnSe-Based Blue-Green Laser Structures*, Wurzburg, p. 13.
Okuyama, H., Itoh, S., Kato, E., Ozawa, M., Nakayama, N., Nakano, K., Ikeda, M., Ishibashi, A., and Mori, Y. (1994). *Electron. Lett.* **30**, 415.
Okuyama, H., Kato, E., Itoh, S., Nakayama, N., Ohata, T., and Ishibashi, A. (1995). *Appl. Phys. Lett.* **66**, 656.
Okuyama, H., Miyajima, T., Morinaga, Y., Hiei, F., Ozawa, M., and Akimoto, K. (1993). *Electr. Lett.* **29**, 766.
Okuyama, H., Nakano, K., Miyajima, T., and Akimoto, K. (1991). *Jpn. J. Appl. Phys.* **30**, L1620.
Ozawa, M., Hiei, F., Ishibashi, A., and Akimoto, K. (1993). *Electr. Lett.* **29**, 503.
Pelekanos, N. T., Ding, J., Hagerott, M., Nurmikko, A. V., Luo, H., Samarth, N., and Furdyna, J. (1992). *Phys. Rev.* **B45**, pp. 6037–6045.
Proceedings of the 6th International Conference on II–VI Compounds, Newport, RI, *J. Cryst. Growth* **138**, 1–4 (1994). North Holland, Amsterdam.
Proceedings of the 6th International Conference on II–VI Compounds, Edinburgh, *J. Cryst. Growth* **Xxx**, yy (1996). North Holland, Amsterdam.
Salokatve, A., Jeon, H., Ding, J., Hovinen, M., Nurmikko, A., Grillo, D. C., Han, J., Li, H., Gunshor, R. L., Hua, C., and Otsuka, N. (1993). *Electr. Lett.* **29**, 2192.
Salokatve, A., private communication.
Samarth, N., Luo, H., Furdyna, J. K., Qadri, S. B., Lee, Y. R., Ramdas, A. K., and Otsuka, N. (1990). *Appl. Phys. Lett.* **54**, 2680.
Suemune, I. (1992). *Jpn. J. Appl. Phys.* **31**, L95.
Taniguchi, S., Hino, T., Itoh, S., Nakano, K., Nakayama, N., Ishibashi, A., and Ikeda, M., (1996). *Electr. Lett.* **32**, 552.
Toda, A., Margalith, T., Imanishi, D., and Ishibashi, A. (1995). *Electr. Lett.* **31**, 235.
Uusimaa, P., Rakennus, K., Salokatve, A., and Pessa, M. (1995). *Appl. Phys. Lett.* **67**, 2197.
Verie, C. (1995). Int. Conf. Semiconductor Heteroepitaxy, Montpellier, France.
Waag, F. Fischer, H. J. Lugauer, J. Litz, J. Laubender, U. Lunz, U. Zehnder, W. Ossau, T. Gerhardt, M. Moller, and G. Landwehr, *J. Appl. Phys.* (in press).
Wang, A., Fischer, F., Lugauer, H. J., Litz, J., Laubender, J., Lunz, U., Zehnder, U., Ossau, W., Gerhardt, T., Moller, M., and Landwehr, G. (in press).
Walecki, W., Nurmikko, A. V., Samarth, N., Luo, H., Furdyna, J. K., and Otsuka, N. (1990). *Appl. Phys. Lett.* **57**, 466.
Walker, C. T., DePuydt, J. M., Haase, M. A., Qiu, J., and Cheng, H. (1993). *Physica* **B185**, 27–35.
Waters, R. G., Bour, D. P., Yellen, S. L., and Ruggieri, N. F. (1990). *IEEE Photonics Techn. Lett.* **2**, 531.
Yu, Z., Boney, C., Hughes, W. C., Rowland, W. H. Jr., Cook, J. W. Jr., Schetzina, J. F., Cabtwell, G., and Harsch, W. C. (1995). *Electr. Lett.* **31**, 1341.
Zhu, X., Littlewood, P. B., Hybertsen, M., and Rice, T. M. (1995). *Phys. Rev. Lett.* **74**, 1633.

CHAPTER 7

Defects and Degradation in Wide Gap II–VI Based Structures and Light-emitting Devices

Supratik Guha

3M CORPORATE RESEARCH LABS,
ST. PAUL, MN 55144*

John Petruzzello

PHILIPS LABORATORIES
PHILIPS ELECTRONICS NORTH AAMERICAN CORPORATION
BRIARCLIFF MANOR, NEW YORK 10510

I. INTRODUCTION	271
II. STRUCTURAL CHARACTERISTICS	272
1. *Single Epilayer Structures*	272
2. *Laser Structures*	288
III. DEGRADATION EFFECTS	298
1. *Experimental Techniques*	299
2. *Degradation in Light-emitting Diodes and Lasers*	301
REFERENCES	317

I. Introduction

Wide gap II–VI materials have made extensive progress over the past 5 years with the accomplishment of several milestones. Injection lasers have now been fabricated that operate in cw mode at room temperature for a few hours. Considering that the first demonstration of injection lasing occurred in 1991 (Haase *et al.*, 1991), the development of the field has certainly been rapid. A principal reason behind this progress has been an improved understanding of the defect structure in these materials, their role in contributing to the degradation process, and finally our ability to reduce their densities in as-grown device structures. Earlier microstructural analyses on ZnSe/GaAs heterostructures were concerned with characterizing the defect types (Brown *et al.*, 1987; Petruzzello *et al.*, 1988) and the heterointerface (Qiu *et al.*, 1990). Material quality was poor, and our

*Present address: IBM T. J. Watson Research Center, Yorktown Heights, New York 10598

understanding of the different defect formation mechanisms was unclear. Much of the earlier attention was toward successfully p-doping ZnSe, which was necessary for the demonstration of a light-emitting diode or injection laser. Following the realization of p-doped ZnSe in 1990 (Park et al., 1990) and the demonstration of a II–VI injection laser, attention started being focused on defect reduction and the role of the defect structure in device degradation. Dislocation densities in high-quality device structures today are routinely obtained in the 10^4 per cm^2 range. This has been achieved by using GaAs substrates with epitaxial GaAs buffer layers, the use of lattice-matched ZnMgSSe and ZnSSe compositions, and careful control of the initial stages of II–VI growth. Considering that 5 years back dislocation densities were typically in the 10^8 per cm^2 range or higher, progress has certainly been made! At the same time, our understanding of the role that defects play in the degradation has also become much clearer. As will be discussed in detail, studies by different groups clearly show that degradation in present devices originates from dislocations that bound stacking faults. These stacking faults in turn are growth faults that originate at the GaAs/II–VI heterovalent interface. Improvements in growth techniques have resulted in our being able to reduce stacking fault densities in device structures and this has then resulted in longer device operating lifetimes.

In Section II we describe our current understanding of the defect structure in ZnSe-based heterostructures, followed by a description of the degradation processes in ZnSe-based lasers and light-emitting diodes in Section III.

II. Structural Characteristics

1. SINGLE EPILAYER STRUCTURES

The starting point for studying epitaxial growth (MBE and MOCVD) toward solid-state blue light emitters was the simple structure of ZnSe epilayers on (001) GaAs substrates. The earliest work was concerned primarily with the optical and electrical characteristics (Yao, 1985, 1986; Park et al., 1987; DePuydt et al., 1988; Shahzad and Cammack, 1990; Marshall and Cammack, 1991) of the epilayers since MBE and MOCVD techniques were expected to reduce impurity levels and allow p-type conductivity. The studies of the structural characteristics (Petruzzello et al., 1988; Guha et al., 1992, 1993) of this system have also received a great deal of attention, especially after the realization of laser diodes. This section is divided into two parts describing the structural properties as a function of initial growth conditions and lattice mismatching effects, respectively. The ZnSe/GaAs interface is very important in determining the resulting defect structure of the epilayer. The defects initiated at this interface are stacking faults and their density ranges from $<10^3$ to $>10^8$ cm^{-2} depending on the initial growth conditions. The structure of the epilayer is also greatly influenced by lattice mismatch. If the values for the mismatch and the epilayer thickness become greater than a critical value, then relaxation of the layer will take place by the nucleation of misfit dislocations (Matthews, 1975). These misfit dislocations are accompanied by threading dislocations and can quickly lead to densities $>10^8$ cm^{-2} in the layer. The lattice mismatch effects are dramatic in the heterostructures used in laser diodes and will be discussed below.

a. Defects: Initial Stages of Growth

There have been a number of studies of the effects of growth parameters, particularly the initial growth mode, on the defect structure of the resulting epilayer (Lilja et al., 1989; Gaines et al., 1993; Guha et al., 1993a). The initial growth mode of the ZnSe can be 2 or 3-dimensional depending on the starting surface of the substrate (with or without buffer layer and Ga or As terminated), the initial exposure (Zn or Se) and temperature of the substrate. The dimensionality is determined qualitatively by the condition of the RHEED pattern (streakiness) during II–VI initiation. In general, the lowest defect densities are obtained when growth is started with a GaAs buffer layer with an As-stabilized surface.

The surface of the GaAs with and without a buffer layer has been studied with scanning tunneling microscopy (STM) (Pashley and Li, 1994). A GaAs substrate that has been desorbed under an As flux (to preserve the 2×4 surface reconstruction) is shown in the large area STM image of Figure 1a. For direct comparison the STM image of Figure 1b is a desorbed substrate with a GaAs buffer layer of 0.4-μm thickness grown on top. It is clear that the desorbed substrate has many more atomic planes exposed than the buffer layer (i.e., the surface is rougher). The buffer layer displays smooth (001) terraces that are several hundred Ångströms across separated by single height steps. The desorbed substrate shows a high density of bright spots across the image but it was not possible to determine whether these are contamination sites or just roughness in the GaAs. The lowest densities of stacking fault defects in ZnSe epilayers grown on GaAs with and without buffer layers are $\approx 10^3$ and $\approx 10^6$, respectively. The exact mechanism of stacking fault nucleation is not understood but it is clear that starting with a buffer layer surface is advantageous for low defect density growth.

In addition to the buffer layer, other initial conditions influence the defect density in ZnSe epilayers such as the initial exposure of the GaAs to Zn and/or Se fluxes and substrate temperature. The structural quality of the epilayers is found to be highest when the substrate is first exposed to a Zn

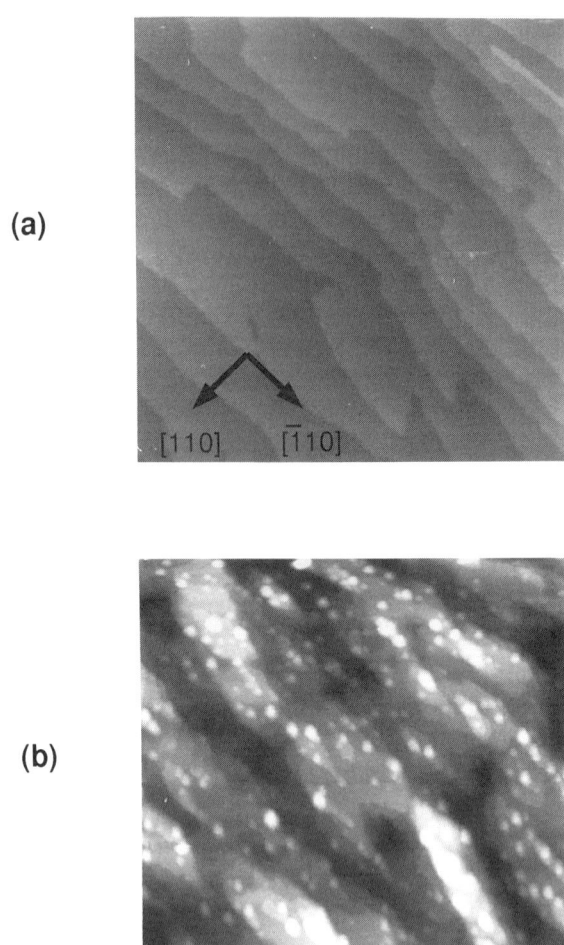

FIG. 1. STM images of (a) an oxide desorbed (001) GaAs substrate and of (b) an MBE grown GaAs buffer layer showing an area of 0.75 × 0.75 μm (from Pashley and Li, 1994).

flux independent of the substrate temperature (200 or 500°C). The only defects found in the thin pseudomorphic Zn-started epilayers (on GaAs buffer layers) are stacking faults at a density of $<10^4 \mathrm{cm}^{-2}$. These stacking faults have a habit plane of {111} and are bounded by partial dislocations along $\langle 110\rangle$'s that are 60° to one another. The faults can be alone or grouped with one, two, or three other fault planes around the same point in the interface. The case where all four {111} planes contain faults around a "nucleation site" at the interface is shown in Figure 2. The fault planes form

7 Defects and Degradation

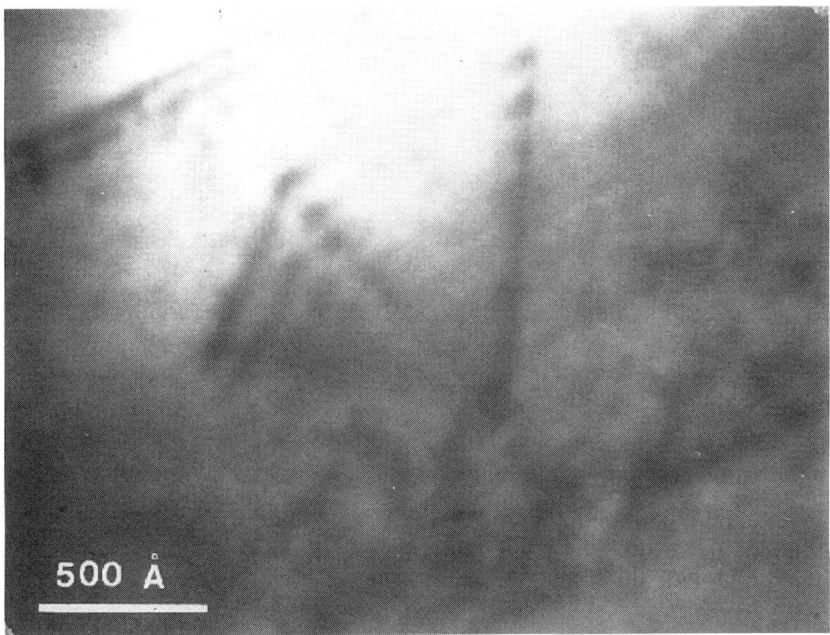

FIG. 2. TEM image of a stacking fault pyramid starting at the GaAs–ZnSe interface and extending to the surface of the ZnSe epilayer (from Petruzzello *et al.*, 1988).

a pyramid, with one apex at the substrate–film interface. There is a distinct difference in the quality of ZnSe epilayers when an initial substrate exposure with Se is done. The substrate temperature at the beginning of Se exposure has a large effect on defects in the resulting layer. If exposure begins at the growth temperature (200°C) then Se-started epilayers are very similar to their Zn-started counterparts. If exposure to Se begins at higher substrate temperatures (>300°C), then a significantly higher density of defects is observed in the resulting epilayers. These epilayers display stacking fault densities of $>10^8$ cm^{-2} (more than four orders of magnitude greater than equivalent Zn-started growths). Representative transmission electron microscopy (TEM) images of Zn- and Se-started epilayers are shown in Figure 3. These initial stages of growth were also studied with STM. An initial Zn exposure was seen to have no effect on the GaAs 2 × 4 reconstructed surface, whereas a Se-stabilized surface (2 × 1) occurs immediately upon exposure. It was found that a highly ordered 2 × 1 Se terminated surface forms if the substrate temperature is >300°C, Figure 4a, but this surface is not like the 2 × 1 surface on ZnSe. The difference in the surfaces is due to electron counting considerations where the Se on GaAs must involve

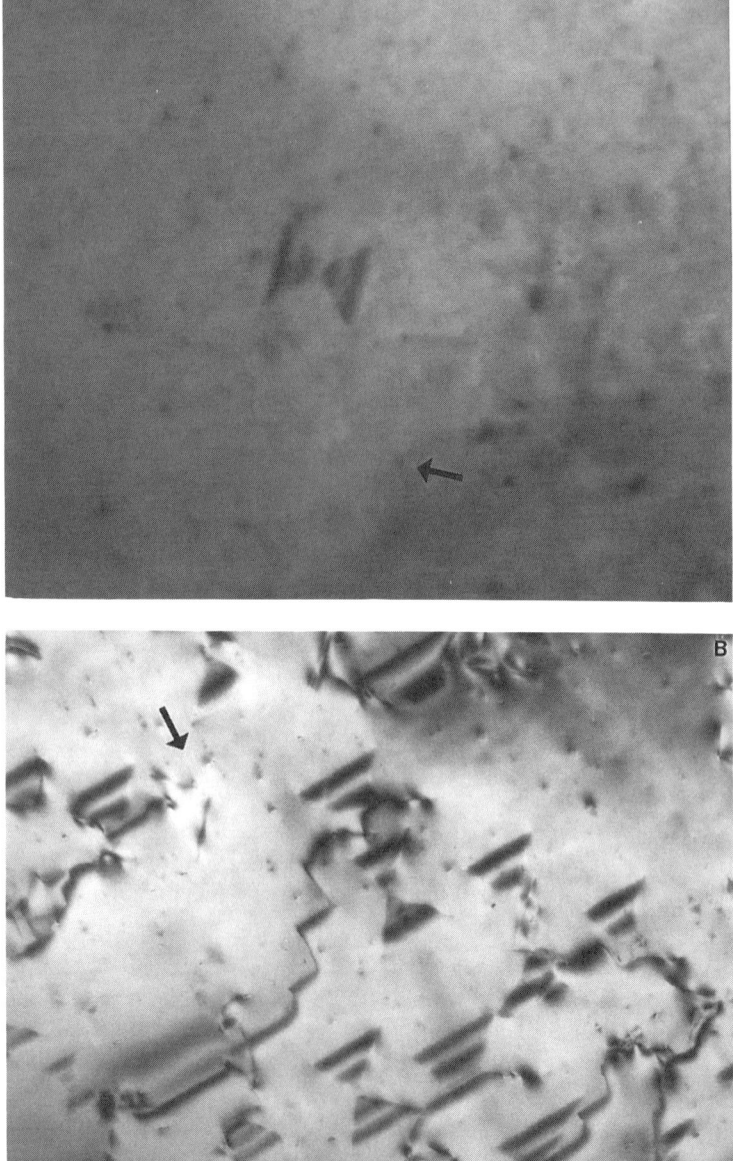

FIG. 3. TEM planar views of (a) a Zn-started and (b) a Se-started ZnSe epilayer showing the difference in stacking fault densities between the two initial growth conditions (from Gaines et al., 1993).

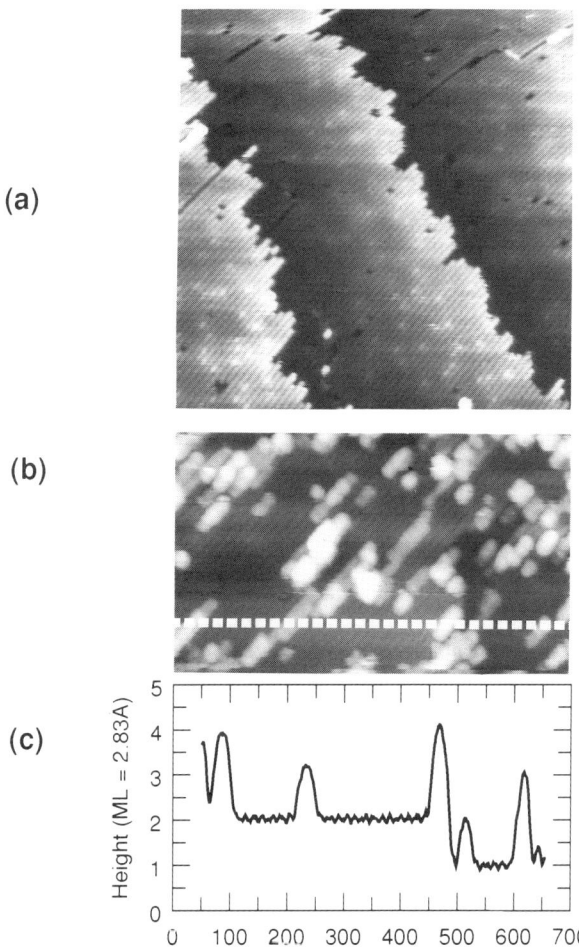

FIG. 4. STM images showing the growth of ZnSe on the Se-stabilized (001) GaAs surface where (a) is the surface prior to ZnSe deposition, (b) is after depositing the equivalent of two monolayers of ZnSe, and (c) is a line profile through the image shown in (b) (from Pashley and Li, 1994).

transfer of electrons from the surface atoms to the bulk to preserve charge neutrality. This difference leads to a more stable surface configuration and makes subsequent growth occur in a 3D mode (ZnSe growth occurs by the formation of islands). An STM image of the equivalent of two monolayers of ZnSe deposited onto the Se-stabilized surface is shown in Figure 4b. The islands of ZnSe are in most cases two or more monolayers high as shown in the line profile of Figure 4c. Thin ≈ 1000 Å epilayers of ZnSe were deposited on both the Zn- and Se-started growths described above and again TEM images revealed that the Se-started epilayers contain orders of magnitude greater densities of stacking faults than those started with Zn. The conclusion that can be drawn from studies of the initial growth stages is that a stable or nonreactive surface is formed when a group VI (Se or S) beam is exposed to the GaAs surface at temperatures < 300°C and that subsequent growth on this surface leads to high densities of stacking fault defects. Therefore it is possible that the background (residual) group VI pressure in the growth chamber could be causing the low density of stacking faults observed in the Zn-started epilayers.

b. Defects: Lattice Mismatch Effects

The defect structure of ZnSe epilayers on GaAs substrates is greatly influenced by the difference in equilibrium lattice constants (lattice mismatch) of the two materials. This heteroepitaxial system has a lattice mismatch, f, of -2.68 and -3.01×10^{-3} (negative value indicates compressive stress on epilayer) at 25 and 350°C (approximate growth temperature), respectively, where f, is given by

$$f = (a_{GaAs} - a_{ZnSe})/a_{GaAs}$$

The difference in the lattice mismatch as a function of temperature arises from the difference in thermal expansion between ZnSe and GaAs.

The growth characteristics of lattice-mismatched systems has been well documented (Matthews, 1975) but a brief summary will be given in the following. The early stages of growth are pseudomorphic (complete registration of atomic planes across the interface) as the epilayer conforms to the lattice constant of the substrate. This causes a tetragonal distortion in the epilayer (biaxial compressive strain, $\varepsilon_{\parallel}$, parallel to the interface plane and tensile strain, $\varepsilon_{\perp}$, parallel to the surface normal). As the epilayer thickness increases, the strain energy builds up. The strain energy increases until a critical thickness is reached where the introduction of misfit dislocations can lower the total energy of the system, with the growth being semicoherent thereafter. After their nucleation, the misfit dislocations accommodate part of the misfit strain with the remainder still accommodated by elastic strain. As the epilayer thickness increases, the density of misfit dislocations increases and they accommodate a larger portion of the lattice mismatch until the epilayer is almost completely relaxed. Along with the increase in density

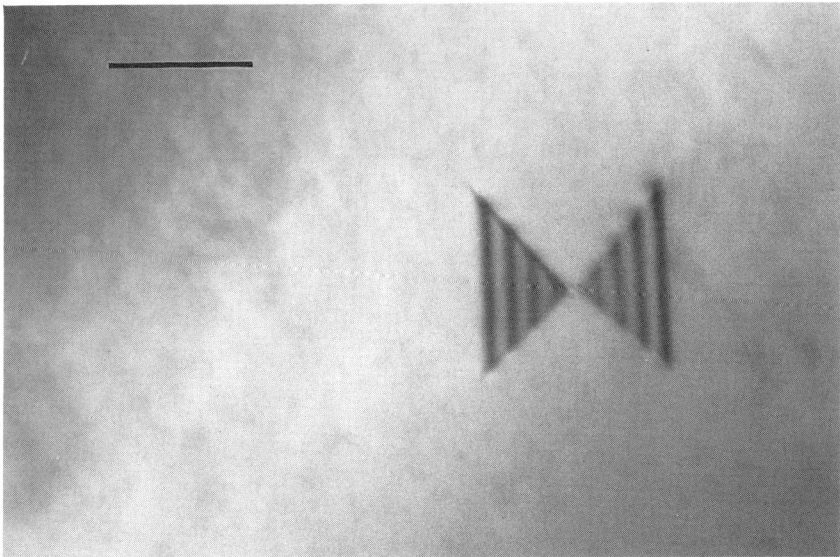

FIG. 5. TEM image of a typical area in a pseudomorphic epilayer of ZnSe where the predominant defect is the stacking fault.

of misfit dislocations is an increase in the number of threading dislocations (threading dislocations terminate misfit dislocations that do not extend to wafer sides).

Pseudomorphic Growth. Plan view transmission electron microscopy (TEM) observations of the epilayers ≤150 nm thick show a defect structure that consists of stacking faults bounded by partial dislocations. Figure 5 is a bright field micrograph of a pseudomorphic epilayer. The stacking faults display the usual black–white fringe contrast. The faults lie on {111} planes and the dislocations are along $\langle 110 \rangle$ directions and are of the Frank partial type with Burgers vector, $b = a/3[111]$. These defects extend from a point near the epilayer–substrate interface to the surface. It was also determined from the Burgers vector analysis that the faults occur predominantly in pairs where points near the interface are close or actually joined. The faults lie preferentially on two {111} planes with the same polarity. This result is not surprising because of the different nature of the {111} planes in the zincblende structure (alternating between cation and anion species). The density of the faults can range from 10^5 to $>10^8$ cm^{-2} depending on the growth conditions as discussed previously.

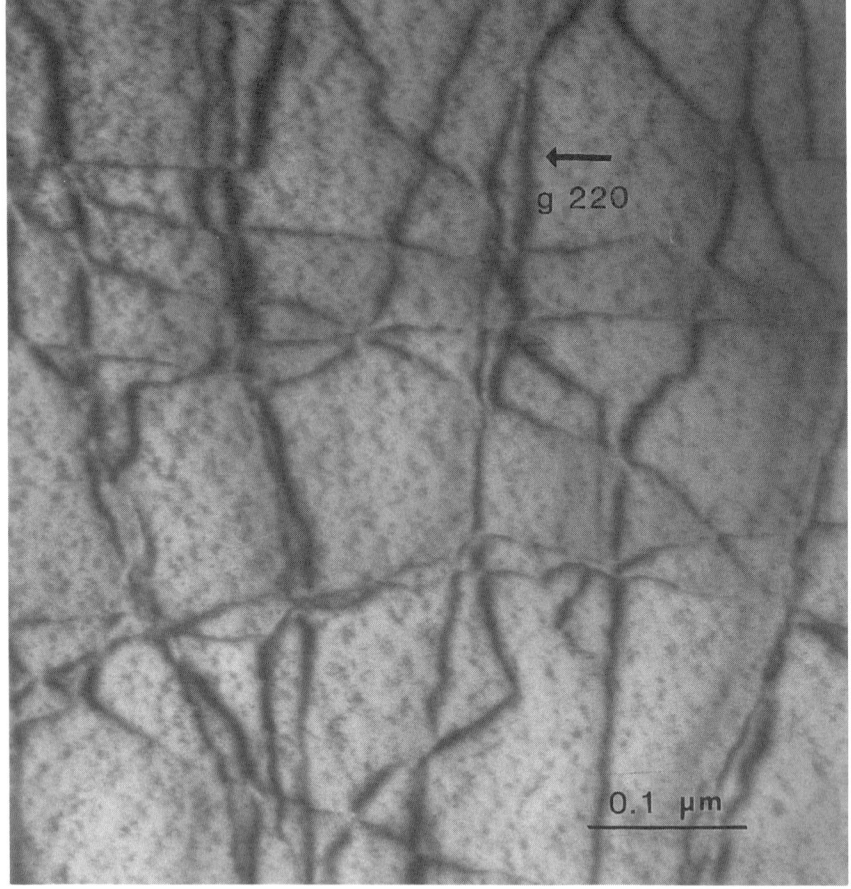

FIG. 6. TEM image of a partially relaxed epilayer (from Petruzzello et al., 1988).

Relaxed Epilayers. The critical thickness for the formation of misfit dislocations was estimated to be about 150 nm by a number of researchers. A typical planar view for a partially relaxed epilayer (≈ 300 nm thick) is shown in Figure 6 where the misfit dislocations lie along $\langle 110 \rangle$ directions in the $\{001\}$ interface plane. The dislocations are predominantly (about 95%) of the 60° type (i.e., $b = a/2[110]$ and a 60° angle between the Burgers vector and dislocation line). It has been found that some of the 60° type dislocations are dissociated into two partials (30° and 90°) with a small separation (<5 nm). The remainder of the dislocations are Lomer edge type with Burger's vectors in the plane of the interface. The Lomer edge type is not a primary dislocation in the zinc-blende structure but arises from the

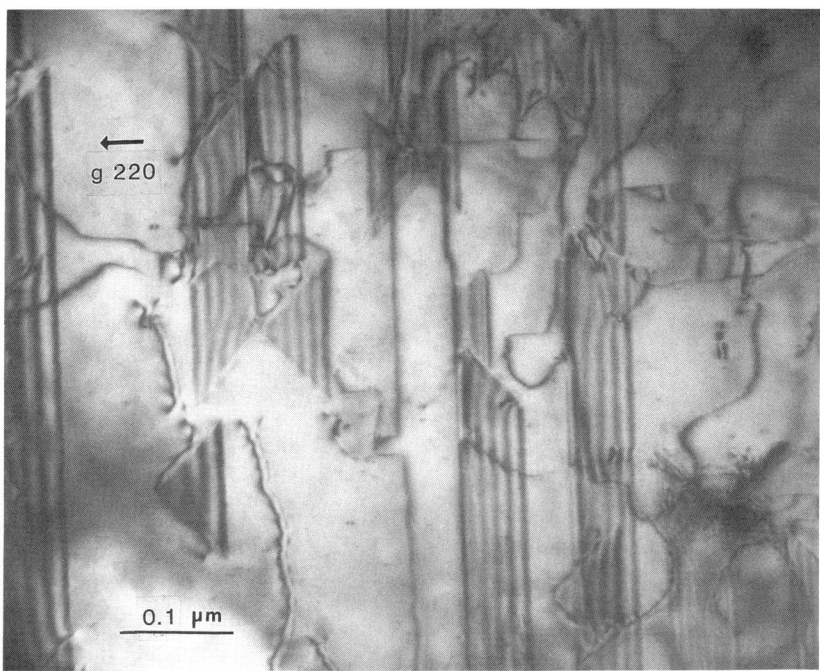

FIG. 7. The elongation of stacking faults into partial misfit dislocations (a) and the formation of perfect misfit dislocations from the faults (b) (from Petruzzello et al., 1988).

interaction of two 60° type dislocations. Many of the misfit dislocations contain zigzags in their line direction which alternates between two perpendicular $\langle 110 \rangle$ directions lying in the (001) plane and is seen in Figure 6. This configuration of misfit dislocations arises from cross slip of the threading segment on $\{111\}$ planes as it glides out toward the edge of the wafer.

The stacking fault defects found in the pseudomorphic epilayers are still present but at a lower density in the partially relaxed epilayers. Some of the faults were observed to elongate along the interface as shown in Figure 7a and a few faults were found at the end of misfit dislocations as shown in Figure 7b.

As the thickness of the epilayer is increased ($150 < t < 500$ nm), so does the density of the misfit dislocations. An irregular network of misfit dislocations is formed with some of the dislocations still present along $\langle 110 \rangle$ directions. The zigzag pattern found in the misfit dislocations of the thinner epilayers is reduced in density and many of the misfit dislocations appear to have a smooth curvature in their line directions. The ratio of 60° to edge type dislocations is decreasing slightly as the thickness increases. Only a very low density of stacking faults remain for this epilayer thickness range.

The epilayers thicker than 500 nm exhibit a further increase in density of misfit dislocations and decreasing ratio of 60° to edge type, as a function of thickness. The density of stacking faults in these epilayers has become too low to observe with the TEM.

Nucleation of Misfit Dislocations. There are three most likely candidates for dislocation sources in the ZnSe/GaAs system. These are the threading dislocations present in the substrate, Frank type partial dislocations (stacking faults), and surface Frank-Read sources. For this system the predominant mechanism for misfit dislocation nucleation appears to be surface sources. This mechanism is supported by the fact that the majority of misfit dislocations in layers just exceeding the critical thickness are of the 60° type and they are the only type obtainable from glide along {111} planes from the surface to the interface. The other sources of misfit dislocations mentioned above are not as dominant as the surface as deduced from their relative densities. The typical density of threading dislocations in the GaAs wafers used is about 10^3 cm^{-2} which is about seven orders of magnitude too small to account for the density of misfit dislocations. The last source of misfit dislocations to be considered are partial dislocations (stacking faults) of the Frank type present in the thinner films where the dislocation reaction is

$$1/3a[111] = 1/2a[110] + 1/6a[112]$$

where the first product is a perfect misfit dislocation and the second is a Shockley type partial that glides to the surface and escapes. It was shown that these defects appear to elongate and then disappear leaving behind a misfit dislocation as the epilayer thickness increases. However, the density of these stacking faults is at least two orders of magnitude smaller than that of the misfit dislocations.

The concentration of Lomer edge type misfit dislocations observed is due to interactions between 60° dislocations as described by Ahearn and Laird (1977). As the density of 60° dislocations increases with thickness so does the probability of an interaction. Therefore the ratio of edge to 60° dislocations also increases with thickness. These edge type dislocations are more efficient at relaxing the misfit strain because their Burgers' vector is in the same direction as the stress whereas the 60° dislocation Burgers' vector is inclined 45° to the stress axis. This will be discussed further in the next section.

Relaxation of Lattice Mismatch. As mentioned previously, the lattice mismatch is manifested as elastic strain in the epilayer when the system is pseudomorphic (free of misfit dislocations). After the nucleation of misfit dislocations the mismatch is accommodated by a combination of the elastic strain and misfit dislocations in such a way that

$$f = \varepsilon + \delta$$

where ε is the elastic strain and δ is the amount of mismatch accommodated by misfit dislocations. The quantity δ is related to the Burgers vector and average separation, d, between misfit dislocations by

$$\delta = b_{\parallel}/d$$

where $b_{\parallel}$ is the component of the Burgers vector parallel to the interface plane. For the 60° type dislocation the resolved component of the Burger vector in the plane of the interface is $a/4[110] = 0.2$ nm, while the resolved component of the Burgers vector for edge type dislocations is $a/2[110] = 0.4$ nm. The effective Burgers vector b_{eff} for a particular layer can then be written as

$$b_{\text{eff}} = b_{60}(x) + b_{\text{edge}}(1 - x)$$

where x is the fraction of 60° type dislocations. The average spacing between dislocations, d, along with the values for δ, b_{eff}, $\varepsilon_{\parallel}$ are shown in Table I. The elastic strain values are shown graphically in Figure 8 as determined with photoluminescence (PL), x-ray diffraction (XRD) and TEM and the agreement between techniques is very good.

TABLE I

SUMMARY OF THE EXPERIMENTAL VALUES FOR THE ZnSe EPILAYERS AS A FUNCTION OF THICKNESS, FROM TEM AND XRD RESULTS

h(nm)	d(nm)	$r\%^a$	b_{eff}	$\delta (\times 10^3)$	$\varepsilon_{\parallel} (\times 10^3)$
50	—	—	—	0.0	−2.68
87	—	—	—	0.0	−2.68
180	575	95	2.10	0.37	−2.31
310	128	90	2.20	1.72	−0.96
550	93	85	2.28	2.45	−0.23
880	86	81	2.34	2.72	0.04
1400	83	77	2.42	3.03	0.35
4900	79	77	2.42	3.06	0.38

[a] Percentage of 60° type dislocations.

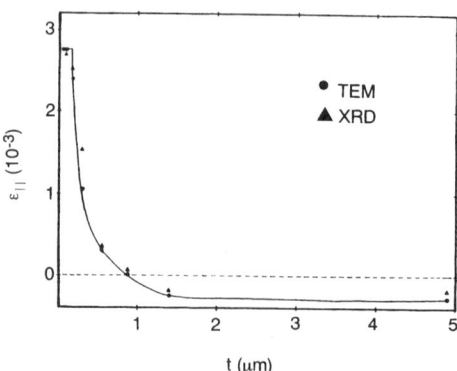

FIG. 8. Plot of elastic strain vs. thickness for ZnSe epilayers as determined by TEM and XRD (from Petruzzello et al., 1988).

It appears that in the early stages of growth the epilayer is pseudomorphic and elastic strain in the layer takes the value of the lattice mismatch ($\varepsilon_| = f$). This regime of growth ends at about 150 nm (critical thickness, h_c) where misfit dislocations appear. From 150 to 500 nm of growth the density of dislocations increases rapidly and consequently the elastic strain in the epilayer decreases accordingly. Above 500 nm the relaxation rate slows until the epilayer is almost completely relaxed after about 900 nm of growth. Further growth above 900 nm results in a reverse of the sign of $\varepsilon_|$. An explanation of this anomaly comes from the difference in thermal expansion of ZnSe and GaAs. The mismatch increases with increasing temperatures and at the growth temperature (350°C) is -3.01×10^{-3}. The relaxation is nearly complete at the growth temperature and therefore upon cooling to room temperature the layer is under biaxial tension.

Effect of N-doping on the Relaxation of ZnSe on GaAs. The previous sections have described the structural characteristics of undoped ZnSe epilayers. To produce laser structures it is necessary to introduce both p-type and n-type dopants (acceptors and donors, respectively) in the cladding regions of the device for carrier injection and recombination in the active region. The incorporation of N in the ZnSe lattice has a dramatic effect on the arrangement and density of misfit dislocations which in turn impact the relaxation of the compressive strain due to the lattice mismatch with the substrate. The micrographs of Figure 9a and b illustrate this effect. The images are planar views of the substrate–epitaxial layer interfaces taken to reveal the misfit dislocation structure for typical (Fig. 9a) undoped and (Fig. 9b) N-doped ([N] = 1×10^{18} cm^{-3}) ZnSe layers. The micrograph of

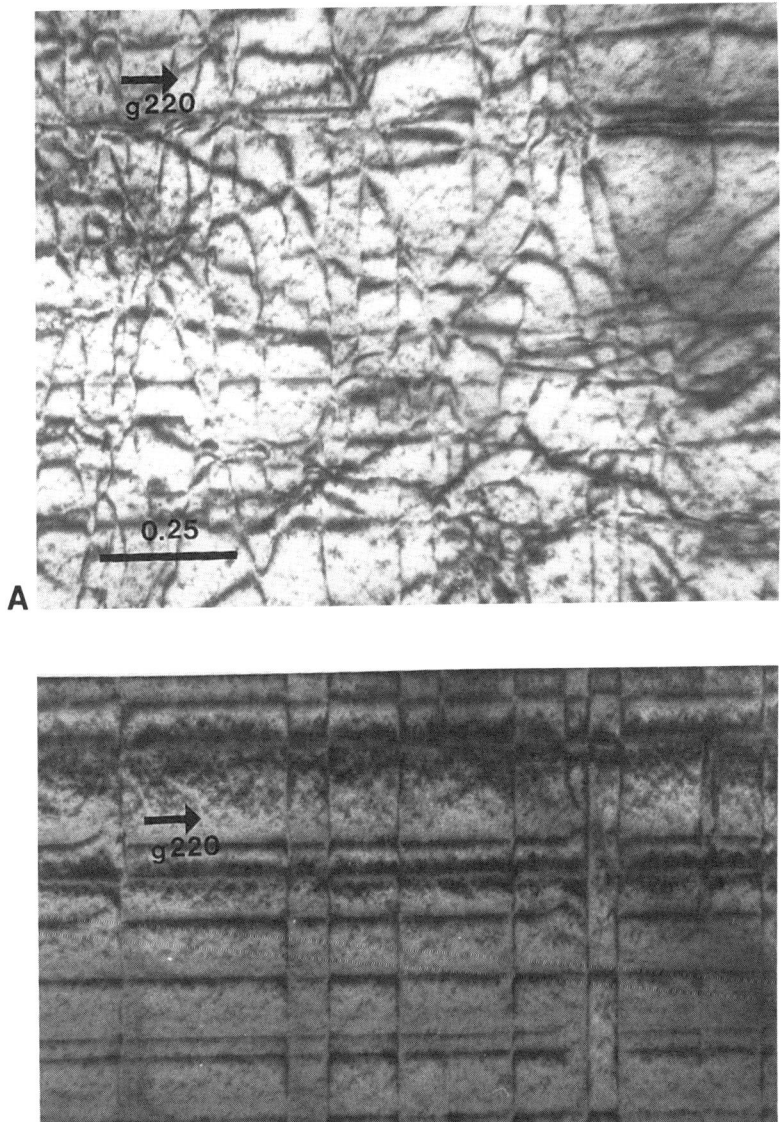

FIG. 9. TEM planar views of the misfit dislocation arrays at the ZnSe–GaAs interface for the (a) undoped and (b) N-doped epilayers of 1 μm thickness (from Petruzzello et al., 1993).

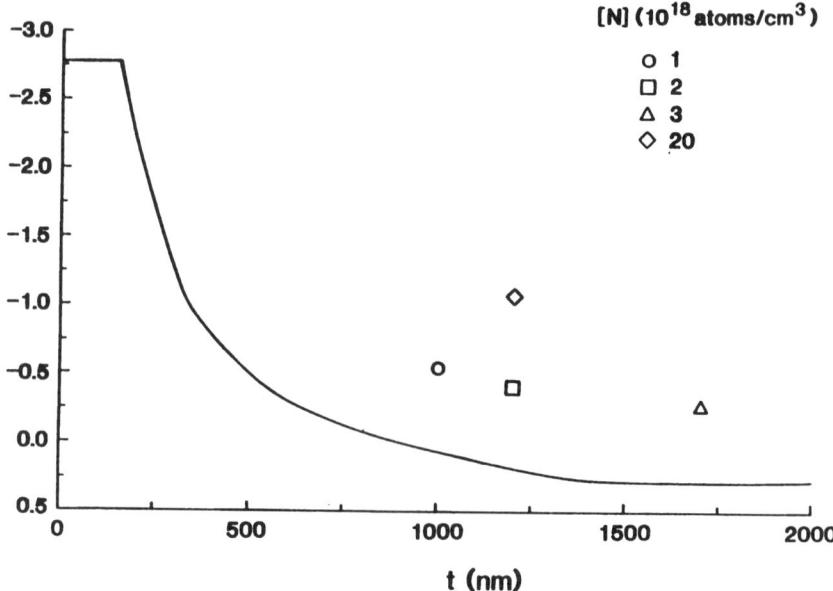

FIG. 10. Plot of elastic strain vs. thickness comparing the undoped and doped epilayers (from Petruzzello et al., 1993).

Figure 9a shows the typical irregular misfit dislocation array for an undoped epitaxial layer. In addition to the misfit dislocations with line directions along the $\langle 110 \rangle$ (usual habit direction found for 60° and edge type dislocations in the zinc-blende structure), undoped layers exhibit a high density of dislocations with other line directions. This system has been described in detail in the previous section. The deviation of the misfit dislocations from the $\langle 011 \rangle$ direction is a consequence of cross slip of the threading segment between $\{111\}$ planes as it glides to produce the misfit dislocation. For N-doped ZnSe, the misfit dislocation array becomes a regular rectangular grid as seen in Figure 10b. In this sample only a very low density of the misfit dislocations deviate from the $\langle 110 \rangle$ direction relative to the undoped case. The explanation for the observance of the regular array of misfit dislocations in the doped layers is that almost all of the misfit dislocations are dissociated into two partial dislocations as discussed earlier and cross slip of this defect configuration is impossible.

The other notable effect of the N-doping observed in the micrographs of Figure 9 is that the density of misfit dislocations decreases significantly for similar thickness layers. The average separation of the misfit dislocations in the undoped layer (Fig. 9a) is about 85 nm and in the 1×10^{18} cm^{-3} layer

(Fig. 9b) is about 125 nm. In both the undoped and N-doped layers the amount of relaxation afforded per misfit dislocation is about the same because the ratio of 60° to edge dislocations is similar. The relaxation of the compressive strain in the N-doped layers is therefore less for a given layer thickness, as the incorporated N acts to "harden" the ZnSe lattice against dislocation movement. Suppressed dislocation motion due to impurity atoms has been analyzed theoretically by Ehrenreich and Hirth (1985). This difference in relaxation of the N-doped layers gives rise to residual strain in the ZnSe.

Figure 10 displays a plot of strain parallel to the interface, $\varepsilon_\parallel$ as a function of thickness, t. The solid line represents data from the study on undoped ZnSe. Undoped layers are pseudomorphic (strain equals lattice mismatch) up to a thickness of about 150 nm and then relaxed by misfit dislocation formation as thickness is increased. Above 900 nm in thickness the layers exhibited a change in strain from compressive to tensile which arose from differences in the thermal expansion between ZnSe and GaAs. The undoped control layer of this study is in agreement with the earlier data but the N-doped layers fall well above the solid line. Note that the magnitude of the residual strain in the different layers cannot be correlated with the total N concentration because the thicknesses of the layers are also changing as

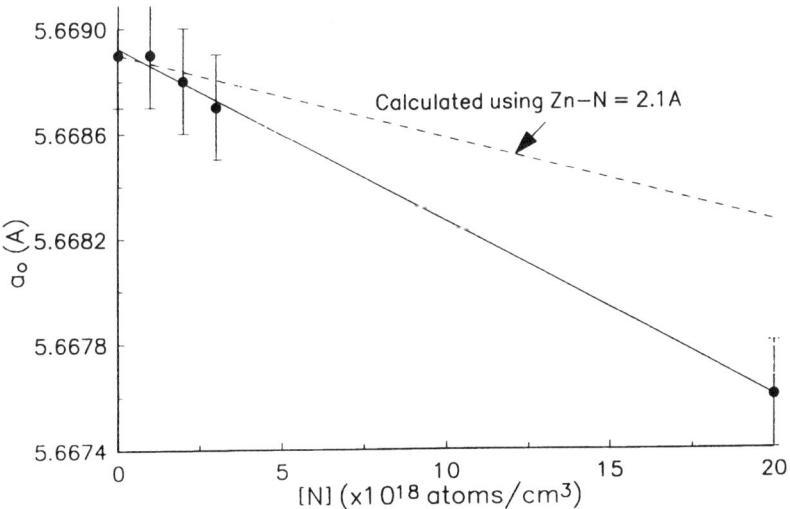

FIG. 11. Plot of the lattice constant vs. N concentration. The dotted line is calculated from theoretical Zn–N bond length and the solid line a fit to the data (from Petruzzello et al., 1993).

shown in Figure 10. However, for the two samples with equal thickness, the layer containing a higher [N] had a larger amount of residual strain.

Another striking effect of the N incorporation is that the lattice constant is decreasing as a function of [N]. A plot of the measured lattice constant (after deconvolution of strain effects) vs. N concentration is shown in Figure 11. The dotted line is calculated using a Zn–N bond length of 2.1 Å in place of the 2.455 Å bond length of Zn–Se (the Zn–N bond length comes from a first principles pseudopotential calculation of N in ZnSe) and the solid line is a fit to the data. The decrease in lattice parameter determined experimentally is larger than the calculated value. A possible explanation for this observation is the generation of point defects, namely vacancies, in the layer with increasing N incorporation. Point defect formation was also observed in ZnSe doped with Ga and O by Miyajima et al. (1991).

2. LASER STRUCTURES

The improvements in laser operating characteristics obtained using $Zn_{1-x}Mg_xS_ySe_{1-y}$ cladding layers are obtained from increased optical confinement, increased electrical confinement, and from improved structural properties. In this part, we stress the improvements in structural properties, and correlate them with laser performance. Defects present in the laser structures include misfit dislocations, threading dislocations, and stacking faults.

a. Ternary Lasers

The structure of the (SCH) in Figure 12a contains a ZnSe guiding region of about 1 μm in thickness. The lattice mismatch between GaAs and ZnSe is about 0.27%, resulting in a critical layer thickness (the thickness at which misfit dislocations start forming and the layer starts relaxing) of about 150 nm as discussed earlier. Thus, relaxation occurs at the interfaces between the guiding and cladding regions. As a result, networks of misfit dislocations are present in both guiding/cladding layer interfaces. Threading dislocations arise from the misfit dislocations and thread through the active region at a density of 10^8 to 10^9 cm^{-2}. Figure 13 is an image of the same structure as in Figure 12a, but with the sample rotated by about 40° about an axis lying parallel to the layer interfaces and perpendicular to the electron beam (perpendicular to the normal of the page). In this image the individual misfit dislocations in the interfaces can be resolved. The average separation between misfit dislocations in the lower cladding/guiding interface is about

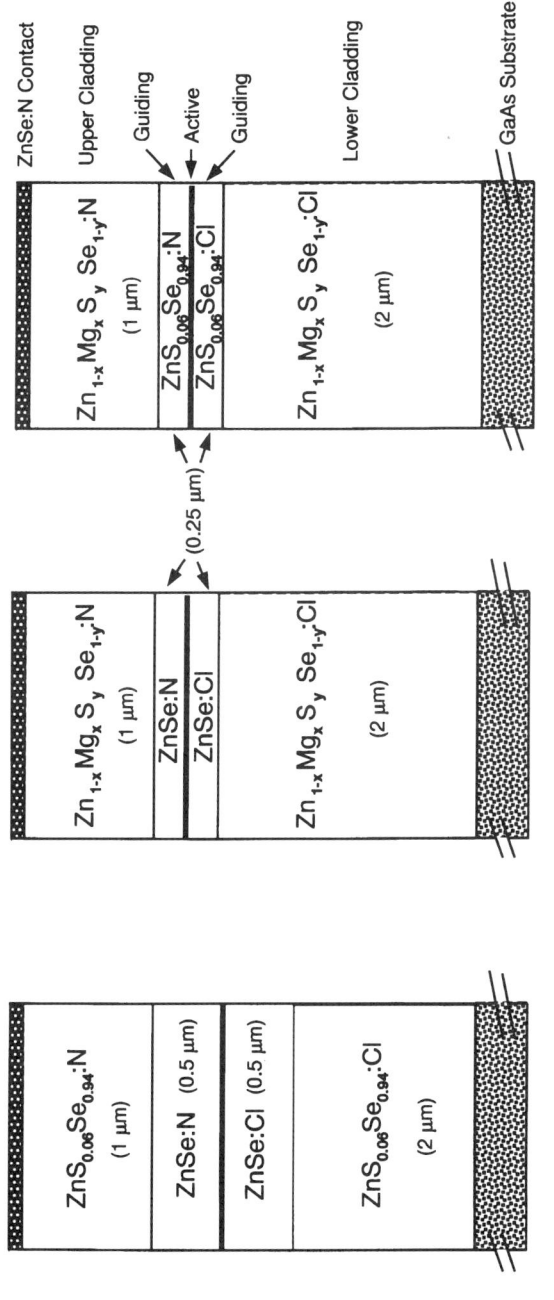

FIG. 12. Schematics of the (a) ternary and (b) and (c) quaternary based SCH lasers (from Petruzzello et al., 1995).

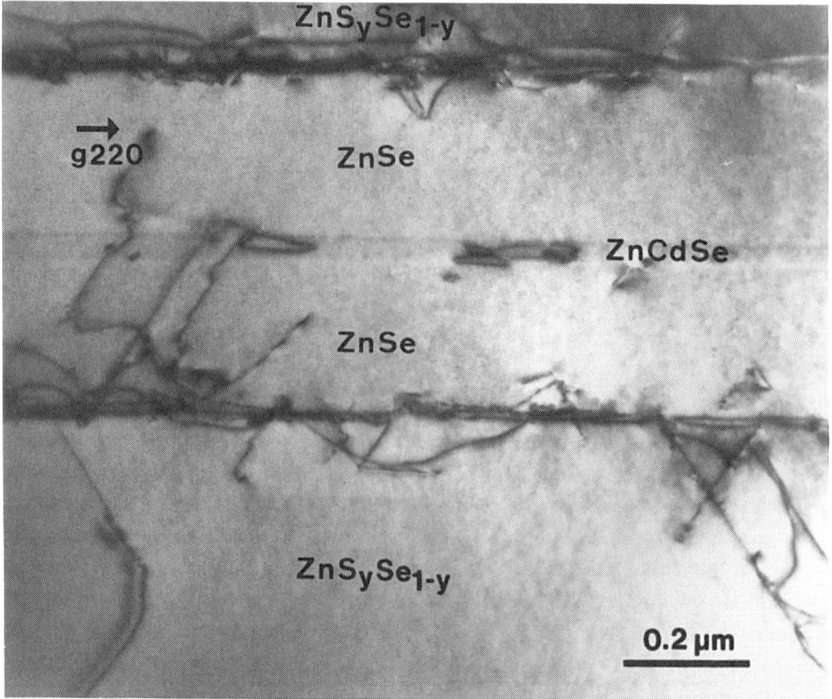

Fig. 13. TEM image of the ternary based SCH as in Figure 12a (from Petruzzello et al., 1995).

75 nm, corresponding to complete relaxation of the ZnSe layer (2.7×10^{-3}). The separation at the upper interface, however, is about 165 nm, corresponding to a lattice relaxation of only 1.2×10^{-3}, calculated by the method described in the previous part. The upper cladding layer thickness is 1 μm, equal to the thickness of the guiding region. Thus, complete relaxation might be expected at this interface also. The explanation for the incomplete relaxation at the upper guiding/cladding interface seems from the nitrogen doping in the upper layers. It was discussed previously that the density of misfit dislocations is lower for a N-doped layer than in an undoped layer of equal thickness.

b. Quaternary Lasers

In this section we compare the structural properties of several quaternary lasers, emphasizing the effects of structural quality on the laser performance.

The lasers were studied using cross-section TEM and 2-dimensional high-resolution x-ray diffraction (2D-HRXRD).

ZnSe Guiding Region. TEM micrographs (Fig. 14a) of a laser with quaternary cladding layers, and a ZnSe guiding region, reveal no dislocations at the interface between the GaAs substrate and the lower quaternary cladding layer, confirming that the lower cladding layer is pseudomorphic with the substrate. Misfit dislocations are present at the lower guiding/cladding interface, but not at the upper interface. The relaxation at the lower guiding/cladding interface is estimated to be about 2.4×10^{-3}, or about 90% of the mismatch. As a result of the partial relaxation of the lower guiding/cladding interface, the density of dislocations threading through the active region is about 10^7 cm^{-2}, more than an order of magnitude below the typical defect density for the ternary laser structure.

The threshold current density of this laser was 1300 A/cm^2 at room temperature, and 500 A/cm^2 at 80 K. Thus, its performance is considerably poorer than that of the fully pseudomorphic laser containing $ZnS_{0.06}Se_{0.94}$ cladding layers, described above. However, its threshold current density is lower and its lifetime is longer at room temperature than for ternary lasers, indicating that the quaternary cladding layers yield device improvements even for nonpseudomorphic structures.

Additional information is obtained from the 2D-HRXRD mapping of the region of reciprocal space near the (224) reflection (Fig. 14b). The figure is a contour plot, with $K_|$ the reciprocal space direction parallel to the interface planes, and $K_\perp$ the reciprocal space direction perpendicular to the interface planes. Several peaks are observed, and are labeled with the corresponding layers in the structure. The peaks were identified based on the location of misfit dislocations, as determined by TEM. A perfectly lattice-matched layer would produce a peak overlying the GaAs substrate peak. A pseudomorphic layer is tetragonally distorted, and produces a peak with the same in-plane lattice parameter, $a_|$ (and the same $K_|$), but with a different $a_\perp$ and $K_\perp$ than the substrate. A relaxed or partially relaxed layer will have a different $a_|$ and a different $a_\perp$ than the substrate. For the laser structure of Figure 14b, the substrate and lower cladding layers have equal $K_|$, but differ in $K_\perp$. Thus, the substrate and lower cladding are pseudomorphic, in agreement with the TEM observation that no misfit dislocations were present at the substrate interface. The perpendicular lattice parameter, $a_\perp$, is 5.6616 Å. The equilibrium lattice constant, a_0, calculated using the ZnSe Poisson's ratio of 0.375, is 5.6573 Å. Thus, the lattice mismatch ($a_{substrate} - a_{epilayer}/a_{substrate}$) with the substrate is -6.3×10^{-4}. This is also the residual strain in the layer, since no relaxation has occurred. The guiding and upper cladding layer peaks (Fig. 14b) are displaced from the substrate peak in both the $K_|$ and $K_\perp$

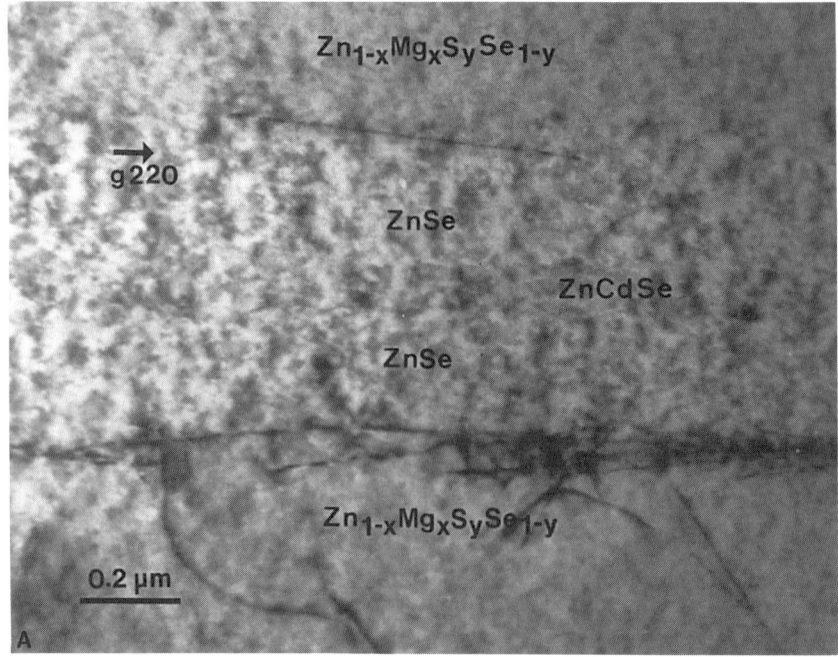

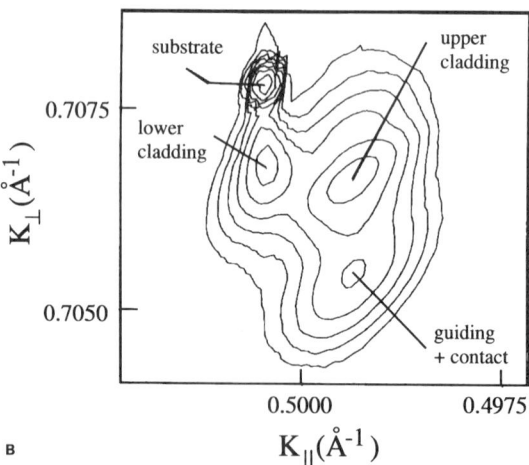

FIG. 14. TEM image (a) of the quaternary based SCH with ZnSe guiding region as in Figure 12b and 2-dimensional x-ray map (b) of the quaternary based SCH with ZnSe guiding region (from Petruzzello et al., 1995).

directions, because of the relaxation at the lower guiding/cladding interface. $a_\parallel$ and $a_\perp$ for the ZnSe guiding region are 5.6721 and 5.6660 Å, respectively. The value of $a_\parallel$, calculated from the measured TEM dislocation density, is 5.6670 Å, in good agreement with the x-ray result. The values of $a_\parallel$ and $a_\perp$, for the upper cladding layer, are 5.6629 and 5.6660 Å, respectively. The calculated equilibrium lattice constant is 5.6646 Å, yielding a lattice mismatch of 2.6×10^{-4} with $a_\parallel$ of the ZnSe guiding region below it. This small lattice mismatch is insufficient to generate misfit dislocations at the upper guiding/cladding interface for a 1-μm-thick cladding layer. The mismatch of the upper cladding layer to the GaAs substrate, however, is 1.9×10^{-3}, considerably larger than the mismatch between the GaAs and the lower cladding layer, and indicates that the composition of the two quaternaries is different.

ZnS_zSe_{1-z} Guiding Region. A TEM micrograph (Fig. 15a) of the quaternary laser indicates the improved structural quality of this laser relative to the ternary laser. The performance of this laser is considerably better than the performance of the previously described, partially relaxed lasers. Figure 15a is a bright-field micrograph containing all of the II/VI layers. All interfaces were found to be free of misfit dislocations, demonstrating that all layers are pseudomorphic with the GaAs substrate. No dislocations were observed to thread through the active region over the transparent part of the sample (approximately 50×0.5 μm). Thus, we estimate the density of defects threading through the active region to be less than 4×10^6 cm^{-2}. A few stacking faults were observed, starting at the upper cladding/guiding interface and propagating to the surface. The origin of these faults is unclear at present.

A 2D-HRXRD map (Fig. 15b) of the region of reciprocal space about the (224) reflection shows several peaks that lie along a line parallel to the [001] direction (the normal to the interfaces). Therefore, the value of $a_\parallel$ equals that of the substrate (5.6537 Å) for all layers, in agreement with the TEM observation that no misfit dislocations are present in any of the interfaces. The value of $a_\perp$ varies, indicating different amounts of strain in the different layers. Because no relaxation has occurred, identification of the layers producing the peaks is more complicated. However, based on a comparison of measured intensities and widths of the various peaks with simulations done with the known layer thicknesses, we assign them to the individual layers as labeled on Figure 15b. The positions of layers can also be deduced by measuring rocking curves for reflections with shallow and high incidence angles with respect to the sample surface. For low incidence angles, diffraction from the upper layers will appear enhanced relative to diffraction

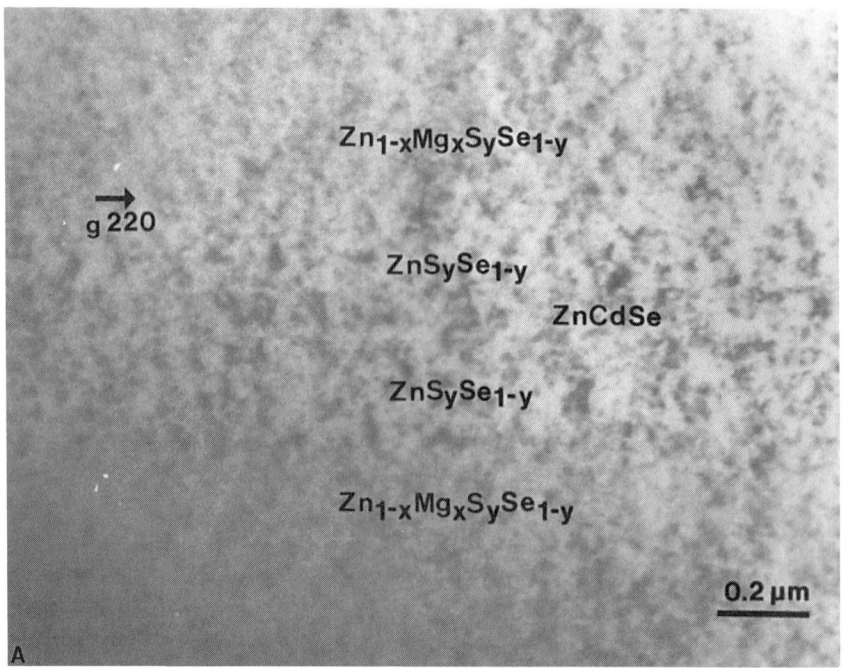

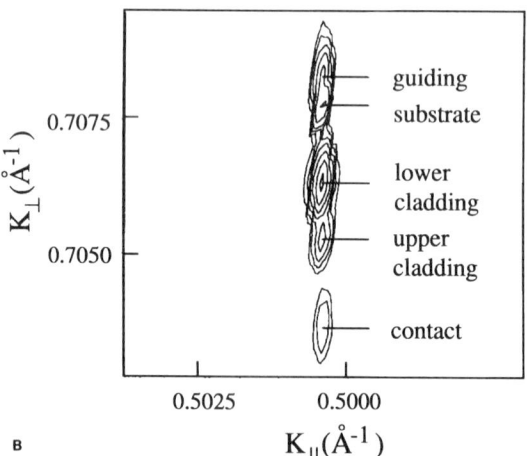

FIG. 15. TEM image (a) of the quaternary based SCH with ZnSSe guiding region as in Figure 12c and 2-dimensional x-ray map (b) of the quaternary based SCH with ZnSSe guiding region (from Petruzzello et al., 1995).

from lower layers, permitting relative positions of the layers to be determined.

From the data of Figure 15b, the lower cladding layer has $a_\perp = 5.6652$ A, leading to an equilibrium lattice parameter, a_0, of 5.6588 A and a lattice mismatch with the substrate of -9.0×10^{-4}. The ZnS_zSe_{1-z} peak is located on the opposite side of the substrate peak, and partially offsets the strain in the lower cladding layer. The value of a_0 is 5.6521, and the lattice mismatch with the substrate is 2.9×10^{-4}. The sulfur content, calculated using Vegard's law with $a_{ZnS} = 5.409$ A, is $z = 0.065$. The peak at lowest K_1 is from the 0.1-μm-thick ZnSe p-type contact layer. The peak labeled "upper cladding layer" has $a_0 = 5.6624$ A and a lattice mismatch with the substrate of -1.5×10^{-3}. The x-ray simulations show that this peak is produced by a layer about 0.25 μm thick, indicating that this peak may represent only part of the upper cladding. The remaining part of the upper cladding peak probably overlies the lower cladding layer peak. Thus, the composition of the quarternary abruptly changed during growth of the upper cladding layer. We speculate that this change is a result of rearrangement of source material in the ZnS furnace, causing a sudden change of flux.

From purely structural considerations, the most stable laser structure would be achieved by perfect lattice matching of every layer. We observe that, if the lattice mismatch is kept below about 1×10^{-3}, then layers of about 2-μm thickness may be grown without misfit dislocations. However, such layers are metastable, and could become unstable during laser operation, resulting in formation of dislocations and degradation of the device.

The equilibrium model of Matthews (1975) predicts a critical layer thickness of 0.16 μm for a lattice mismatch of 1×10^{-3}. This model does not account for the activation energy required to nucleate the misfit dislocations from the surface, and thus represents a lower bound on the critical layer thickness. Maree et al. (1985) considered this problem for the case of half-loop nucleation from the surface of zinc-blende materials. They estimated the critical thickness to be about 10 μm for a lattice mismatch of 1×10^{-3}. Their experimental data for mismatch of 1×10^{-3} falls between the two models. Thus, the observation that pseudomorphic layers of $Zn_{1-x}Mg_xS_ySe_{1-y}$ can be grown up to 2 μm thick with this mismatch is understandable.

c. *Effects of Stacking Faults on Laser Performance*

$Zn_{1-x}Mg_xS_ySe_{1-y}$-containing lasers may be grown with all layers pseudomorphic with the substrate, as described above. However, growth of the quaternary is accompanied by a variable density of stacking faults and threading dislocations which probably arise from the faults. The presence of these defects impacts the laser performance. In this section, we compare the structure and performance of three quaternary lasers with defect density (threading through the active region) varying from $<4 \times 10^6 \text{ cm}^{-2}$ to $6 \times 10^8 \text{ cm}^{-2}$. The lasers with the lowest defect density yielded the best operating characteristics. As expected, reduction of the defect density, well

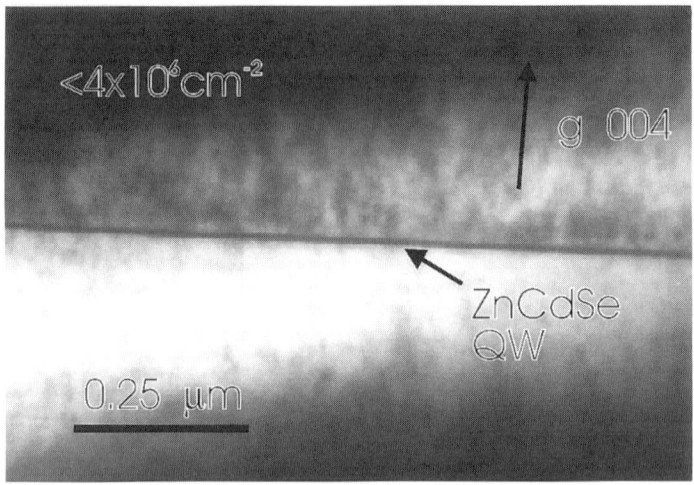

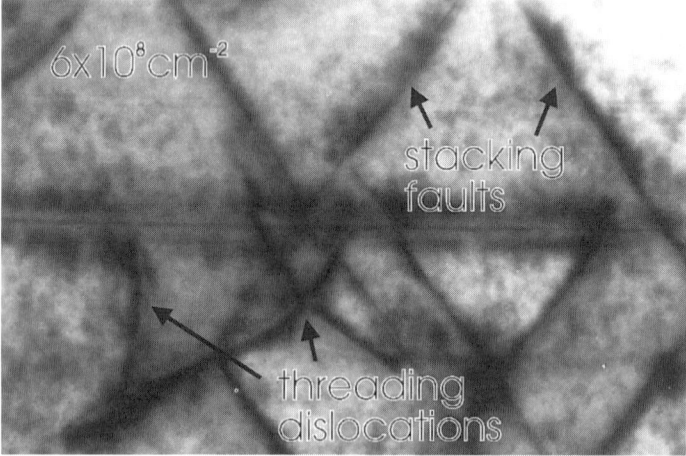

FIG. 16. TEM images of the active regions in two pseudomorphic SCH lasers with (a) a low density and (b) a high density of stacking fault defects (from Petruzzello et al., 1995).

known to be a critical factor in determining lifetime of III/V lasers, is imperative for II/VI lasers as well.

TEM micrographs (Fig. 16a and b) strongly contrast the active regions of the laser structures with lowest and highest defect densities, $<4 \times 10^6\,\mathrm{cm}^{-2}$ and $6 \times 10^8\,\mathrm{cm}^{-2}$, respectively. Stacking faults are the majority defect in Figure 16, and appear as straight dark lines lying in (111) planes. The defect density in Figure 16b is comparable to the defect density in ternary laser structures, although the majority defects in the ternary lasers are threading dislocations.

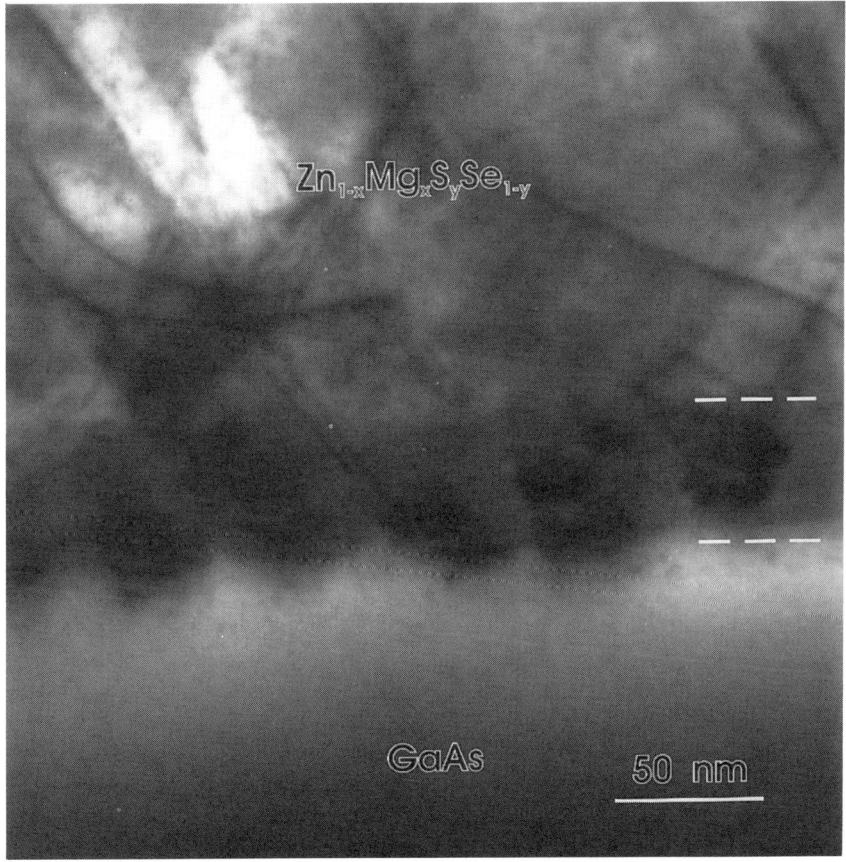

FIG. 17. TEM image showing the dislocation loop contrast in the vicinity of the stacking fault nucleation site (from Petruzzello *et al.*, 1995).

The stacking faults originate primarily from the interface between the GaAs substrate and the lower cladding layer, and are present throughout the entire structure. Higher resolution TEM imaging, at high tilt angles, of the GaAs interface (Fig. 17), reveals regions of darker contrast in the interface plane, that appear to be dislocation loops. Stacking faults emanate from these loops. The origin and detailed structure of the loop/stacking fault combination is unclear, but appears to be related to details of the substrate preparation, oxide desorption, and/or initiation of growth. Although the majority of stacking faults start at the GaAs interface, we have also observed faults beginning at points higher in the epitaxial layer.

The defects have a significant effect on the threshold current densities of lasers. Comparison of SCH lasers made from molecular beam epitaxy (MBE) growth with varying defect densities shows that those structures with the highest defect density also have the highest threshold current densities. The three lasers are identical in structure, except for the variation in defect density and a small variation in Cd content (about 2%, based on the photoluminescence peak wavelengths). The room-temperature threshold current density of the most defected device is a factor of four higher than that of the least defected device. The value of T_0 is degraded by the presence of defects. T_0 is lowest for the structure with highest defect density. In addition, the differential external efficiency depends on the defect density, with the highest efficiency occurring for the lowest defect density. The above observations may be explained by the defects acting as nonradiative recombination centers.

III. Degradation Effects

The initial thrust of research following the first demonstration of injection lasing in 1991 (Haase et al., 1991) was in the direction of development of new materials, reducing threshold current densities, and trying to attain room temperature cw operation. Around 1992, electroluminescence imaging observations of light-emitting diode (LED) structures at 3M clearly showed the formation of dark nonradiative recombination regions that expanded during device operation. Analysis of these degraded LEDs by Guha et al. (1993b) showed that the degradation was microstructural and activated by preexisting threading defects such as stacking faults that were grown in during epitaxial deposition. This form of degradation is currently the major impediment for long lived devices and a large part of the research effort has since focused on understanding the degradation mechanism, and reducing preexisting stacking faults in the epitaxial structures. Significant progress has been made over the past few years in reducing the densities of these stacking faults from 10^7 per cm^2 to about 10^5 per cm^2 or lower (numbers quoted are ones that can be routinely obtained). This improvement has been the major cause for room temperature cw lifetimes increasing recently to a few hours as demonstrated by groups at Sony and a combined 3M and Philips group. While these are encouraging results, a 10,000-hr cw lifetime is usually taken to be a reasonable number at which these lasers will begin to be considered as marketable: consequently the technology is still in its infancy and much needs to be developed.

Once the degradation induced by preexisting defects, which we will refer to as *extrinsic* microstructural degradation, is eliminated, other degradation modes that are likely masked at present will have to be addressed. Degradation initiated at facets and processing damage induced failure are likely to become important. While at present the degradation is related to events in the device active region, contact degradation from currently used ZnSeTe contacts may also be anticipated for lifetimes greater than about 5 to 10 hr.

Beginning with a brief description of the experimental techniques used to study the degradation mechanisms, we discuss the various degradation mechanisms operative in wide gap II–VI lasers and light emitting diodes, our understanding of the degradation process, and degradation modes that we expect to become important in the future, as the device lifetimes increase.

1. EXPERIMENTAL TECHNIQUES

Most of the relevant degradation studies so far have been microstructural imaging based and this has been appropriate since the degradation is crystal defect related. Transmission electron, cathodoluminescence, electroluminescence, and photoluminescence microscopy techniques have been the four techniques that, used in complementary fashion, have yielded useful information.

a. Transmission Electron Microscopy

Plan view transmission electron microscopy using standard diffraction imaging techniques has been extensively used for characterization of degraded devices by different groups (Guha *et al.*, 1993; Hua *et al.*, 1994; Tomiya *et al.*, 1995). Due to the availability of selective etches that preferentially remove the GaAs substrate, specimens of degraded devices can be made with large electron transparent areas (~ 1 mm) for TEM observation.

It is well known that ion milling can induce subsurface damage in the form of small dislocation loops in ZnSe and related wide gap II–VI compounds. Electron beam irradiation of ZnSe can also produce such subsurface damage as has been reported earlier by Stacy and Fitzpatrick (1978), and such damage along with sample contamination appears to be further exacerbated in compounds that contain S and Mg. Such damage does not usually interfere with conclusions made from conventional diffraction analysis, though it should be kept in mind that very small defects such as small dislocation loops that may be present can be masked by the ion milling and/or electron beam damage.

b. Cathodoluminescence (CL) Microscopy

This is a very powerful technique that has become increasingly important for studying degradation and defects when their densities are low enough (below 10^5 per cm^2) so that transmission electron microscopy measurements become unreliable. Cathodoluminescence microscopy (CL) is a useful technique for rapidly monitoring defect densities in the 10^3 to 10^7 per cm^2 range. As will be shown in this article, it is also very useful for monitoring the propagation of nonradiative degradation features in degraded devices. Our studies have been mostly done using panchromatic CL where all of the light is collected through an optical microscope system (inserted into the [SEM] chamber) and detected using a photomultiplier tube that is most sensitive in the blue (490 mn) region of the spectrum. There has not been much recent work on spectrally discriminative CL of ZnSe-based compounds (Rich *et al.*, 1994), though potentially it could be very useful for studying device structures since images formed with emission wavelengths of the clad, guide, and active regions could provide information about defect behavior in these different layers. When cathodoluminesce from within a few thousand angstroms of the surface is the main contribution to the image, as in a single II–VI layer where the collected luminescence is limited by the absorption depth, the resolution is limited by the minority carrier diffusion length and is about 0.1 μm for ZnSe. If the layer of interest is buried, for instance a quantum well a micron below the surface, the resolution is poorer and limited by the lateral spread of the incident electron beam. For situations such as this it is preferable to etch or ion mill down to within a few thousand angstroms of the layer of interest for the highest spatial resolution.

c. Photoluminescence Imaging (PLI)

This is a technique complementary to CL, and unlike CL, the excitation is done with incident photons instead of electrons. A quantum well sample is typically needed to provide enough signal to obtain clear images at room temperature. Typical samples used are therefore ZnMgSSe/ZnCdSe/ZnMgSSe quantum wells. For the incident beam, the 351 and 364 nm lines of an Ar laser suffice to excite electron hole pairs in both the clad and quantum well layers, while for selective excitation of only the quantum well, the 448 nm line of the laser may be used. The advantage of PL imaging is the ease of use, quick turnaround, and importantly, the ability to study optical degradation of samples in real time. The disadvantage is a somewhat poorer spatial resolution compared to CL, since the image resolution is limited by the optics of the imaging microscope. The incident power in typical PL imaging is about 80 W/cm^2, and estimates from photoluminescence emission peak shifts indicate temperature rises of below about 50°C, thus minimizing artefacts introduced from thermal effects. Studies of degradation features observed in photoluminescence and electroluminescence imaging indicate that they are similar, implying that optical degradation and electrical degradation of the active region proceed similarly (Haugen *et al.*, 1995).

d. Electroluminescence (EL) Imaging

As shown in the schematic of Figure 18, in EL microscopy, a laser or LED is operated electrically and the electroluminescence collected from the top and used to form a spatial image of the device active region. This is facilitated by using a thin (100 Å) Au pad as the top electrical contact so that it is transparent to the emitted light. Since actual devices may be tested and the spatial degradation observed in real time, this is the most direct means of studying device degradation. As in PLI, the spatial resolution is limited by the optics of the imaging microscope.

2. Degradation in Light-emitting Diodes and Lasers

a. Description of Physical Features

The microstructural degradation features observed in both lasers and LEDs appear to be the same, and typical evolution of the different degradation features is shown in the sequence of EL micrographs of Figure 18 for a LED. The current density during testing was increased from 10 A/cm^2 to 100 A/cm^2. The following features, in their order of appearance, are observed. At turn on, the initial nonradiative defect structure consists of a distribution of faint dark spots and dark lines parallel to the $\langle 011 \rangle$ direction in the $\{100\}$ interface plane and reflects the preexisting defect

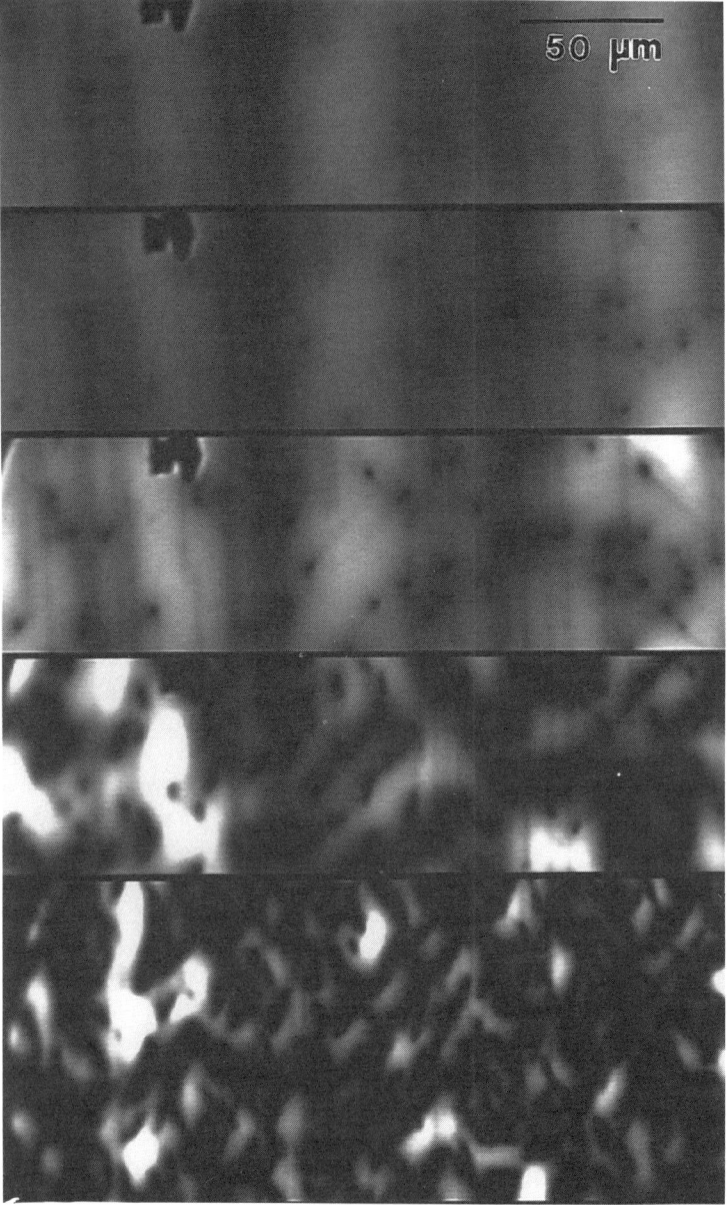

FIG. 18. Electroluminescence micrographs showing the various stages of degradation in a light-emitting diode (from Guha et al., 1993b).

structure. The dark spot defects are typical of as-grown structures and are also observed by both CL and PL imaging. A positive correlation of their densities with the densities of stacking faults as measured by TEM in a large number of samples indicates these dark spots to be grown in stacking faults in the device structure. The dark lines along $\langle 011 \rangle$ are misfit dislocations that lie in the ZnCdSe/ZnSSe interface. They are a consequence of lattice relaxation of the quantum well and can be avoided by using thinner quantum wells.

At the first stage of degradation, the faint dark spots become darker (Fig. 18b) indicating that they act as stronger nonradiative recombination centers. In the second stage, these dark spots eject out dark lines along the $\langle 001 \rangle$ directions (Fig. 18c). In the third and final stage, these lines break up into geometrical shaped dark (and therefore nonradiative) patches that cover the entire device active region (Fig. 18e). The patches are spatially bound either by misfit dislocations, or end abruptly along high index directions that make angles of 7° to 15° with $\langle 011 \rangle$. The reason for the former is understood and will be discussed later, while the latter is not clearly understood at this time.

It is interesting to note the manner in which $\langle 100 \rangle$ dark line defects (DLDs) emerge from the dark spots which are the preexisting stacking faults. As seen by both electroluminescence and photoluminescence imaging, $\langle 100 \rangle$ DLD formation is preceded by the appearance of small mobile dark spot defects, and an example is shown in the sequence of Figure 19, taken by photoluminescence imaging. Observations of the motion of numerous such mobile defects indicate that they have the following properties:

1. The mobile dark spot defects (DSD) originate from the preexisting stacking faults and move along exact $\langle 100 \rangle$ directions at velocities of approximately 5 μm per second (the velocity varies somewhat depending upon the degree of excitation).
2. They may draw behind them small $\langle 100 \rangle$ dark line defect segments, and then detach from these segments and move away. The $\langle 100 \rangle$ dark line defects thus formed, then grow throughout the active region of the device.
3. They can be arrested or may reflect off strain centers such as other defects, indicating strain field interactions.
4. They can also originate from $\langle 100 \rangle$ DLDs that have formed.

The strain field interaction observed suggests that the mobile DSDs may be small dislocation loops. Their rapid motion along exact $\langle 100 \rangle$ over long ($\sim 50 \, \mu$m) distances is indicative of glide along this direction. The only type of dislocation in the zinc-blende lattice that would explain this observation are loops of the $\{100\}/\langle 100 \rangle$ type which can glide along the $\langle 100 \rangle$ glide

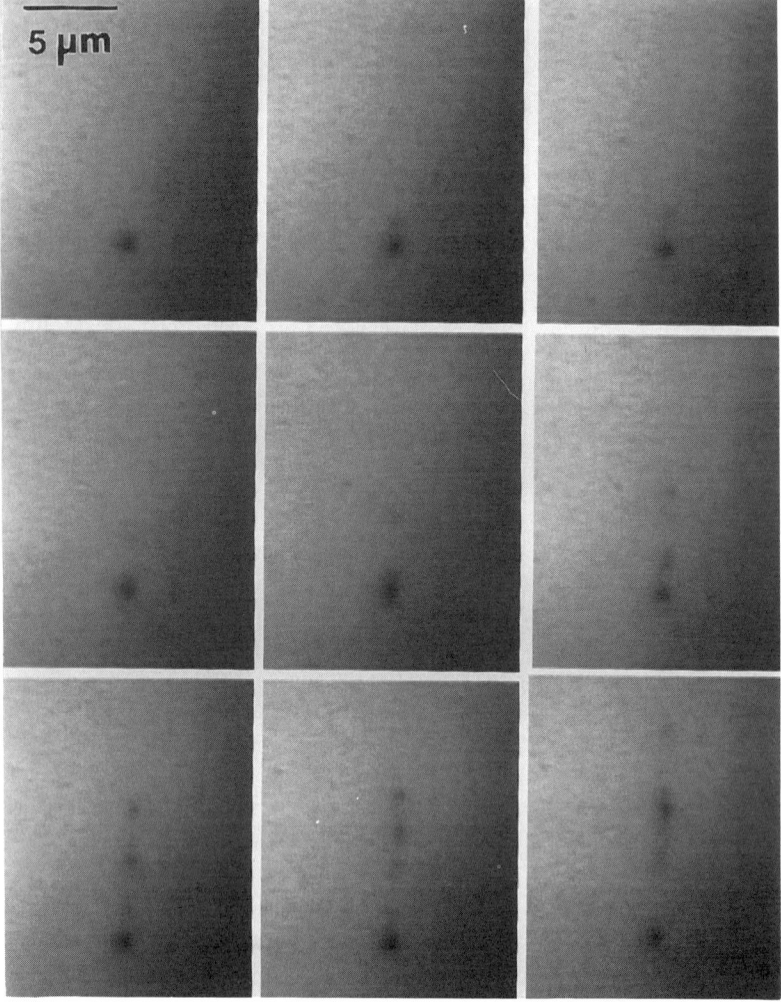

FIG. 19. Sequence of images taken by photoluminescence imaging, showing mobile dark spot defects originating from a preexisting stacking fault.

cylinder (Hirth and Loethe, 1982). Such dislocation loops lie on the $\{100\}$ plane and have a Burgers vector normal to that plane along $\langle 100 \rangle$. They are stable for the zinc-blende system (Hirth and Loethe, 1982), and have been observed earlier in degraded InGaAsP lasers by Chu and Nakahara (1990). Indeed, recent TEM observations in optically degraded samples have confirmed their presence in degraded ZnCdSe quantum well structures

(U'Ren et al., 1995). Reflection off other strain centers so that a [100] oriented defect is reflected to the [010] direction, for instance, is however not possible for such a loop which would be constrained to move along only one ⟨100⟩ direction. One can reconcile this, however, by noting that the observation of apparent reflection may be a case of one mobile defect generating a new one perpendicular to it upon impact with a stationary defect.

The appearance of the mobile defects and the subsequent appearance of the ⟨100⟩ DLDs depend on the starting stacking fault density in the sample. When the starting stacking fault density is higher than about 5×10^7 per cm^2, so that the stacking fault spacing is roughly about 1.5 μm, the degradation occurs by darkening of the dark spot defects until the non-radiative regions corresponding to them overlap (Fig. 20). No ⟨100⟩ dark line defects form in this case, presumably because of the close proximity of the strain fields of the as-grown defects, indicating that the DLDs require large regions of high-quality dislocation-free material to successfully propagate. This can be explained in terms of the strain field interaction of the

FIG. 20. Degradation in a LED structure containing a high ($>5 \times 10^7 \, cm^{-2}$) starting defect density.

mobile loops with the threading dislocations bounding the stacking fault. The nucleated dislocation loop experiences a force (the Peach-Kohler force) exerted on it due to the strain field of the preexisting threading dislocation bounding the stacking fault. When the stacking fault density is low and the force is repulsive, it causes the loop to be ejected and glide along the $\langle 100 \rangle$ direction. When the stacking fault density is high, and the strain fields of the partials bounding the faults start overlapping, the nucleated loop now also feels the repulsion effects of the neighboring stacking faults, and is thus locked in and does not glide. Degradation then proceeds by the darkening and radial extension of the dark spot defects and no $\langle 100 \rangle$ DLDs are formed.

Transmission electron microscopy studies on degraded LEDs and lasers have been carried out by different groups (Guha et al., 1993; Guha et al., 1994; Hua et al., 1994; Tomiya et al., 1995; Hovinen et al., 1995) to deduce the microstructure of the degraded defects. Success at trying to observe the $\langle 100 \rangle$ dark line defects has been varied. Guha et al. (1994) have attempted to observe $\langle 100 \rangle$ DLDs by TEM in specimens made from degraded LEDs. While the specimens, as expected, readily exhibited the $\langle 100 \rangle$ DLDs in CL imaging, observations of the same areas by TEM under different diffraction conditions failed to reveal the defects. The failure to observe $\langle 100 \rangle$ DLDs has also been reported by Hovinen et al. (1995) and the TEM micrographs of degraded regions published by Tomiya et al. (1995) also do not show features that correspond to the sharp $\langle 100 \rangle$ DLDs that can be seen in their EL micrographs as well. On the other hand, there have been two reports (U'Ren et al., 1995; Kuo et al., 1995) of the observation of DLDs in TEM in optically degraded quantum well samples. U'Ren et al. have determined that the $\langle 100 \rangle$ dark line defects consist of small loops that exhibit diffraction contrast consistent with that expected from loops with Burgers vectors of the type $a/2\langle 100 \rangle$. If the $\langle 100 \rangle$ dark line defects constitute an array of small point defect clusters or very small dislocation loops, it would be hard to observe them via TEM since the strain contrast would not be adequate. Their observation would also be further complicated by masking effects from the ion beam/electron beam radiation damage that creates a distribution of subsurface defects. This explains the failure by some groups to observe the $\langle 100 \rangle$ DLDs. On the other hand, if these point defects clusters or small loops are optically active, and are spaced at distances of the order of the minority carrier diffusion lengths ($\sim 0.1\,\mu$m) or less, $\langle 100 \rangle$ DLDs would readily be observed by CL, EL, or PL imaging. These results thus indicate that the $\langle 100 \rangle$ dark lines are probably aggregates of very small dislocation loops or point defect clusters. Depending on the sample (as will be discussed later) and with increased degradation, at some point the small loops may grow to form larger ones in which case the DLDs would be observable by TEM as was perhaps the case for U'Ren et al. (1995) and Kuo et al. (1995). Careful ion milling experiments followed by CL observations (Guha et al., 1994) established that the ⟨100⟩ DLDs reside in the quantum well of the device.

The final product of the degradation process, the dark patches, on the other hand can clearly be observed by transmission electron microscopy and consist of dense networks of dislocations that act as nonradiative recombination centers. The first transmission electron microscopy study of these defects was carried out on degraded LEDs by Guha et al. (1993b) and subsequently Hua et al. (1994) and Tomiya et al. (1995) have published results of such degradation patches observed in degraded laser stripes. Figure 21 shows plan view TEM orthogonal ⟨022⟩ bright field images of a defect patch showing that it consists of a dense network of two mutually orthogonal sets of dislocations aligned along 011 directions. Assuming that the dislocations were of the perfect type, diffraction studies indicated that the Burgers vector is parallel to the ⟨011⟩ directions in the {100} interface plane. The local densities of these defects are in the 10^{10} per cm^2 range, and electron stereomicroscopy analysis shows them to be coplanar with the misfit dislocations that lie along the quantum well interfaces, thus implying their presence in the quantum well. Guha et al. (1993b) reported dislocation networks originating out of both stacking faults as well as misfit dislocations, and situations regarding the two are shown in Figure 22 where defects originate from a stacking fault and an arrowhead-shaped dislocation patch originates from a misfit dislocation. Note that while the arrowhead has a mean direction close to ⟨100⟩, it flares out and therefore does not correspond to the long and relatively sharp ⟨100⟩ DLDs described earlier. It is likely, though, that this patch originated out of a DLD and hence it maintains that average direction. Hua et al. (1994) have published beautiful micrographs of these defect patches originating from stacking faults; an example is shown in Figure 23a. As shown in the schematic of Figure 23b, a grown-in stacking fault bound by Shockley partials has nucleated degradation defects at the quantum well. Tomiya et al. (1995) also observe dislocations appearing out of stacking faults (bound by Frank partials in this case), forming similar arrowhead shapes, though in contrast to the findings of Guha et al. (1993) they observe defects with Burgers vectors of the ⟨011⟩ type that are inclined at 45° to the interface plane.

As had been mentioned earlier and as can be seen from the TEM images, the patches end along arbitrary directions that makes angles of between 7° and 15° with the 011 directions (Guha et al., 1993; Hovinen et al., 1995). The reason for this is not understood. In the presence of misfit dislocations, the patches also end abruptly at the misfit dislocations. This is likely a consequence of the interactions between the strain fields of the defects in the

FIG. 21. Plan view bright field ($\langle 022 \rangle$) type TEM images of dislocations arising due to degradation (from Guha et al., 1993b).

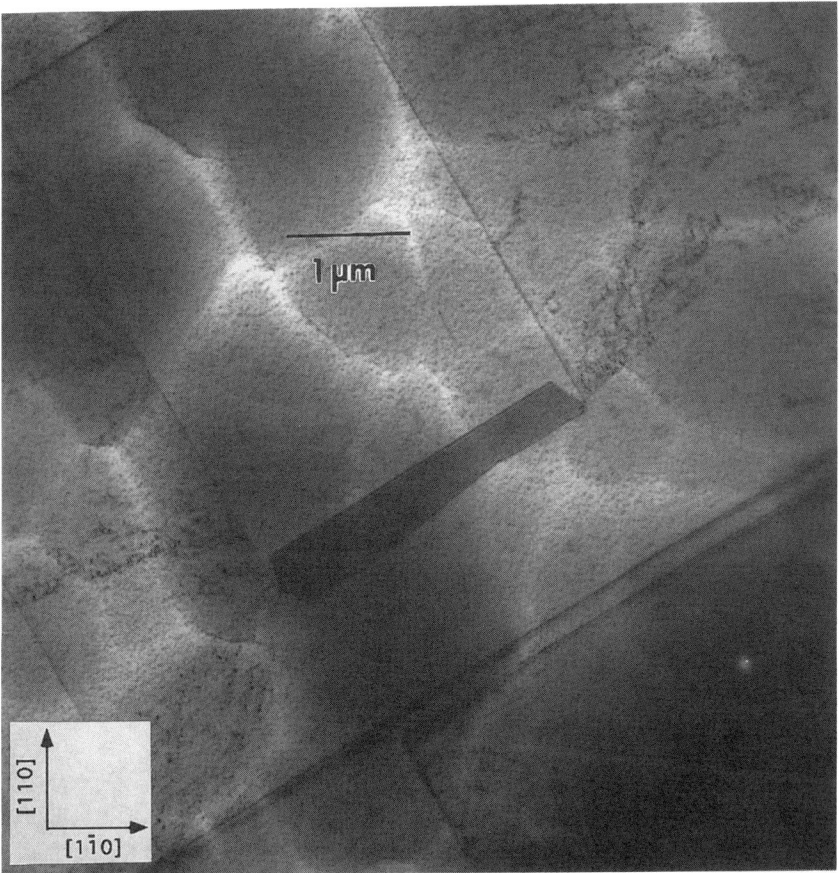

FIG. 22. Plan view bright field TEM images showing degradation defects originating from a stacking fault.

patches and the strain fields of the misfit dislocations. The misfit dislocations are of the 60° mixed type and glide on the {111} plane. Consider the normal component of the stress σ_{yy} of the misfit dislocation, that is perpendicular to the dislocation line direction and lies in a plane parallel to the slip plane. This component of the stress changes from positive to negative across the slip plane. Hence in regions of the quantum well below this slip plane, σ_{yy} is positive, while in regions above the slip plane it is negative. Owing to their own stress fields and depending on its nature, the dislocations in the patches will thus be attracted by regions in the well on one side of the dislocation and repelled by the other side.

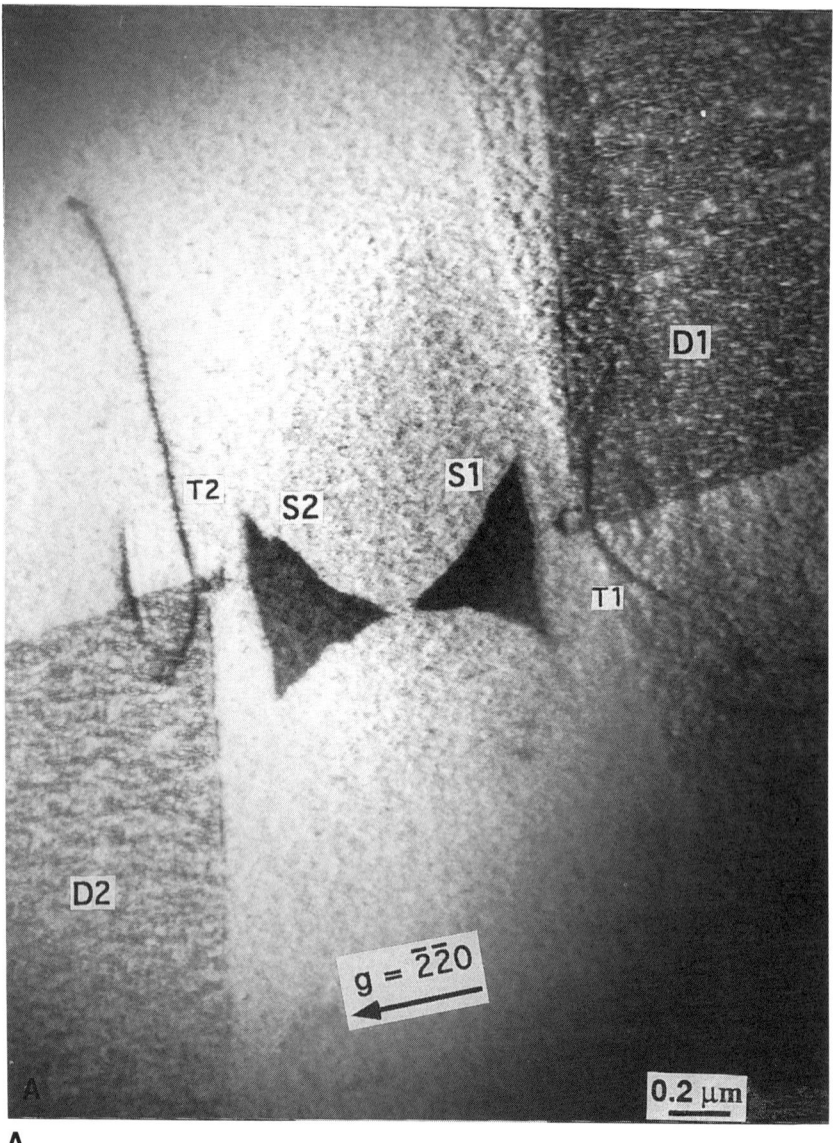

A

Fig. 23. (a) Bright field TEM image showing dislocations due to degradation originating from a stacking fault bound by Shockley partials; (b) schematic explaining the dislocation formation (from Hua et al., 1994).

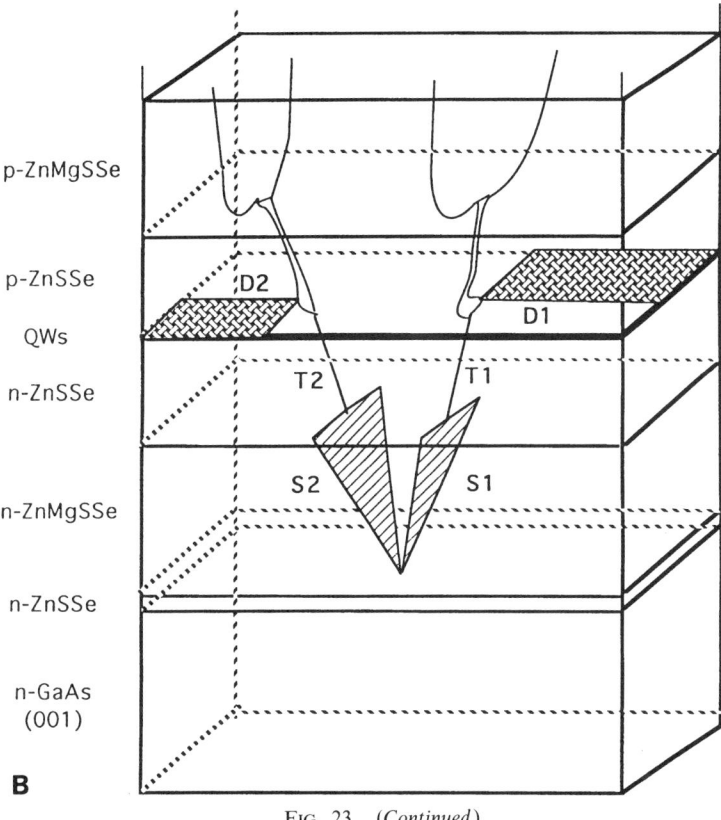

FIG. 23. (Continued)

The development of the features described above remains the cause for failure even for our best lasers that have lasted in the 20 to 70 minute range when operated cw at room temperature. Figure 24 shows a panchromatic CL image of a typical failed index-guided laser. Note that the cause of failure has been a cluster of $\langle 100 \rangle$ DLDs that spread out a distance of about 20 μm along the length of the laser, while the remaining portion of the laser appears unharmed in terms of localized degradation. Observations of different degraded laser stripes indicate that the failed stripes contain one such large patch of degraded defects, presumably initiated from a preexisting stacking fault. The lifetimes of the lasers thus still appear limited by the preexisting defect structure: the fact that large portions of the stripe remain unharmed is encouraging and indicates that a significant gain in lifetimes would be expected once the preexisting defects are eliminated.

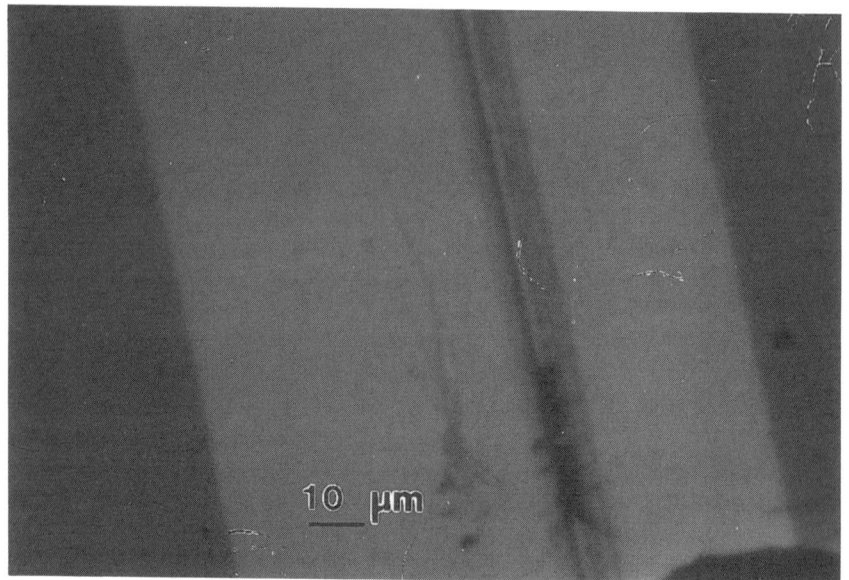

FIG. 24. Cathodoluminescence image of a failed stripe laser showing the formation of $\langle 100 \rangle$ dark line defects.

b. The Degradation Process

As mentioned before, the dominant degradation mode is driven by preexisting stacking faults and recent improvements in cw lifetimes have resulted from the ability to reduce as-grown preexisting stacking fault densities. The following is an explanation, consistent with observations so far, of the process by which we believe the degradation process occurs.

The observations of the microstructural features described earlier indicate the formation of point defects which can form small clusters or small dislocation loops. Barring the formation of Frenkel defects (vacany interstitial pairs), or localized lattice instabilities associated with atomic relaxations (Park and Chadi), point defects need to be transported to the active region (the quantum well) to provide a net increase in the density of degradation-induced defects. A ready way of this occurring is provided by the preexisting stacking faults that thread through the device structure. The partial dislocations bounding the stacking fault can act as channels for pipe diffusion that transport point defects into the quantum well from the guide/clad layers, the substrate, or the surface, which would act as sources for point defects. These preexisting defects thus mediate the transport of point defects to the quantum well. This speculation is indeed consistent with the observation that the first stage of degradation involves the darkening (in EL or PL imaging) of the dark spot defects that correspond to the preexisting stacking faults. This indicates that they act as stronger nonradiative recombination centers and implies a buildup of point defect concentrations in the vicinity of the region where the threading dislocation intersects the quantum well. If the diffusivity of point defects along the dislocation core is fast enough, then the threading dislocation can be viewed as a means for a ready supply of point defects to the quantum well. Consider initially that there is a concentration of point defects along the threading dislocation. Nonradiative recombination at the threading dislocation core that intersects the quantum well will result in a higher diffusivity (for in-plane motion within the quantum well) of point defects neighboring the dislocation due to nonradiative recombination enhanced motion (discussed later). Consequently these defects will spread outward and their concentrations around the dislocation core will be replenished by pipe diffusion. This will result in a gradual accumulation of point defects around the threading dislocation cores in the quantum well.

A high enough concentration of such point defects can then result in their coalescing to form small dislocation loops. When the concentration is high enough, small loops nucleate on the $\{100\}$ plane with Burgers vectors along $\langle 100 \rangle$. These loops then glide readily along the $\langle 100 \rangle$ glide cylinder and are the mobile dark spot defects observed in the EL and PL imaging as described before. These mobile loops can sweep point defects along their motion, and can also leave behind other small loops and point defect clusters along their slip traces along $\langle 100 \rangle$. These defects left behind are then naturally aligned along the $\langle 100 \rangle$ direction and constitute the $\langle 100 \rangle$ dark line defects.

Once these initial $\langle 100 \rangle$ DLDs are formed, their subsequent progress and growth is slower and is due to nonradiative recombination enhanced diffusive motion (as opposed to glide as in the mobile DSDs), a subject studied extensively in the context of degradation in III–V injection lasers (Lang and Kimmerling, 1974; Weeks *et al.*, 1975). Trapping of a carrier at a point defect whose energy level (E_{tr}) resides within the bandgap would result in a release of energy (E) in the form of localized lattice vibrations which would locally enhance the mobility of the defect. An estimate of the enhancement in diffusion rates that results may be made in the following way (Guha *et al.*, 1994). The diffusion coefficient of the defect is proportional to the rate at which it hops to the adjacent site, $\Omega = v_0 \exp(-Q/kT)$. We may interpret v_0 as the jump attempt frequency and Q as the activation energy for a successful jump. After capture of a carrier, the defect is in an energized state and sees an activation energy of $(Q - E)$ for a time τ, the typical time scale on which the defect transfers energy to the lattice. If R is the nonradiative recombination rate, then the ratio of the effective hopping rate ω' due to nonradiative recombination to ω is given by:

$$\omega'/\omega = R\tau \exp(E/kT) \qquad (1)$$

This assumes that $\exp[E/kT] \gg 1$, the fraction of time spent by the defect in the excited state is small, and that the attempt frequencies v_0 and v_1 are the same in both cases. A more detailed analysis by Weeks et al. (1975) includes a factor of $[(Q - E)/Q]^{(S-1)}$ on the right-hand side of Eq (1), where S is the number of effective oscillators in the defect, with 8–10 being a reasonable number. Estimates for the case of ZnSe by Guha et al. (1994) indicate that an enhancement in the diffusion coefficients by a factor of 10^{22} is possible under typical operating conditions and can account for the degradation effects observed.

The growth of the $\langle 100 \rangle$ DLD thus occurs by nonradiative recombination enhanced motion and is sustained by the supply of point defects to the quantum well via the high diffusivity paths along the preexisting threading dislocations. Since the $\langle 100 \rangle$ DLDs constitute an atmosphere of point defects, they can now act as sources for new mobile dark DSDs which can in turn initiate $\langle 100 \rangle$ DLDs that are perpendicular to the first one. With continued device operation, when the point defect concentration along the DLDs is high enough, macroscopic dislocation networks can nucleate. The threshold at which nucleation of such dislocations can occur will depend on the extent to which the quantum well is strained. Nucleation of the dislocation networks along $\langle 100 \rangle$ and their subsequent spreading out then result in the final product of the degradation—the dark, geometrically shaped patches that consist of dense networks of dislocations. The entire process then originates from the preexisting threading dislocations that bound the stacking faults and the development of the degradation according to this model would depend on the supply of point defects transported by these threading defects.

c. Other Degradation Mechanisms

Intrinsic Degradation. This section considers degradation behavior that may become significant once stacking fault related degradation is eliminated by reducing the stacking fault densities in the as-grown material. There has been considerable speculation that since the wide-gap compounds are "soft" materials and the photon energies are comparable to the bond energies, they would have a tendency toward self destruction under optical fields. This implies the generation of defects in the active region in the absence of any extrinsic sources such as stacking faults or free surfaces (such as facets) and we will refer to such degradation as intrinsic degradation. If a material self-destructs by the spontaneous generation of point defects, then such a process (if readily active) would suggest a catastrophic degradation and would preclude the observation of cw lasing at room temperature of a few hours. The cross-sections for such processes are thus lower, though so far there have been no serious estimates of the probabilities of such processes and no direct experiments reported that clearly demonstrate the significance and extent of such intrinsic degradation behavior. Quantitative photoluminescence imaging experiments done on quantum well samples with low stacking fault densities are suspect because the range of influence of a stacking fault in providing defects (such as mobile DSDs) can be quite large and is not well known. Similar optical experiment on patterned small mesas would eliminate the uncertainty arising from the effect of distant stacking faults, but would introduce the effects of defects arising from free surfaces. So while this is clearly a question with technological relevance, answering it experimentally is nontrivial.

Let us consider ways by which defects can be generated intrinsically in the crystal. Let us first consider defect generation in a perfect crystal. Provided there is no source for point defects, there are two ways by which a point defect can be created. The first is by the formation of a Frenkel defect by which an atom moves from a substitutional to an interstitial site and creates a vacancy–interstitial pair. This cannot be initiated by a passing photon since it cannot provide the momentum transfer required for this. The second is by the breaking of bonds and the formation of localized metastable lattice instabilities that are electrically active as described by Park and Chadi (1995). While their calculations show that such defects may be stable for p-doped ZnSe, the energetics of and the barriers to the formation of such defects in strained ZnCdSe quantum wells are not known. Consider now defect generation mediated by the presence of a point defect that acts a nonradiative recombination center. Nonradiative recombination enhanced local lattice vibrations at such point defects can result in the formation of Frenkel pairs or localized lattice instabilities in the vicinity of the point defect. This would thus be another mechanism for the multiplication of point defects within the active region. The above represent the pathways by which point defects may be created intrinsically. The description though is at best speculative since no published estimates are available as to the barriers for the formation of such defects in the material systems of our interest.

Ishibashi (1995) has recently attempted to interpret laser lifetime data as a function of preexisting defect density to extract information regarding possibilities for intrinsic degradation that have been discussed above. He describes the time-dependent defect density as:

$$N(t) = N_0(t/t_0)^\gamma \exp(E_a/kT) \qquad (2)$$

where N_0 is the initial defect density and t_0 is a constant. At a critical defect density of N_c at time t_c when the laser dies,

$$\ln(t_c/t_0) = \text{const.} - (1/\gamma) \log(N_0) \qquad (3)$$

If the defect formation is of a self-proliferating and therefore catastrophic nature leading to the device failure, then γ would be expected to be greater than unity. However, fits to data of laser diode lifetime vs. defect density by Ishibashi indicated that $\gamma \sim 0.5$. This clearly implies that the laser diodes have not failed due to the above mentioned intrinsic reasons. The data thus clearly shows that the effects of such intrinsic degradation are not felt at lifetime levels of about an hour and any such effects would thus be significant for timescales much larger than that. Since the lifetimes are limited by the preexisting defects, the data however cannot provide any quantitative information regarding these timescales.

Contact Degradation and Failure at Facets. Both of these are issues that may become important in the future. As of now they have not been well studied and not much information is available.

The past few years have resulted in the development of the graded ZnSe/ZnTe contact that has considerably reduced operating voltages of lasers to about 4 volts and improved electrical characteristics. Though they are not a lifetime limiting step at this point, contact-related lifetimes will probably be significant at the 5- to 10-hr lifetime level. Microstructural details of contact failure have not been studied in detail and it is little understood. The as-grown contact layers have high dislocation densities in the $> 10^8$ per cm^2 range and the glide of these dislocations into the active region is one possibility. Diffusion and redistribution of the metastable nitrogen p-dopant within the ZnSe/ZnTe layers and segregation at dislocations is another possibility.

Degradation arising from facets has not been the common cause for device failure so far, but again, its contribution is unknown regarding long-term operation. Degradation has been observed in our studies originating from facets, but in those cases was probably related to uneven cleaving resulting in stress concentrations. For uniformly cleaved facets, the presence or absence of surface states that act as nonradiative recombination sites can result in the propagation of degradation features. Since a facet can be an infinite source of vacancies, nonradiative recombination enhanced motion of defect centered around such surface states can result in degradation in the active layer much in the same fashion as described for stacking faults.

REFERENCES

Brown, P. D., Jones, A. P. C., Russell, G. J., Woods, J., Cockayne, B., and Wright, P. J. (1987). *Microscopy of Semiconducting Material 1987*. IOP, London, 123.
Chu, S. N. G., and Nakahara, S. (1990). *Appl. Phys. Lett.* **56**, 434.
DePuydt, J. M., Potts, J. E., and Mohapatra, S. K. (1988). *J. Appl. Phys.* **52**, 4756.
Ehrenreich, M., and Hirth, J. P. (1985). *Appl. Phys. Lett.* **46**, 668.
Gaines, J. M., Petruzzello, J., and Greenberg, B. L. (1993). *J. Appl. Phys.* **73**, 2835.
Guha, S., Munekata, H., LeGoues, F. K., and Cheng, L. L. (1992). *Appl. Phys. Lett.* **60**, 3220.
Guha, S., Munekata, H., and Cheng, L. L. (1993). *J. Appl. Phys.* **73**, 2294.
Guha, S., Munekata, H., Cheng, L. L., and Tang, W. C. (1993). *J. Cryst. Growth* **127**, 308.
Guha, S., DePuydt, J. M., Haase, M. A., Qiu, J., and Cheng, H. (1993). *Appl. Phys. Lett.* **63**, 3107.
Guha, S., Cheng, H., Haase, M. A., DePuydt, J. M., Qiu, J., Wu, B. J., and Hofler, G. E. (1994). *Appl. Phys. Lett.* **65**, 801.
Haase, M. A., Qiu, J., DePuydt, J. M., and Cheng, H. (1991). *Appl. Phys. Lett.* **59**, 1273.
Haugen, G. M., Guha, S., Cheng, H., DePuydt, J. M. Haase, M. A., Hofer, G. E., Qiu, J., and Wu, B. J. (1995). *Appl. Phys. Lett.* **66**, 358.
Hirth, J. P., and Loethe, J. (1982). *Theory of Dislocations*. Wiley, New York.
Hua, G. C., Otsuka, N., Grillo, D. C., Fan, Y., Han, J., Ringle, M. D., Gunshor, R. L., Hovinen, M., and Nurmikko, A. V. (1994). *Appl. Phys. Lett.* **65**, 1331.
Ishibashi, A., (1995). *IEEE J. Selected Topics in Quantum Electron.* **1**, 741.
Kuo, L. H., Salamanca Riba, L., Wu, B. J., Haugen, G. M., DePuydt, J. M., Hofler, G., Cheng, H., (1995). *J. Vac. Sci. Technol* **B13**, 1694.
Lang, D. V., and Kimmerling, L. C. (1974). *Phys. Rev. Lett.* **33**, 489.
Lilja, J., Keskinen, J., Hovinen, M., and Pessa, M. (1989). *J. Vac. Sci. Technol.* **B7**, 593.
Maree, P. M. J., Olthof, R. I. J., Franken, J. W. M., van der Veen, J. F., Bulle-Lieuwma, C. W. T., Viegers, M. P. A., and Zalm, P. C. (1985). *J. Appl. Phys.* **58**, 3097.
Marshall, T., and Cammack, D. C. (1991). *J. Appl. Phys.* **69**, 4149.
Matthews, J. W. (1975). In *Epitaxial Growth*, ed. J. W. Matthews. Academic Press, New York, 559.
Miyajima, T., Okuyama, H., Akimoto, K., Mori, Y., Wei, L., and Tanigawa, S. (1991). *Appl. Phys. Lett.* **59**, 1482.
Neumark, G. F., and Yi, G. J. (1994). *The Encyclopaedia of Advanced Materials Vol. 4*. Pergamon, New York, 2383.
Park, C. H., and Chadi, D. J., unpublished.
Park, R. M., Kleiman, J., and Mar, H. A. (1987). *SPIE* vol. 796 *Growth of Compound Semiconductors*, 86.
Park, R. M., Troffer, M. B., Rouleasu, C. M., DePuydt, J. M., and Haase, M. A. (1990). *Appl. Phys. Lett.* **57**, 2127.
Pashley, M. D., and Li, D. (1994). *Mat. Sci. and Eng. B*.
Petruzzello, J., Greenberg, B. L., Cammack, D. C., and Dalby, R. (1988). *J. Appl. Phys.* **63**, 2299.

Petruzzello, J., Gaines, J., Van der Sluis, P., Olego, D. and Ponzoni, C. (1993). *Appl. Phys. Lett.* **62**, 1496.
Petruzzello, J. M., Herko, J., Gaines, S., Marshall, T. (1995). *Phys. Stat. Sol.* **B187**, 297.
Qiu, J., Qian, Q. D., Kobayashi, M., Gunshor, R. L., Menke, D. R., Li, D., and Otsuka, N. (1990). *J. Vac. Sci. Technol.* **B8**, 701.
Shahzad, K., and Cammack, D. C. (1990). *Appl. Phys. Lett.* **56**, 180.
Stacy, W. T., and Fitzpatrick, B. J. (1978). *J. Appl. Phys.* **49**, 4765.
Tomiya, S., Morita, E., Ukita, M., Okuyama, H., Itoh, S., Nakano, K., and Ishibashi, A. (1995). *Appl. Phys. Lett.* **66**, 1208.
U'Ren, G. D., Haugen, G. M., Baude, P. F., Haase, K. K. Law, M. A., Miller, T. J., and Wu, B. J. (1995) *Appl. Phys. Lett.* **67**, 3862
Weeks, J. D., Tully, J. C., and Kimmerling, L. C. (1975). *Phys. Rev.* **B12**, 3286.
Yao, T. (1985). In *The Technology and Physics of Molecular Beam Epitaxy*, ed. E. H. C. Parker. Plenum, London, 313.
Yao, T. (1986). *J. Appl. Phys.* **25**, L544.

Index

A

Absorption spectra, 169–170, 187, 208
Alloy(s), 2, 5, 13, 71
Annealed p-type layer, 76
Ax center. *See also* Charged carrier transport
  optical cross-section, 50
  thermal capture barrier, 51

B

Band gap renormalization, 204, 217
Band offset, 22, 25, 30, 32, 178, 181, 184, 228, 232, 234, 247
Be, 230–231, 253
BeTe, 232, 239
BeTe/ZnSe, 239
Bohr radius, 185–186, 189, 264
Buffer layers, 15, 22, 99–100, 231, 273

C

Carrier confinement, 247
CdTe/CdMnTe, 232
Capacitance-voltage characteristics, 24, 40, 76, 108, 112
CBE. *See* Chemical beam epitaxy
Chemical potential, 127–132, 145
Chemical beam epitaxy (CBE), 83
Charged carrier transport
  acceptor activation energy as a function of bandgap, 48
  activation energy, 40, 42
  AX centers, 47

Hall measurements, 34, 101
hydrogenic, 43
hydrogenic effective-mass model, 50
persistent photoconductivity, 48
temperature dependent Hall measurements, 42, 45, 47–48
use of Zn(Se,Te) graded bandgap contact, 40–41
van der Pauw, 40, 101
Chlorine, 102. *See also* ZnSe:Cl
Compensating Defects, 69
Complex formation, 146
Compositional modulation, 245
Cracking, 17, 19, 95, 98, 110, 114
Critical thickness, 16, 280. *See also* Lattice matching
CW laser operation, lifetime, 28, 242, 299

D

Deep levels
  AX center, 47–53, 149
  breaking or rearranging of bonds, 123
  compensating defects, 69
  deep level emission, 96, 123
  DX center, 48, 125, 147–148, 152
  persistent photoconductivity (PPC), 48
Defects
  compensating defects, 69
  cracks, 17, 19
  critical thickness, 16
  cross-hatched pattern, 17
  $\langle 100 \rangle$ dark line defects, 306, 311, 313
  dark line defects, 254–255, 297, 303
  dark spot defects, 303, 313
  defect density, 29, 255

Defects (*continued*)
  defect diffusion coefficient, 313
  dislocations, 19, 29, 272
  dislocations climb, 20
  dislocation dipoles, 19–20
  dislocation loops, 298
  dislocation networks, 20
  extended defects, 29, 256
  formation energy, 126–127, 130–131, 140
  misfit dislocations, 14, 157, 273, 278, 280–282, 284–288, 293, 295, 303, 307, 309
  native defects, 60, 122, 125, 132, 135–138, 153–156
  point defects, 20, 122, 124, 133, 253, 306, 313
  stacking faults, 14, 18, 19, 29, 253–255, 272–282, 295–298, 305–313
  threading dislocations, 19, 29, 256, 282, 288, 291, 297, 306
  time dependent defect density, 316
Defect etch, bromine-methanol, 29
Deformation potential, 165, 184
Degradation
  degradation mechanisms, 19, 252–256, 272, 298–299, 301–317
  stacking faults and laser lifetime, 295–298
  of LEDs, 255, 298
Diamagnetic shift, 219
Differential quantum efficiency, 243, 253
Diffusion coefficient of defects, 313–314
Doping,
  activation energy. *See* Charged carrier transport
  amphoteric doping, 230
  compensation, 69–70, 124–125, 140–141, 153–156, 158
  effect of nitrogen doping on misfit dislocation density, 284
  electrical passivation, 98
  Fermi level, 123
  formation of complexes, 125, 150
  Li, 74, 124, 130, 140–141, 144–145
  Li interstitials, 124, 144–145
  nitrogen doping, 74, 99,105, 148–149, 158
  n-type doping, 72, 151
  oxygen in ZnSe, 151
  P and As in ZnSe, 147
  passivation, 76. *See also* Hydrogen passivation

p-doping of ZnTe, 31, 158
self-activated centers, 137–138
self-compensation, 38, 45, 105, 122
shrinking of the lattice constant with N-doping, 158
sodium in ZnSe, 147
solid solubility, 47
solubility limit, 125, 154, 156, 158
using GSMBE, 109

E

Elastic strain, 284
Electrical contacts to n-ZnSe
  In contacts, 101
  Ti/Au contacts, 28
Electrical contacts to p-ZnSe
  As, 39
  BeTe/ZnSe, 239, 252
  contacts to p-ZnSe, 30–31, 239
  HgSe, 37
  hole injection, 33
  Li, 39, 74
  N2, NH3, N2 plasma, 39, 105
  Na, 39, 147
  ohmic contacts, 31, 40
  palladium, 237
  Pd/Pt/Au contact, 237
  resonant tunneling, 37, 239
  specific contact resistance, 35–37, 237, 239
  transmission-line model (TLM), 35
  Zn(Se,Te) graded bandgap contact, 32–37, 40, 235, 252, 316
Electroluminescence, 20, 301
Electron-hole plasma, 211, 232, 247, 256
Epilayers. *See* Buffer layers
Epitaxial growth methods
  gas source molecular beam epitaxy (GSMBE), 83
  molecular beam epitaxy (MBE), 3
  nonequilibrium growth, 4, 114
Excitons-electron/hole correlation
  2D excitons in quantum wells, 257
  biexcitons, 217, 256, 258–260
  bleaching, 202–203, 208, 211, 215, 219
  exciton binding energy, 165–167, 173, 175–176, 180–185, 202, 257

INDEX

exciton lifetime, 196
exciton-phonon coupling, 163, 173–174, 195
excitonic gain. *See* Gain and gain mechanisms
free exciton, 198
localization, 170, 196–198, 201, 216, 219
modeling of excitons in quantum wells, 164–168
nonlinear excitonic properties, 202
significance of LO phonon energy, 173
thermal stability in quantum wells, 171, 176

F

Fabry-Perot
oscillation in optical emission spectra, 244
thin film interference effects in optical pyrometer readings, 9
Fermi level, 123, 132, 139, 144
Frank type partial dislocations, 282, 307. *See also* Defects

G

GaAs
epilayers. *See* Buffer layers
substrates, 5, 232, 234, 293
Ga_2Se_3, 25
GaN, 228, 230–231
GaN/InGaN, 228
Gain and gain mechanisms
excitonic processes in QW lasers, 256–266
gain, 247
gain spectra, 262, 266
gain spectra in SCH laser diodes, 244
Mott transition, 256
Phase space filling (PSF), 204
Gas source molecular beam epitaxy (GSMBE)
advantages and disadvantages, 97–98
AsH_3, 98
hydride sources, 97–98
passivation during doping, 110–112
PH_3, 98
SeH_2, 98

growth of ZnSe:N, 105, 108–109
Growth by MBE, 4
Growth mechanism, 67

H

H_2, 76, 98
H_2Se, 110
Heavy-hole (HH), 164, 173, 177, 184, 201, 208, 222, 258, 264, 266
Heterovalent interface
charge inversion layers, 26
electrical transport, 26
energy band diagrams, 26–27
interface states, 22–25
interfacial bonding, 23
interfacial compound, 25
MIS experiments, 23, 25
valence mismatch, 22
variations in band offset, 26
Hydride sources, 97–98
Hydrogen passivation, 62, 76, 98, 108, 113–115, 126

I

InGaAs, 231, 254
InGaP, 231
InN, 230
Interface: II-VI/III-V
InGaN/GaN LED, 245
interdiffusion, 60
interstitial, 124
ionicity, 230
stacking fault nucleation, 19, 272

L

Lasers
dislocations in ternary lasers, 288
distributed feedback (DFB) lasers, 249
excitonic lasing, 258
laser diodes (LDs), 59, 227, 228, 229, 234, 262
lasing threshold, 201, 242
pseudomorphic structures, 236
threshold current densities, 241, 243, 247, 291

Lattice matching
  lattice mismatch and critical thickness, 273, 278, 295
  lattice mismatch and relaxation, 282–288
  of II-VI structures to GaAs, 5, 14, 229
  of ZnSe to InGaAs, 231
  using BeTe, 239   ZnCdSe mismatched to ZnSe, 232
LEDs. See Light-emitting diodes
Lifetime broadening, 173, 213
Light-emitting diodes (LEDs), 59
Light-hole (LH), 164, 173, 177, 180, 184, 208
LO phonons
  assisted recombination, 181
  interaction with excitons, 195   LO phonon energy, 163, 173, 176, 187, 217–218, 258
  replicas, 106

M

Magneto-optical spectroscopy, 234
Many-body effects, 263, 266
Many-body nonlinearities, 202
MBE. See Molecular Beam Epitaxy
Metal-insulator-semiconductor (MIS) structures, inverse layers, 26
Metalorganic chemical vapor deposition (MOCVD), 38
Metalorganic vapor-phase epitaxy (MOVPE)
  growth by MOVPE, 59
  growth mechanism, 67
  low temperature growth, 62
  n-doping, 72
  nonequilibrium conditions, 65
  p-doping, 74
  photo-assisted MOVPE, 65, 71
  photodecomposition and photo-excitation, 65, 72
  premature reactions, 60
  purity of source precursors, 64
  source precursors, 60
Metalorganic molecular beam epitaxy (MOMBE),
  advantages and disadvantages, 85
  control of flux, 98
  cracking, 86, 98
  electron beam effects, 91, 96
  growth rate enhancement, 93, 95–97
  growth rate suppression, 94–96
  growth rates, 89–90
  photo-assisted growth, 86, 91–97
  precursors, 86
  surface site blockage by ethyl radicals, 90
  thermal decomposition, 86
Metastable structures, 5
MgSe, MgS, 15
MnTe, 4
Molecular beam epitaxy (MBE)
  2 dimensional, layer by layer nucleation and growth, 28, 29, 273
  3 dimensional island nucleation and growth, 28, 273, 277
  measurement of substrate temperature (pyrometer oscillations), 7–12
  migration enhanced epitaxy (MEE), 29
  nonequilibrium growth, 4, 38, 156
  selenium stabilized surface, 21
  source fluxes measured with quartz crystal microbalance, 6
  sticking coefficients, 6, 11, 105
MOVPE. See Metalorganic vapor-phase epitaxy
MOMBE. See Metalorganic molecular beam epitaxy
MOMBE and MOVPE sources
  $((CH_3)_2NLi)$, 74
  (CH_3I), 61
  (CpLi), 74
  (DASe), 61
  (DiPNH), 75
  (Li_3N), 74
  (MASe), 61
  $((MeCp)_2Mg)$, 62
  (n-BuCl), 73
  (t-BuNH2), 75
  (TEG), 90
  (TMG), 97
  DES, 60
  DESe, 60, 86
  DEZn, 60, 86
  DMSe, 60
  DMZn, 60, 86
  DMZn-(NEt3)2, 61
  H_2S, 61
  H_2Se, 61
  NH_3, 74

N

Nonradiative recombination, 20, 298, 313

O

Optical absorption,
  absorption strength, 167–168
  interband transitions, 203
  spectra, 169, 173, 187, 190, 208
Output characteristics, 243

P

Photo-assisted growth, 65
Photoluminescence (PL)
  acceptor binding energy, 106
  chlorine doped ZnSe, 104
  donor-to-acceptor pair (DAP) emission, 75, 106
  excitonic bands, 170
  free exciton feature, 101
  imaging of defects, 300–301
  LO phonon replicas, 106
  MOMBE ZnSe characterization, 95–96
  MQWs, 187
  nitrogen doped ZnSe, 105–106
  temperature-dependent PL, 20
  thermally activated nonradiative recombination centers, 20
  time resolved PL, 72, 197
  under hydrostatic pressure, 181
Phototransport processes, 189
Photocurrent spectroscopy, 190, 193
Polariton, 186, 196, 201
Polarity, 230
Pseudomorphic, 14, 234, 240, 278–281, 284–287, 291
Purity of precursors, 63–64

Q

Quantum confined Stark effect, 202
Quantum efficiency, 247
Quantum wells
  electron subbands, 170
  electronic confinement, 228
  multi-quantum well (MQW), 14, 175, 177, 189, 198, 201, 219, 220

quantum confinement, 163
quantum size effect, 170
QW LED, 77–78, 229
  Type I, 232
  Type II, 232

R

Rabi splitting, 266
Radio frequency (RF) plasma source, 99
Reflection high energy electron diffraction (RHEED)
  GaAs surface reconstruction c(4 × 4), (2 × 4), (4 × 6), (4 × 3), 23, 27, 275
  on ZnSe, c(2 × 2) and (2 × 1), 29, 88, 275
  streakiness, 273
  surface reconstruction patterns, 23

S

Scanning tunneling microscopy (STM), 29, 273, 275, 277
Secondary ion mass spectroscopy (SIMS), 45, 100, 102, 106, 108, 126, 158
Shockley partial dislocations and stacking faults, 18, 19, 282, 307. *See also* Defects
Source precursors, 60
Spontaneous emission, 262
Stacking faults. *See* Defects
Stimulated emission, 198, 215, 218–219, 244, 258–259, 262
Strain
  compressive, 15, 17–19, 278, 284
  critical thickness, 16, 278
  tensile, 15, 17–19, 278
Substrate temperature, 7
Superlattice, 170, 181, 184–185, 217, 220, 239
Surface morphology, 17, 29, 89
Surface stoichiometry, 6, 101, 135

T

Thermodynamic equilibrium, 127–128, 131–132, 156
Transmission electron microscopy (TEM)
  bright field image, 19

Transmission electron microscopy (*continued*)
   cross-sectional observation, 19, 297
   of degraded LEDs and lasers, 254, 306–312
   plan view TEM, 19, 299
   showing compositional modulation, 245
   stereo-microscopy, 19
Transmission measurements, 170

U

Uniaxial strain, 232

V

Vertical cavity surface emitting lasers
   distributed Bragg reflectors, 250
   surface emitting, 249
   VCSEL, 249–252

X

X-ray diffraction
   2D HRXRD mapping, 291–293
   dislocation density, 283
   FWHM, 15
   multiple narrow peaks and drifting lattice constant, 8–11
   X-ray rocking curves, 6, 170
X-ray photoelectron spectroscopy (XPS), 25

Z

ZnCdSe, 232, 234
ZnCdSe/ZnSe and ZnCdSe/ZnSSe quantum wells, 234–236
ZnBeMgSe, 230
ZnMgSSe quaternary, 5, 14, 99, 101–102, 105, 230, 234, 240–243, 288
ZnMnSe, 232
ZnSe:Cl, 101–103
ZnSSe, 234
ZnTe
   p-doping using N, 31, 34–35, 237
   Zn(Se,Te), 31–32, 137

Contents of Volumes in This Series

Volume 1 **Physics of III–V Compounds**

C. Hilsum, Some Key Features of III–V Compounds
Franco Bassani, Methods of Band Calculations Applicable to III–V Compounds
E. O. Kane, The k-p Method
V. L. Bonch-Bruevich, Effect of Heavy Doping on the Semiconductor Band Structure
Donald Long, Energy Band Structures of Mixed Crystals of III–V Compounds
Laura M. Roth and Petros N. Argyres, Magnetic Quantum Effects
S. M. Puri and T. H. Geballe, Thermomagnetic Effects in the Quantum Region
W. M. Becker, Band Characteristics near Principal Minima from Magnetoresistance
E. H. Putley, Freeze-Out Effects, Hot Electron Effects, and Submillimeter Photoconductivity in InSb
H. Weiss, Magnetoresistance
Betsy Ancker-Johnson, Plasma in Semiconductors and Semimetals

Volume 2 **Physics of III–V Compounds**

M. G. Holland, Thermal Conductivity
S. I. Novkova, Thermal Expansion
U. Piesbergen, Heat Capacity and Debye Temperatures
G. Giesecke, Lattice Constants
J. R. Drabble, Elastic Properties
A. U. Mac. Rae and G. W. Gobeli, Low Energy Diffraction Studies
Robert Lee Mieher, Nuclear Magnetic Resonance
Bernard Goldstein, Electron Paramagnetic Resonance
T. S. Moss, Photoconduction in III–V Compounds
E. Antončik and J. Tauc, Quantum Efficiency of the Internal Photoelectric Effect in InSb
G. W. Gobeli and F. G. Allen, Photoelectric Threshold and Work Function
P. S. Pershan, Nonlinear Optics in III–V Compounds
M. Gershenzon, Radiative Recombination in the III–V Compounds
Frank Stern, Stimulated Emission in Semiconductors

Volume 3 Optical of Properties III–V Compounds

Marvin Hass, Lattice Reflection
William G. Spitzer, Multiphonon Lattice Absorption
D. L. Stierwalt and R. F. Potter, Emittance Studies
H. R. Philipp and H. Ehrenveich, Ultraviolet Optical Properties
Manuel Cardona, Optical Absorption above the Fundamental Edge
Earnest J. Johnson, Absorption near the Fundamental Edge
John O. Dimmock, Introduction to the Theory of Exciton States in Semiconductors
B. Lax and J. G. Mavroides, Interband Magnetooptical Effects
H. Y. Fan, Effects of Free Carries on Optical Properties
Edward D. Palik and George B. Wright, Free-Carrier Magnetooptical Effects
Richard H. Bube, Photoelectronic Analysis
B. O. Seraphin and H. E. Bennett, Optical Constants

Volume 4 Physics of III–V Compounds

N. A. Goryunova, A. S. Borschevskii, and D. N. Tretiakov, Hardness
N. N. Sirota, Heats of Formation and Temperatures and Heats of Fusion of Compounds $A^{III}B^{V}$
Don L. Kendall, Diffusion
A. G. Chynoweth, Charge Multiplication Phenomena
Robert W. Keyes, The Effects of Hydrostatic Pressure on the Properties of III–V Semiconductors
L. W. Aukerman, Radiation Effects
N. A. Goryunova, F. P. Kesamanly, and D. N. Nasledov, Phenomena in Solid Solutions
R. T. Bate, Electrical Properties of Nonuniform Crystals

Volume 5 Infrared Detectors

Henry Levinstein, Characterization of Infrared Detectors
Paul W. Kruse, Indium Antimonide Photoconductive and Photoelectromagnetic Detectors
M. B. Prince, Narrowband Self-Filtering Detectors
Ivars Melngalis and T. C. Harman, Single-Crystal Lead-Tin Chalcogenides
Donald Long and Joseph L. Schmidt, Mercury-Cadmium Telluride and Closely Related Alloys
E. H. Putley, The Pyroelectric Detector
Normal B. Stevens, Radiation Thermopiles
R. J. Keyes and T. M. Quist, Low Level Coherent and Incoherent Detection in the Infrared
M. C. Teich, Coherent Detection in the Infrared
F. R. Arams, E. W. Sard, B. J. Peyton, and F. P. Pace, Infrared Heterodyne Detection with Gigahertz IF Response
H. S. Sommers, Jr., Macrowave-Based Photoconductive Detector
Robert Sehr and Rainer Zuleeg, Imaging and Display

Volume 6 Injection Phenomena

Murray A. Lampert and Ronald B. Schilling, Current Injection in Solids: The Regional Approximation Method
Richard Williams, Injection by Internal Photoemission
Allen M. Barnett, Current Filament Formation

R. Baron and J. W. Mayer, Double Injection in Semiconductors
W. Ruppel, The Photoconductor-Metal Contact

Volume 7 Application and Devices
PART A

John A. Copeland and Stephen Knight, Applications Utilizing Bulk Negative Resistance
F. A. Padovani, The Voltage-Current Characteristics of Metal-Semiconductor Contacts
P. L. Hower, W. W. Hooper, B. R. Cairns, R. D. Fairman, and D. A. Tremere, The GaAs Field-Effect Transistor
Marvin H. White, MOS Transistors
G. R. Antell, Gallium Arsenide Transistors
T. L. Tansley, Heterojunction Properties

PART B

T. Misawa, IMPATT Diodes
H. C. Okean, Tunnel Diodes
Robert B. Campbell and Hung-Chi Chang, Silicon Carbide Junction Devices
R. E. Enstrom, H. Kressel, and L. Krassner, High-Temperature Power Rectifiers of $GaAs_{1-x}P_x$

Volume 8 *Transport and Optical Phenomena*

Richard J. Stirn, Band Structure and Galvanomagnetic Effects in III–V Compounds with Indirect Band Gaps
Roland W. Ure, Jr., Thermoelectric Effects in III–V Compounds
Herbert Piller, Faraday Rotation
H. Barry Bebb and E. W. Williams, Photoluminescence I: Theory
E. W. Williams and H. Barry Bebb, Photoluminescence II: Gallium Arsenide

Volume 9 Modulation Techniques

B. O. Seraphin, Electroreflectance
R. L. Aggarwal, Modulated Interband Magnetooptics
Daniel F. Blossey and Paul Handler, Electroabsorption
Bruno Batz, Thermal and Wavelength Modulation Spectroscopy
Ivar Balslev, Piezopptical Effects
D. E. Aspnes and N. Bottka, Electric-Field Effects on the Dielectric Function of Semiconductors and Insulators

Volume 10 Transport Phenomena

R. L. Rhode, Low-Field Electron Transport
J. D. Wiley, Mobility of Holes in III–V Compounds
C. M. Wolfe and G. E. Stillman, Apparent Mobility Enhancement in Inhomogeneous Crystals
Robert L. Petersen, The Magnetophonon Effect

Volume 11 Solar Cells

Harold J. Hovel, Introduction; Carrier Collection, Spectral Response, and Photocurrent; Solar Cell Electrical Characteristics; Efficiency; Thickness; Other Solar Cell Devices; Radiation Effects; Temperature and Intensity; Solar Cell Technology

Volume 12 Infrared Detectors (II)

W. L. Eiseman, J. D. Merriman, and R. F. Potter, Operational Characteristics of Infrared Photodetectors
Peter R. Bratt, Impurity Germanium and Silicon Infrared Detectors
E. H. Putley, InSb Submillimeter Photoconductive Detectors
G. E. Stillman, C. M. Wolfe, and J. O. Dimmock, Far-Infrared Photoconductivity in High Purity GaAs
G. E. Stillman and C. M. Wolfe, Avalanche Photodiodes
P. L. Richards, The Josephson Junction as a Detector of Microwave and Far-Infrared Radiation
E. H. Putley, The Pyroelectric Detector—An Update

Volume 13 Cadmium Telluride

Kenneth Zanio, Materials Preparation; Physics; Defects; Applications

Volume 14 Lasers, Junctions, Transport

N. Holonyak, Jr. and M. H. Lee, Photopumped III–V Semiconductor Lasers
Henry Kressel and Jerome K. Butler, Heterojunction Laser Diodes
A. Van der Ziel, Space-Charge-Limited Solid-State Diodes
Peter J. Price, Monte Carlo Calculation of Electron Transport in Solids

Volume 15 Contacts, Junctions, Emitters

B. L. Sharma, Ohmic Contacts to III–V Compound Semiconductors
Allen Nussbaum, The Theory of Semiconducting Junctions
John S. Escher, NEA Semiconductor Photoemitters

Volume 16 Defects, (HgCd)Se, (HgCd)Te

Henry Kressel, The Effect of Crystal Defects on Optoelectronic Devices
C. R. Whitsett, J. G. Broerman, and C. J. Summers, Crystal Growth and Properties of $Hg_{1-x}Cd_xSe$ alloys
M. H. Weiler, Magnetooptical Properties of $Hg_{1-x}Cd_xTe$ Alloys
Paul W. Kruse and John G. Ready, Nonlinear Optical Effects in $Hg_{1-x}Cd_xTe$

Volume 17 CW Processing of Silicon and Other Semiconductors

James F. Gibbons, Beam Processing of Silicon
Arto Lietoila, Richard B. Gold, James F. Gibbons, and Lee A. Christel, Temperature Distributions and Solid Phase Reaction Rates Produced by Scanning CW Beams

Arto Leitoila and James F. Gibbons, Applications of CW Beam Processing to Ion Implanted Crystalline Silicon
N. M. Johnson, Electronic Defects in CW Transient Thermal Processed Silicon
K. F. Lee, T. J. Stultz, and James F. Gibbons, Beam Recrystallized Polycrystalline Silicon: Properties, Applications, and Techniques
T. Shibata, A. Wakita, T. W. Sigman, and James F. Gibbons, Metal-Silicon Reactions and Silicide
Yves I. Nissim and James F. Gibbons, CW Beam Processing of Gallium Arsenide

Volume 18 Mercury Cadmium Telluride

Paul W. Kruse, The Emergence of $(Hg_{1-x}Cd_x)Te$ as a Modern Infrared Sensitive Material
H. E. Hirsch, S. C. Liang, and A. G. White, Preparation of High-Purity Cadmium, Mercury, and Tellurium
W. F. H. Micklethwaite, The Crystal Growth of Cadmium Mercury Telluride
Paul E. Petersen, Auger Recombination in Mercury Cadmium Telluride
R. M. Broudy and V. J. Mazurczyk, (HgCd)Te Photoconductive Detectors
M. B. Reine, A. K. Soad, and T. J. Tredwell, Photovoltaic Infrared Detectors
M. A. Kinch, Metal-Insulator-Semiconductor Infrared Detectors

Volume 19 Deep Levels, GaAs, Alloys, Photochemistry

G. F. Neumark and K. Kosai, Deep Levels in Wide Band-Gap III–V Semiconductors
David C. Look, The Electrical and Photoelectronic Properties of Semi-Insulating GaAs
R. F. Brebrick, Ching-Hua Su, and Pok-Kai Liao, Associated Solution Model for Ga–In–Sb and Hg–Cd–Te
Yu. Ya. Gurevich and Yu. V. Pleskon, Photoelectrochemistry of Semiconductors

Volume 20 Semi-Insulating GaAs

R. N. Thomas, H. M. Hobgood, G. W. Eldridge, D. L. Barrett, T. T. Braggins, L. B. Ta, and S. K. Wang, High-Purity LEC Growth and Direct Implantation of GaAs for Monolithic Microwave Circuits
C. A. Stolte, Ion Implantation and Materials for GaAs Integrated Circuits
C. G. Kirkpatrick, R. T. Chen, D. E. Holmes, P. M. Asbeck, K. R. Elliott, R. D. Fairman, and J. R. Oliver, LEC GaAs for Integrated Circuit Applications
J. S. Blakemore and S. Rahimi, Models for Mid-Gap Centers in Gallium Arsenide

Volume 21 Hydrogenated Amorphous Silicon Part A

Jacques I. Pankove, Introduction
Masataka Hirose, Glow Discharge; Chemical Vapor Deposition
Yoshiyuki Uchida, dc Glow Discharge
T. D. Moustakas, Sputtering
Isao Yamada, Ionized-Cluster Beam Deposition
Bruce A. Scott, Homogeneous Chemical Vapor Deposition
Frank J. Kampas, Chemical Reactions in Plasma Deposition

Paul A. Longeway, Plasma Kinetics
Herbert A. Weakliem, Diagnostics of Silane Glow Discharges Using Probes and Mass Spectroscopy
Lester Gluttman, Relation between the Atomic and the Electronic Structures
A. Chenevas-Paule, Experiment Determination of Structure
S. Minomura, Pressure Effects on the Local Atomic Structure
David Adler, Defects and Density of Localized States

Part B

Jacques I. Pankove, Introduction
G. D. Cody, The Optical Absorption Edge of a-Si:H
Nabil M. Amer and Warren B. Jackson, Optical Properties of Defect States in a-Si:H
P. J. Zanzucchi, The Vibrational Spectra of a-Si:H
Yoshihiro Hamakawa, Electroreflectance and Electroabsorption
Jeffrey S. Lannin, Raman Scattering of Amorphous Si, Ge, and Their Alloys
R. A. Street, Luminescence in a-Si:H
Richard S. Crandall, Photoconductivity
J. Tauc, Time-Resolved Spectroscopy of Electronic Relaxation Processes
P. E. Vanier, IR-Induced Quenching and Enhancement of Photoconductivity and Photoluminescence
H. Schade, Irradiation-Induced Metastable Effects
L. Ley, Photoelectron Emission Studies

Part C

Jacques I. Pankove, Introduction
J. David Cohen, Density of States from Junction Measurements in Hydrogenated Amorphous Silicon
P. C. Taylor, Magnetic Resonance Measurements in a-Si:H
K. Morigaki, Optically Detected Magnetic Resonance
J. Dresner, Carrier Mobility in a-Si:H
T. Tiedje, Information about band-Tail States from Time-of-Flight Experiments
Arnold R. Moore, Diffusion Length in Undoped a-Si:H
W. Beyer and J. Overhof, Doping Effects in a-Si:H
H. Fritzche, Electronic Properties of Surfaces in a-Si:H
C. R. Wronski, The Staebler-Wronski Effect
R. J. Nemanich, Schottky Barriers on a-Si:H
B. Abeles and T. Tiedje, Amorphous Semiconductor Superlattices

Part D

Jacques I. Pankove, Introduction
D. E. Carlson, Solar Cells
G. A. Swartz, Closed-Form Solution of I–V Characteristic for a-Si:H Solar Cells
Isamu Shimizu, Electrophotography
Sachio Ishioka, Image Pickup Tubes
P. G. LeComber and W. E. Spear, The Development of the a-Si:H Field-Effect Transistor and Its Possible Applications
D. G. Ast, a-Si:H FET-Addressed LCD Panel
S. Kaneko, Solid-State Image Sensor

Masakiyo Matsumura, Charge-Coupled Devices
M. A. Bosch, Optical Recording
A. D'Amico and G. Fortunato, Ambient Sensors
Hiroshi Kukimoto, Amorphous Light-Emitting Devices
Robert J. Phelan, Jr., Fast Detectors and Modulators
Jacques I. Pankove, Hybrid Structures
P. G. LeComber, A. E. Owen, W. E. Spear, J. Hajto, and W. K. Choi, Electronic Switching in Amorphous Silicon Junction Devices

Volume 22 Lightwave Communications Technology
Part A

Kazuo Nakajima, The Liquid-Phase Epitaxial Growth of IngaAsp
W. T. Tsang, Molecular Beam Epitaxy for III–V Compound Semiconductors
G. B. Stringfellow, Organometallic Vapor-Phase Epitaxial Growth of III–V Semiconductors
G. Beuchet, Halide and Chloride Transport Vapor-Phase Deposition of InGaAsP and GaAs
Manijeh Razeghi, Low-Pressure Metallo-Organic Chemical Vapor Deposition of $Ga_xIn_{1-x}As P_{1-y}$ Alloys
P. M. Petroff, Defects in III–V Compounds Semiconductors

Part B

J. P. van der Ziel, Mode Locking of Semiconductor Lasers
Kam Y. Lau and Ammon Yariv, High-Frequency Current Modulation of Semiconductor Injection Lasers
Charles H. Henry, Spectral Properties of Semiconductor Lasers
Yasuharu Suematsu, Katsumi Kishino, Shigehisa Arai, and Fumio Koyama, Dynamic Single-Mode Semiconductor Lasers with a Distributed Reflector
W. T. Tsang, The Cleaved-Coupled-Cavity (C^3) Laser

Part C

R. J. Nelson and N. K. Dutta, Review of InGaAsP InP Laser Structures and Comparison of Their Performance
N. Chinone and M. Nakamura, Mode-Stabilized Semiconductor Lasers for 0.7–0.8- and 1.1–1.6-μm Regions
Yoshiki Horikoshi, Semiconductor Lasers with Wavelength Exceeding 2 μm
B. A. Dean and M. Dixon, The Functional Reliability of Semiconductor Lasers as Optical Transmitters
R. H. Saul, T. P. Lee, and C. A. Burus, Light-Emitting Device Design
C. L. Zipfel, Light-Emitting Diode-Reliability
Tien Pei Lee and Tingye Li, LED-Based Multimode Lightwave Systems
Kinichiro Ogawa, Semiconductor Noise-Mode Partition Noise

Part D

Federico Capasso, The Physics of Avalanche Photodiodes
T. P. Pearsall and M. A. Pollack, Compound Semiconductor Photodiodes

Takao Kaneda, Silicon and Germanium Avalanche Photodiodes
S. R. Forrest, Sensitivity of Avalanche Photodetector Receivers for High-Bit Long-Wavelength Optical Communication Systems
J. C. Campbell, Phototransistors for Lightwave Communications

Part E

Shyh Wang, Principles and Characteristics of Integratable Active and Passive Optical Devices
Shlomo Margalit and Amnon Yariv, Integrated Electronic and Photonic Devices
Takaoki Mukai, Yoshihisa Yamamoto, and Tatsuya Kimura, Optical Amplification by Semiconductor Lasers

Volume 23 Pulsed Laser Processing of Semiconductors

R. F. Wood, C. W. White, and R. T. Young, Laser Processing of Semiconductors: An Overview
C. W. White, Segregation, Solution Trapping, and Supersaturated Alloys
G. E. Jellison, Jr., Optical and Electrical Properties of Pulsed Laser-Annealed Silicon
R. F. Wood and G. E. Jellison, Jr., Melting Model of Pulsed Laser Processing
R. F. Wood and F. W. Young, Jr., Nonequilibrium Solidification Following Pulsed Laser Melting
D. H. Lowndes and G. E. Jellison, Jr., Time-Resolved Measurements During Pulsed Laser Irradiation of Silicon
D. M. Zebner, Surface Studies of Pulsed Laser Irradiated Semiconductors
D. H. Lowndes, Pulsed Beam Processing of Gallium Arsenide
R. B. James, Pulsed CO_2 Laser Annealing of Semiconductors
R. T. Young and R. F. Wood, Applications of Pulsed Laser Processing

Volume 24 Applications of Multiquantum Wells, Selective Doping, and Superlattices

W. Weisbuch, Fundamental Properties of III–V Semiconductor Two-Dimensional Quantized Structures: The Basis for Optical and Electronic Device Applications
H. Morkoc and H. Unlu, Factors Affecting the Performance of (Al, Ga)As/GaAs and (Al, Ga)As/InGaAs Modulation-Doped Field-Effect Transistors: Microwave and Digital Applications
N. T. Linh, Two-Dimensional Electron Gas FETs: Microwave Applications
M. Abe et al., Ultra-High-Speed HEMT Integrated Circuits
D. S. Chemla, D. A. B. Miller, and P. W. Smith, Nonlinear Optical Properties of Multiple Quantum Well Structures for Optical Signal Processing
F. Capasso, Graded-Gap and Superlattice Devices by Band-Gap Engineering
W. T. Tsang, Quantum Confinement Heterostructure Semiconductor Lasers
G. C. Osbourn et al., Principles and Applications of Semiconductor Strained-Layer Superlattices

Volume 25 Diluted Magnetic Semiconductors

W. Giriat and J. K. Furdyna, Crystal Structure, Composition, and Materials Preparation of Diluted Magnetic Semiconductors

W. M. Becker, Band Structure and Optical Properties of Wide-Gap $A^{II}_{1-x}Mn_xB^{VI}$ Alloys at Zero Magnetic Field
Saul Oseroff and Pieter H. Keesom, Magnetic Properties: Macroscopic Studies
T. Giebultowicz and T. M. Holden, Neutron Scattering Studies of the Magnetic Structure and Dynamics of Diluted Magnetic Semiconductors
J. Kossut, Band Structure and Quantum Transport Phenomena in Narrow-Gap Diluted Magnetic Semiconductors
C. Riquaux, Magnetooptical Properties of Large-Gap Diluted Magnetic Semiconductors
J. A. Gaj, Magnetooptical Properties of Large-Gap Diluted Magnetic Semiconductors
J. Mycielski, Shallow Acceptors in Diluted Magnetic Semiconductors: Splitting, Boil-off, Giant Negative Magnetoresistance
A. K. Ramadas and R. Rodrique, Raman Scattering in Diluted Magnetic Semiconductors
P. A. Wolff, Theory of Bound Magnetic Polarons in Semimagnetic Semiconductors

Volume 26 III–V Compound Semiconductors and Semiconductor Properties of Superionic Materials

Zou Yuanxi, III–V Compounds
H. V. Winston, A. T. Hunter, H. Kimura, and R. E. Lee, InAs-Alloyed GaAs Substrates for Direct Implantation
P. K. Bhattachary and S. Dhar, Deep Levels in III–V Compound Semiconductors Grown by MBE
Yu. Yu. Gurevich and A. K. Ivano-Shits, Semiconductor Properties of Superionic Materials

Volume 27 High Conducting Quasi-One-Dimensional Organic Crystals

E. M. Conwell, Introduction to Highly Conducting Quasi-One-Dimensional Organic Crystals
I. A. Howard, A Reference Guide to the Conducting Quasi-One-Dimensional Organic Molecular Crystals
J. P. Pouquet, Structural Instabilities
E. M. Conwell, Transport Properties
C. S. Jacobsen, Optical Properties
J. C. Scott, Magnetic Properties
L. Zuppiroli, Irradiation Effects: Perfect Crystals and Real Crystals

Volume 28 Measurement of High-Speed Signals in Solid State Devices

J. Frey and D. Ioannou, Materials and Devices for High-Speed and Optoelectronic Applications
H. Schumacher and E. Strid, Electronic Wafer Probing Techniques
D. H. Auston, Picosecond Photoconductivity: High-Speed Measurements of Devices and Materials
J. A. Valdmanis, Electro-Optic Measurement Techniques for Picosecond Materials, Devices, and Integrated Circuits
J. M. Wiesenfeld and R. K. Jain, Direct Optical Probing of Integrated Circuits and High-Speed Devices
G. Plows, Electron-Beam Probing
A. M. Weiner and R. B. Marcus, Photoemissive Probing

Volume 29 Very High Speed Integrated Circuits: Gallium Arsenide LSI

M. Kuzuhara and T. Nazaki, Active Layer Formation by Ion Implantation
H. Hasimoto, Focused Ion Beam Implantation Technology
T. Nozaki and A. Higashisaka, Device Fabrication Process Technology
M. Ino and T. Takada, GaAs LSI Circuit Design
H. Hirayama, M. Ohmori, and K. Yamasaki, GaAs LSI Fabrication and Performance

Volume 30 Very High Speed Integrated Circuits: Heterostructure

H. Watanabe, T. Mizutani, and A. Usui, Fundamentals of Epitaxial Growth and Atomic Layer Epitaxy
S. Hiyamizu, Characteristics of Two-Dimensional Electron Gas in III–V Compound Heterostructures Grown by MBE
T. Nakanisi, Metalorganic Vapor Phase Epitaxy for High-Quality Active Layers
T. Nimura, High Electron Mobility Transistor and LSI Applications
T. Sugeta and T. Ishibashi, Hetero-Bipolar Transistor and LSI Application
H. Matsueda, T. Tanaka, and M. Nakamura, Optoelectronic Integrated Circuits

Volume 31 Indium Phosphide: Crystal Growth and Characterization

J. P. Farges, Growth of Dislocation-free InP
M. J. McCollum and G. E. Stillman, High Purity InP Grown by Hydride Vapor Phase Epitaxy
T. Inada and T. Fukuda, Direct Synthesis and Growth of Indium Phosphide by the Liquid Phosphorous Encapsulated Czochralski Method
O. Oda, K. Katagiri, K. Shinohara, S. Katsura, Y. Takahashi, K. Kainosho, K. Kohiro, and R. Hirano, InP Crystal Growth, Substrate Preparation and Evaluation
K. Tada, M. Tatsumi, M. Morioka, T. Araki, and T. Kawase, InP Substrates: Production and Quality Control
M. Razeghi, LP-MOCVD Growth, Characterization, and Application of InP Material
T. A. Kennedy and P. J. Lin-Chung, Stoichiometric Defects in InP

Volume 32 Strained-Layer Superlattices: Physics

T. P. Pearsall, Straied-Layer Superlattices
Fred H. Pollock, Effects of Homogeneous Strain on the Electronic and Vibrational Levels in Semiconductors
J. Y. Marzin, J. M. Gerárd, P. Voisin, nad J. A. Brum, Optical Studies of Strained III–V Heterolayers
R. People and S. A. Jackson, Structurally Induced States from Strained and Confinement
M. Jacobs, Microscopic Phenomena in Ordered Superlattices

Volume 33 Strained-Layer Superlattices: Materials Science and Technology

R. Hull and J. C. Bean, Principles and Concepts of Strained-Layer Epitaxy
William J. Schaff, Paul J. Tasker, Mark C. Foisy, and Lester F. Eastman, Device Applications of Strained-Layer Epitaxy

S. T. Picraux, B. L. Doyle, and J. Y. Tsao, Structure and Characterization of Strained-Layer Superlattices
E. Kasper and F. Schäffler, Group IV Compounds
Dale L. Martin, Molecular Beam Epitaxy of IV–VI Compounds Heterojunction
Robert L. Gunshor, Leslie A. Kolodziejski, Arto V. Nurmikko, and Nobuo Otsuka, Molecular Beam Epitaxy of II–VI Semiconductor Microstructures

Volume 34 Hydrogen in Semiconductors

J. I. Pankove and N. M. Johnson, Introduction to Hydrogen in Semiconductors
C. H. Seager, Hydrogenation Methods
J. I. Pankove, Hydrogenation of Defects in Crystalline Silicon
J. W. Corbett, P. Deák, U. V. Desnica, and S. J. Pearton, Hydrogen Passivation of Damage Centers in Semiconductors
S. J. Pearton, Neutralization of Deep Levels in Silicon
J. I. Pankove, Neutralization of Shallow Acceptors in Silicon
N. M. Johnson, Neutralization of Donor Dopants and Formation of Hydrogen-Induced Defects in n-Type Silicon
M. Stavola and S. J. Pearton, Vibrational Spectroscopy of Hydrogen-Related Defects in Silicon
A. D. Marwick, Hydrogen in Semiconductors: Ion Beam Techniques
C. Herring and N. M. Johnson, Hydrogen Migration and Solubility in Silicon
E. E. Haller, Hydrogen-Related Phenomena in Crystalline Germanium
J. Kakalios, Hydrogen Diffusion in Amorphous Silicon
J. Chevalier, B. Clerjaud, and B. Pajot, Neutralization of Defects and Dopants in III–V Semiconductors
G. G. DeLeo and W. B. Fowler, Computational Studies of Hydrogen-Containing Complexes in Semiconductors
R. F. Kiefl and T. L. Estle, Muonium in Semiconductors
C. G. Van de Walle, Theory of Isolated Interstitial Hydrogen and Muonium in Crystalline Semiconductors

Volume 35 Nanostructured Systems

Mark Reed, Introduction
H. van Houten, C. W. J. Beenakker, and B. J. van Wees, Quantum Point Contacts
G. Timp, When Does a Wire Become an Electron Waveguide?
M. Büttiker, The Quantum Hall Effect in Open Conductors
W. Hansen, J. P. Kotthaus, and U. Merkt, Electrons in Laterally Periodic Nanostructures

Volume 36 The Spectroscopy of Semiconductors

D. Heiman, Spectroscopy of Semiconductors at Low Temperatures and High Magnetic Fields
Arto V. Nurmikko, Transient Spectroscopy by Ultrashort Laser Pulse Techniques
A. K. Ramdas and S. Rodriguez, Piezospectroscopy of Semiconductors
Orest J. Glembocki and Benjamin V. Shanabrook, Photoreflectance Spectroscopy of Microstructures
David G. Seiler, Christopher L. Littler, and Margaret H. Wiler, One- and Two-Photon Magneto-Optical Spectroscopy of InSb and $Hg_{1-x}Cd_xTe$

Volume 37 The Mechanical Properties of Semiconductors

A.-B. Chen, Arden Sher and W. T. Yost, Elastic Constants and Related Properties of Semiconductor Compounds and Their Alloys
David R. Clarke, Fracture of Silicon and Other Semiconductors
Hans Siethoff, The Plasticity of Elemental and Compound Semiconductors
Sivaraman Guruswamy, Katherine T. Faber and John P. Hirth, Mechanical Behavior of Compound Semiconductors
Subbanh Mahajan, Deformation Behavior of Compound Semiconductors
John P. Hirth, Injection of Dislocations into Strained Multilayer Structures
Don Kendall, Charles B. Fledderman, and Kevin J. Molloy, Critical Technologies for the Micromachining of Silicon
Ikuo Matsuba and Kinjii Mokuya, Processing and Semiconductor Thermoelastic Behavior

Volume 38 Imperfections in III/V Materials

Udo Scherz and Matthias Scheffler, Density-Functional Theory of sp-Bonded Defects in III/V Semiconductors
Maria Kaminska and Eicke R. Weber, EL2 Defect in GaAs
David C. Look, Defects Relevant for Compensation in Semi-Insulating GaAs
R. C. Newman, Local Vibrational Mode Spectroscopy of Defects in III/V Compounds
Andrzej M. Hennel, Transition Metals in III/V Compounds
Kevin J. Malloy and Ken Khachaturyan, DX and Related Defects in Semiconductors
V. Swaminathan and Andrew S. Jordan, Dislocations in III/V Compounds
Krzysztof W. Nauka, Deep Level Defects in the Epitaxial III/V Materials

Volume 39 Minority Carriers in III-V Semiconductors: Physics and Applications

Niloy K. Dutta, Radiative Transitions in GaAs and Other III–V Compounds
Richard K. Ahrenkiel, Minority-Carrier Lifetime in III–V Semiconductors
Tomofumi Furuta, High Field Minority Electron Transport in p-GaAs
Mark S. Lundstrom, Minority-Carrier Transport in III–V Semiconductors
Richard A. Abram, Effects of Heavy Doping and High Excitation on the Band Structure of GaAs
David Yevick and Witold Bardyszewski, An Introduction to Non-Equilibrium Many-Body Analyses of Optical Processes in III–V Semiconductors

Volume 40 Epitaxial Microstructures

E. F. Schubert, Delta-Doping of Semiconductors: Electronic, Optical, and Structural Properties of Materials and Devices
A. Gossard, M. Sundaram, and P. Hopkins, Wide Graded Potential Wells
P. Petroff, Direct Growth of Nanometer-Size Quantum Wire Superlattices
E. Kapon, Lateral Patterning of Quantum Well Heterostructures by Growth of Nonplanar Substrates
H. Temkin, D. Gershoni, and M. Panish, Optical Properties of Ga$_{1-x}$In$_x$As/InP Quantum Wells

Volume 41 High Speed Heterostructure Devices

F. Capasso, F. Beltram, S. Sen, A. Pahlevi, and A. Y. Cho, Quantum Electron Devices: Physics and Applications
P. Solomon, D. J. Frank, S. L. Wright, and F. Canora, GaAs-Gate Semiconductor–Insulator–Semiconductor FET
M. H. Hashemi and U. K. Mishra, Unipolar InP-Based Transistors
R. Kiehl, Complementary Heterostructure FET Integrated Circuits
T. Ishibashi, GaAs-Based and InP-Based Heterostructure Bipolar Transistors
H. C. Liu and T. C. L. G. Sollner, High-Frequency-Tunneling Devices
H. Ohnishi, T. More, M. Takatsu, K. Imamura, and N. Yokoyama, Resonant-Tunneling Hot-Electron Transistors and Circuits

Volume 42 Oxygen in Silicon

F. Shimura, Introduction to Oxygen in Silicon
W. Lin, The Incorporation of Oxygen into Silicon Crystals
T. J. Schaffner and D. K. Schroder, Characterization Techniques for Oxygen in Silicon
W. M. Bullis, Oxygen Concentration Measurement
S. M. Hu, Intrinsic Point Defects in Silicon
B. Pajot, Some Atomic Configurations of Oxygen
J. Michel and L. C. Kimerling, Electrical Properties of Oxygen in Silicon
R. C. Newman and R. Jones, Diffusion of Oxygen in Silicon
T. Y. Tan and W. J. Taylor, Mechanisms of Oxygen Precipitation: Some Quantitative Aspects
M. Schrems, Simulation of Oxygen Precipitation
K. Simino and I. Yoneaga, Oxygen Effect on Mechanical Properties
W. Bergholz, Grown-in and Process-Induced Effects
F. Shimura, Intrinsic/Internal Gettering
H. Tsuya, Oxygen Effect on Electronic Device Performance

Volume 43 Semiconductors for Room Temperature Nuclear Detector Applications

R. B. James and T. E. Schlesinger, Introduction and Overview
L. S. Darken and C. E. Cox, High-Purity Germanium Detectors
A. Burger, D. Nason, L. Van den Berg, and M. Schieber, Growth of Mercuric Iodide
X. J. Bao, T. E. Schlesinger, and R. B. James, Electrical Properties of Mercuric Iodide
X. J. Bao, R. B. James, and T. E. Schlesinger, Optical Properties of Red Mercuric Iodide
M. Hage-Ali and P. Siffert, Growth Methods of CdTe Nuclear Detector Materials
M. Hage-Ali and P. Siffert, Characterization of CdTe Nuclear Detector Materials
M. Hage-Ali and P. Siffert, CdTe Nuclear Detectors and Applications
R. B. James, T. E. Schlesinger, J. Lund, and M. Schieber, $Cd_{1-x}Zn_xTe$ Spectrometers for Gamma and X-Ray Applications
D. S. McGregor, J. E. Kammeraad, Gallium Arsenide Radiation Detectors and Spectrometers
J. C. Lund, F. Olschner, and A. Burger, Lead Iodide
M. R. Squillante, and K. S. Shah, Other Materials: Status and Prospects
V. M. Gerrish, Characterization and Quantification of Detector Performance
J. S. Iwanczyk and B. E. Patt, Electronics for X-Ray and Gamma Ray Spectrometers

M. Schieber, R. B. James, and T. E. Schlesinger, Summary and Remaining Issues for Room Temperature Radiation Spectrometers

Volume 44 I-VI Blue/Green Light Emitters: Device Physics and Epitaxial Growth

J. Han and R. L. Gunshor, MBE Growth and Electrical Properties of Wide Bandgap ZnSe-based II–VI Semiconductors

Shizuo Fujita and Shigeo Fujita, Growth and Characterization of ZnSe-based II–VI Semiconductors by MOVPE

Easen Ho and Leslie A. Kolodziejski, Gaseous Source UHV Epitaxy Technologies for Wide Bandgap II–VI Semiconductors

Chris G. Van de Walle, Doping of Wide-Band-Gap II–VI Compounds — Theory

Roberto Cingolani, Optical Properties of Excitons in ZnSe-Based Quantum Well Heterostructures

A. Ishibashi and A. V. Nurmikko, II–VI Diode Lasers: A Current View of Device Performance and Issues

Supratik Guha and John Petruzzello, Defects and Degradation in Wide-Gap II–VI-based Structures and Light Emitting Devices.

ISBN 0-12-752144-5